Biogeography

A Study of Plants in the Ecosphere

Biogeography

A Study of Plants in the Ecosphere

Joy Tivy

Oliver and Boyd

Oliver & Boyd
Tweeddale Court
14 High Street
Edinburgh, EH1 1YL
A Division of Longman Group Limited

ISBN 0 05 001585 0

Set in 'Linotype' 10/12 pt. Baskerville
and printed in the United Kingdom by
W. & G. Baird Limited, Belfast

Contents

Preface

Biogeography is the 'Cinderella' of geography. It is still a relatively neglected and underdeveloped field of study at both school and university, despite the considerable lip-service that has been paid to its significance. Many have drawn attention to its value in providing) within the framework of the early formulated and recently 're-discovered' ecosystem concept (an integrated holistic approach to environmental studies. Some have accorded it the distinction of being the 'vital' link between physical and human geography. Others have enthusiastically advocated (without necessarily demonstrating) its potential as a means of unifying a subject whose peripheral organs often seem to be growing at a rate greater than that of its body! That biogeography is slowly gaining greater recognition is unquestionable. However, there are still all too few British or American universities where, in general physical geography courses, it receives or attracts the same attention as either climatology or geomorphology—and fewer still where it can be pursued at a more advanced Honours level. In the latter cases instruction is often relegated to departments of biology, botany, zoology or pedology. Those who go so far as to profess biogeography are 'rare', indeed some would say 'odd' species. English text-books on the subject, written by geographers, can be counted on one hand. A large proportion of its most elementary data must be distilled from non-geographical and often highly specialised and technical biological sources. And developments in the latter have long outstripped those in biogeography.

There are good reasons if not necessarily valid excuses for this state of affairs. First, because of the range of phenomena with which the biogeographer is faced, his field of study is less amenable to isolation and systematisation than other branches of physical geography, such as climatology, hydrology or geomorphology. Biogeography is not easy to define or delimit precisely; and the distinction between it and the closely related subject of ecology often tends to be more a function of scale and emphasis than of content or method. Second, the majority of teachers and students of geography, whether at school or university, know much less about the nature and scope of biogeography than about practically any other aspect of their subject. Opportunities to combine a study of biology and geography are limited. Unfamiliarity (exacerbated by a high degree of urbanisation) with even the most common plants has tended to create a psychological barrier which makes the average geography student chary of involving himself too deeply in a field which requires the acquisition of a whole lot of new and apparently highly 'technical' terms (among which Latin names seem to be the most daunting!) and concepts. And the continuing emphasis in many, particularly school, texts on 'explanatory' descriptions of the vegetation or soils of the world, often completely divorced from any concrete terms of reference, has done little to overcome this 'built-in' resistance and to stimulate interest in the subject.

This book is the product of a long period of 'experimentation' in the teaching of biogeography at various levels to university students, many of whom had no previous training in biology, or for that matter in any of the physical sciences. Its aim is twofold. The first is to summarise and explain (in a way which it is hoped will be comprehensive and palatable to teachers and students at Sixth Form and University levels) those biological processes and concepts which the author considers basic to the understanding of the principal characteristics of and the complex interrelationships within the 'organic world' or 'biosphere'. In the interests of clarity an attempt has been made to cut a path through the terminological jungle of ecology and to be selective, indeed as sparing as possible, in the use of uncommon and/or exotic plant and animal names. The second is to bring together the bases and aspects of biogeography which at present can only be obtained from a great number of widely scattered sources and, in doing so, to bridge the gap between the earlier traditional zonal study of vegetation and soil and modern developments, particularly in plant ecology. The theme is that of *organic resources*, and the reciprocal relationship between these and *man*. Throughout, the emphasis is on plants, as the primary food producers which form the essential link between man and his physical environment. The first half of the book is concerned with a systematic analysis of the effect of environmental (ecological) factors—climate, soil, biological competition, animals and man on the functioning, evolution and adaptation, and distribution of plants. The second half deals with the nature of vegetation and a consideration of the principal characteristics of the structure and function of the major types of ecosystems—marine, forest, grassland and desert—and the potentialities and problems of their particular organic resources for use by man through time.

It is hoped that this book will go some way to answering the perennial *'cri du coeur'* of the aspiring student of biogeography for a basic text-book which would provide a starting point and a guide to the highly ramified highways and byways of a vast and complex field. In its compilation the author has drawn on and selected from a wide variety of sources. However, it was decided in the interests of readability to omit numbered references in the text. The lists of references are not, nor were they intended to be, comprehensive. They do, however, include all those sources on which each chapter has been based, and they were selected as those most valuable and relevant to further elucidation and a deeper study of the subjects under consideration.

Finally the author would like to acknowledge her indebtedness to those generations of geography students at the Universities of Edinburgh and Glasgow whose lively interest and constructive criticism in the classroom and extreme fortitude in the field were a continuing source of inspiration. Without them, and the constant encouragement of long-suffering friends and colleagues, this book would never have been written.

1
Scope and Development of Biogeography

Biogeography, as the term indicates, is both a biological and a geographical science. Its 'field of study' is the biologically inhabited part of the lithosphere, atmosphere and hydrosphere—or, as it has become known—the *biosphere*. Its subject matter covers the multitudinous forms of plant and animal life which inhabit this relatively shallow but densely populated zone, as well as the complex biological processes which control their activities. The approach to and aim of the subject is geographical in so far as it is primarily concerned with the *distribution* (together with the causes and implications thereof!) of organisms and biological processes. However, although this 'field of study' is shared by, and is common to, both biology and geography it is not the exclusive preserve of either of these two sciences. By its very character, biogeography is situated at, and overlaps, the boundaries of a great number of other disciplines. The geologist, climatologist, pedologist, geomorphologist as well as the botanist, zoologist, geneticist and geographer (to mention but a few!) all 'cultivate' or 'crop', as the case may be, particular parts of this very large and varied field; and in doing so they are, to a greater or lesser extent, essential to, as well as being dependent on, an understanding of biogeography. As a result the approach to—or concept of—biogeography is in large measure determined by the training, interest, and objective of the particular student.

The geographer's interest tends to focus on (to be organised around) the spatial variation of two basic characteristics or processes, rather than on any particular components of the biosphere. The first is the intimate interrelationship between the organic and inorganic elements of the earth's environment; the character of the biosphere is primarily a product of the continual interaction or interchange between the lithosphere and the atmosphere. The second is the reciprocal relationship between man and the biosphere. On the one hand the latter provides the vital link between man and his physical environment; and despite the advances of modern science and technology man is still, whether he realises or likes it, completely dependent on the biosphere for his food. On the other hand, because of an ability, greater than that possessed by any other form of life, to exploit organic resources, he is not only an integral part of the biosphere but is now the ecologically dominant organism in it. It is somewhat ironic that the significance of biogeography should be more fully appreciated by the biologist than by the majority of geographers. It is particularly well expressed by M. G. Lemée, Professor of Ecology at Orsay in his recent book, *Précis de Biogéographie,* when he states 'C'est [la biogéographie] aussi une science géographique, car elle tend à établir les rapports avec des peuplements végétaux et animaux avec les autres grandes phénomènes géographiques, climats, géomorphologie, sols, activités humaines, pour attendre à une vue synthétique des aspects de la surface du Globe. Pour le géographe la connaissance de la partie vivante du paysage intervient comme un élément de première importance de ce complex, car liée aux autres éléments par d'étroites relations mutuelles, elle constitue un indicateur très sensible des caractères du milieu géographique'. Variations in the definition and, hence, the delimitation of the scope of biogeography have, to a considerable extent, depended on whether or not the aim or point of view has been primarily biological or geographical and on the rôle accorded to man in the study of the biosphere. For many biologists man is regarded as an important, if somewhat 'unnatural', element in the environment of plants and animals. The geographer's rather greater concentration on and appreciation of the nature and extent of man's rôle as an ecological factor has tended to distinguish his approach and contribution to biogeography up-to-date. Some would perhaps subscribe to the view of the late Margaret Anderson that the study of 'the biological relations between man, *considered as an animal* (author's italics), and the whole of his animate and inanimate environment' is the essence of biogeography.

In both biological and geographical literature attention has been primarily (though not exclusively) concerned with the study of plant rather than animal geography. There are various reasons for this. Greater mobility combined with the small size and an elusive habit of life of the majority of animal species make a study of their distribution more difficult than that of plants. Also, until fairly recently, zoogeography had not developed to anything like the same extent as plant geography. Other considerations have, however, been responsible for the geographer's preoccupation with plant life. Plants—vegetable matter—both living and decaying, comprise the greatest bulk of the total world *biomass* (i.e., volume of organic material) both above and below the ground surface, and in water bodies. Not only is the animal biomass small in comparison, but most of it is composed of micro-organisms the majority of which live in the soil. Hence plants are the most conspicuous components of the biosphere and, *'en masse'*, form a major landscape element. Further, plants are more *directly* dependent upon and affected by their physical habitat than animals. In comparison to the latter, plants are relatively immobile or at least lacking effective means of independent locomotion. As a result they provide a better index (outward and visible expression of the total environment—physical, biological and human) of the site they occupy; and this is no less true of the isolated tree in the city square than of the cornfield or the tropical jungle! Also, because of their greater biomass, plants exert a greater influence on the character of the atmosphere and soil they occupy. They not only modify the physical habitat, but in doing so they create a particular biological environment which would not otherwise exist. Finally plants provide the primary source of food-energy for all other living organisms—including man. They are basic to the geography of animals and constitute the most important of man's resources.

Biogeography is firmly rooted in the biological sciences on whose data, concepts and methods the geographer is obliged to draw and whose developments have inevitably influenced his particular interest in and approach to the biosphere. Biogeography originated—as did so many of the other closely related but now highly specialised and discrete disciplines of botany, zoology, geology, climatology, etc.—in the early and more catholic field of 'Natural History' or what today are called the 'Earth Sciences'. A growing curiosity about the nature of the earth became organised around the description and classification of 'natural phenomena'. And by the latter half of the eighteenth century, the Swedish botanist, Carl von Linné, (Carolus Linneaus, 1707–1778) had laid the foundations of modern biological taxonomy and nomenclature. The late eighteenth century and the nineteenth century constitute the era of such great explorer-naturalists as Alexander von Humboldt, Edward Forbes, Joseph Hooker, Louis Agassiz, Alfred Wallace and Charles Darwin. As early as 1804, von Humboldt, often described as the 'father of plant geography', had published twenty-six volumes recording his observations on plants and other environmental data collected during the course of his extensive travels in South America. Exploration, motivated by a combination of economic and scientific incentives, led to the accumulation of a growing body of factual data. In addition, far-reaching voyages drew early attention not only to the extent of biological diversity, but to striking variations and anomalies in the distribution of different types of plants and animals. And from the search for the causes of these variations there emerged two interrelated concepts which revolutionised the study of natural history and initiated the development of the biological sciences. One was that of the 'adaption' of organisms to their physical environment, the other that of the process of 'natural selection' of those best fitted or adapted to survive in a given habitat. Together these formed the basis of the Darwinian theory of evolution and the origin of species.

Initially dependent upon the data collected and the concepts formulated by the early naturalists, the subsequent development of biogeography has been distinguished by two distinct, though not completely exclusive lines of investigation—one primarily taxonomic, the other ecological. The first is characteristic of *plant (or phyto) geography* (and its counterpart animal or zoogeography). Although frequently employed in a wider sense, plant geography is, strictly speaking, an accepted branch of botany. It is the botanist's study of the distribution of different 'types of plants' or plant taxa (families, genera, species, etc.) and originated in the early inventories or *Floras* compiled from different parts of the world. Its object is the analysis, and explanation, of the geographical range of particular taxa or floras as a means to a fuller understanding of their origin, evolution and dispersion. Plant geography has made and continues to make important contributions to the elucidation and assessment of the relative importance of the factors which determine floristic distribution; and one of the most important has undoubtedly been the light that it has thrown on the effect of past events on present distributions, including the time of origin of a taxon and the environmental changes that have taken place during the course of its evolution. That aspect of plant geography which has aroused most interest for geographers, and to which they have made a not insignificant contribution, is the effect of man on the range of particular species: the deliberate or accidental spread of particular plants and the consequence for the areas of immigration. Also, the fascinating problem of the origins and dissemination of domesticated species—the cultigens—is an aspect of human/historical geography which has been receiving increasing attention and is an important part of biogeography.

The geographer's approach to biogeography, however, has been influenced and determined more by ecological than taxonomic concepts. Indeed the late Marion Newbigin, the *doyenne* of English biogeographers, regarded its scope and aims as virtually synonymous with those of ecology. The term ecology (from Gr. *oikos* = home) was originally coined by zoologists at the end of the nineteenth century to define 'the study of the reciprocal relations between organisms and their environment'. But it was not the ecological approach alone which influenced the development of biogeography: the understanding of the distribution of taxa (of plant geography) was also dependent on a knowledge of the interrelationships of particular species to their environment or, as it became known, *autecology*. It was, however, the study of the distribution of vegetation rather than floras which became the geographer's prime concern and basic to his concept of biogeography. The nature of vegetation (the sum total of all the plants) is dependent on the relative proportions of the various species, on the particular assemblage, group or 'community' of plants occupying a particular area. Vegetation geography (or ecological plant geography as it has sometimes been called) is then virtually synonymous with that branch of ecology referred to as *synecology*, i.e. the study of (plant) communities in relation to their environment: and the development of biogeography has been influenced to a lesser or greater extent, over time, by ecological studies of the four main attributes of plant communities—their structure, development, composition and function.

The study of vegetation geography, however, developed nearly a century before the recognition and acceptance of ecology as an academically respectable branch of the biological sciences. It, in fact, emerged at the same time and for the same reasons as did plant geography; indeed, initially there was little distinction between the two subjects. Descriptions of the very obvious variations in the world's vegetation cover (which must have made such an impression on the early globe-trotting naturalists) were an integral part of the inventories of natural phenomena that were being accumulated. At an early stage they focused attention on the very obvious significance of climate as an important 'ecological' variable. This, on the one hand, resulted in the concept of 'life-zones' as originally defined by C. H. Merriam in 1894. These were, primarily, 'thermal' climatic zones each distinguished by a characteristic flora and fauna. They were later to find quantitative expression in 'the growing season' as a means of defining either vegetation or agricultural limits. On the other hand, von Humboldt's successors—Alphonse de Candolle, A. Grisebach, and O. Drude—had, during the course of the nineteenth century, been putting increasing emphasis on the variation, particularly in the form and structure of the *major* types of vegetation, from one part of the world to another. Attempts to relate the distribution of vegetation and explain its morphological features in terms of adaption to environmental, and more especially climatic, conditions, culminated in the publication of A. W. F. Schimper's classic survey of world vegetation, *Plant Geography on a Physiological Basis.*

The English translation of this major work appeared in 1903, and it soon became the main reference source on which many subsequent descriptions and explanations of world vegetation were (and indeed still are!) based. The, by now, presumed causal relationships between climate and vegetation distributions were beginning to have a profound influence on the study of all the natural sciences. It is not without significance that one of the earliest classifications of world climates was proposed by a biologist, W. Köppen, in 1918; it was, in fact, an attempt to establish the climatic parameters which coincided with the boundaries between major types of vegetation. Further, vegetation-climate relationships profoundly influenced geographical thinking of the time. They were the basis of the major *'Natural Regions of the World'* set out by A. J. Herbertson in 1905. Indeed the assumption that climate was the major or 'master' ecological factor early became, and to a marked extent has remained, an unchallenged tenet in both ecological and biological studies.

Until the turn of the century concepts of vegetation were essentially static. The gradual growth of ecological work, however, began to direct attention towards the significance of other factors in the determination of the nature of vegetation. The importance of the 'time element' gained increasing recognition. The pioneer work of the American ecologists inspired originally by Henry Cowles—on the *development* of vegetation on a sand-dune complex in Michigan—is a *landmark* in the history of ecology. It demonstrated the process of *succession* in the establishment of a vegetation cover and the essentially 'dynamic' nature of both the physical habitat and its associated biological communities. The genetic approach to the study of vegetation was firmly established by the subsequent writings of Frederick E. Clements, one of Cowles' most distinguished pupils. He finally elaborated the concept of the *climax*—expressed in the form of the dominant plants and determined (in his opinion) by climate—as the ultimate terminal stage in vegetation development. These trends in ecology closely paralleled the contemporaneous, but largely independent developments in geomorphology and pedology, which had been influenced by the Darwinian theory of organic evolution. The concept of landforms as the product of processes operating over a period of time had already gained acceptance among geologists. The 'cycle of erosion' as reflected in the varying stages in the development of landforms from 'youth through maturity to old age', and culminating in the *peneplain,* was initially propounded by William Morris Davis in 1899. Concurrently the importance of climate and vegetation as factors in soil formation were attracting attention in Russia. This resulted in the concept of the *zonal soil* and the genetic classification of major soil groups whose profiles reflected, predominantly, the overriding influence of the climatic regime and associated type of vegetation under which they had developed. Attempts, particularly in America, to apply the Davisian concept of 'stage' to soil studies, led to the equation of the 'zonal' soil with the 'mature' soil—the final terminal phase in soil formation when the resulting profile has attained a state of dynamic equilibrium under the prevailing climatic conditions.

Theoretically the attainment of soil maturity would be accompanied by a similar stage in the development of associated landforms and the final establishment of the climax vegetation. The latter would then be dominated by that form of plant life best adapted to (and hence representing the most 'complete' expression of) the prevailing climatic conditions. In contrast to geomorphology, pedology and climatology, biogeographical studies have been dominated to a much greater extent and for a considerably longer period by the 'zonal' approach and the concept of the climatic 'climax'. Biogeography early became, and tended to remain, synonymous with a study of types of world vegetation and related soil zones. In both biological and geographical studies its ultimate aim was the description and explanation of the 'natural' or 'potential' climatic climax vegetation. Even contemporary textbooks published under the title of 'Biogeography' are global in scope and retain an emphasis on 'natural or semi-natural' vegetation considered in a broad climatic context. To an even greater extent the treatment of biogeography in school texts is heavily biased towards explanatory descriptions of climatically defined types of world vegetation and their associated soils.

Biogeography has in fact failed to keep pace with the increasingly rapid advances that have taken place in ecology within the last fifty years or so. There have been, it is true, a growing number of smaller-scale regional and local vegetation studies, produced by biologists and to a lesser extent by geographers. However, probably largely as a result of the 'dominance' of geomorphology in schools of physical geography, the emphasis on the genesis and status of vegetation has been retained in biogeography longer than might otherwise have been the case. It was further reinforced by the development of Quaternary geology together with associated techniques of pollen analysis and methods of dating 'fossil' organic matter. And many 'schools of biogeography' have become strongly oriented towards palynological studies with the aim of reconstructing former environmental conditions and tracing the development of vegetation since early post-glacial times. As such, biogeography became the 'hand-maiden' of historical geography.

At the same time, however, biological work was being concentrated more and more on detailed autecological and synecological field studies. These entailed the refinement of methods of description and the development of controlled field and laboratory experiments designed to test empirically-based hypotheses. As a result many of the early formulated generalisations about the significance of plant forms, the nature of vegetation and its ecological relationships, required drastic reassessment. Also, attention was being increasingly directed towards two aspects of vegetation which had hitherto received much less attention from either ecologists or geographers. One was the study of the composition and inter-specific relationships of plant communities—that branch of ecology which became known as *plant (or phyto) sociology*. The other was the more recent investigation of ecological 'energetics': of the processes involved in the ecological relationships between organisms and their environment.

While the American-British school of ecology, under the influence of Frederick Clements and A. G. Tansley respectively, was pursuing studies of the habitat and status of vegetation communities, European botanists were more concerned with the floristic composition (the sociological attributes) of plant communities together with inter- and intra-specific associations and relationships. The detailed and precise data demanded by these studies stimulated the development of more quantitative, and hence standardised methods of sampling and describing vegetation. The 'quadrat' (a rectangular area of a size appropriate to the sampling of a particular type of vegetation cover) became the 'unit area' of study, the species-list provided the basic data. From the nineteen-thirties onwards the Zurich-Montpellier school of ecology exerted (particularly after the publication of J. Braun-Blanquet's text-book on plant sociology) a major influence on phyto-sociological concepts and methods. These were centred on the recognition and description of 'associations'—stands of vegetation characterised by the possession of common sociological features distinctive enough for them to be grouped together as a 'community-type'. The 'association' became the basic unit in the classification of vegetation and the definition of 'the association' a long-standing source of contention among ecologists.

Since the last war, however, the emphasis in ecological studies has been shifting not only from extensive to intensive study but from vegetation taxonomy to detailed work on ecological relationships and processes. Such trends are, on the one hand, the outcome of new and more efficient methods of collecting and analysing field data; on the other, of a growing awareness of the importance of understanding the way in which the whole biosphere functions. However, despite major technical developments, the nature and interaction of ecological variables is such that not all biological phenomena are amenable to the same degree of either objectivity, accuracy, or standardisation of measurement as can be achieved in the physical sciences. And the last two decades, in particular, have seen the rapidly increasing use of: (1) statistical analyses designed to assess the significance and 'margin of error' of the results of ecological investigations; (2) mechanical aids in the collection, storage and processing of data, which have been accompanied by a tremendous increase in the volume of data available; and (3) methods other than the traditional *hierarchical classification* of vegetation as a means of analysing causal relationships between species or between species and their environment. Of these the most important is that of *ordination*; this is an attempt (by those who maintain that variation in vegetation is continuous) to arrange sampled stands of vegetation in sequences or an order which can be related to either environmental or community variations.

Finally the growing awareness during the last three decades of the rapid depletion of organic material and the increasing modification of the biosphere by man has highlighted the seriousness and complexity of existing ecological problems. The significance of man as a universal and long-established ecological variable is, at last, being fully appreciated by biologists. 'Over-cropping', exacerbated by direct and indirect habitat modification (not least in recent years by the pollution of air, water and soil) has sparked off biological chain-reactions which man has, as yet, been unable to control completely. The resulting imbalance has manifested itself in drastic fluctuations of populations, particularly of animals and pathogenic organisms. Increasing human population combined with technical developments continue to intensify pressures on the biosphere and to increase the seriousness of these problems. The need to maintain both the diversity and amount of organic production and, where possible, to increase food production is becoming even more urgent.

The solution of such problems and the efficient management of organic resources is dependent on a better understanding of ecological functions and processes and of the nature of ecological interrelationships. And this need has undoubtedly stimulated a revival of interest in the *ecosystem* as a functioning unit. Neither the term nor the concept are new. The term 'ecosystem' was originally coined by A. G. Tansley in the nineteen-thirties to express the sum total of organisms and their physical habitat and the concept has long been the basis of ecological studies. Until quite recently, however, concentration and specialisation on particular components or aspects of the biosphere has tended to divert attention from the necessity of relating the former to the whole interacting system. Recent population studies by zoologists such as Charles Elton and V. C. Wynne-Edwards combined with quantitative studies of energy-flow and nutrient circulation in the biosphere have helped to reinstate the ecosystem as the fundamental integrating concept in ecological studies.

The 're-discovery' of the ecosystem concept is also beginning to influence and re-invigorate biogeographical thinking. In the first place it provides a much more satisfactory 'conceptual framework' to which the component parts of the biosphere, plants *and* animals, soil *and* atmosphere, can be related and within which their interactions can be understood. It is not limited by the restrictions of scale; the ecosystem concept can be applied to any self-regulating ecological system from the whole biosphere (or ecosphere) to an individual acorn! It provides a standardised basis—in terms of 'energy equivalence'—for the comparison of structurally dissimilar organic communities and/or physical habitats. Finally, man is, and can be treated as, an integral part of the ecosystem—indeed the most important ecological factor—rather than as an unnatural biological accident.

References

ALLRED, B. W. and CLEMENTS, E. S. (Eds). 1949. *Dynamics of vegetation.* Selections from writings of Frederick E. Clements, Ph.D. H. H. Wilson Co., New York.

ANDERSON, M. S. 1951. *A geography of living things.* English Universities Press, London.

BARROWS, H. H. 1923. Geography as human ecology. *Ann. Ass. Am. Geogr.*, **13**: 1–14.

BIROT, P. 1960. *Cours de biogéographie.* C.U.D., Paris.

BIROT, P. 1965. *Formations végétales du globe.* S.E.D.E.S., Paris.

BRAUN-BLANQUET, J. 1932. *Plant sociology.* Translated by H. S. Conrad and G. D. Fuller. McGaw-Hill, New York.

CAIN, S. A. 1944. *Foundations of plant geography.* Harper, New York.

CHEVALIER, A. and CUENOT, L. 1927. *Biogéographie*, Vol. III in *Traité de géographie physique*, (Ed. de Martonne) 4th Ed. Armand Colin, Paris.

CLEMENTS, F. E. 1916. *Plant succession.* Carnegie Institute of Washington.

COWLES, H. C. 1899. The ecological relations of the vegetation on the sand dunes of Lake Michigan, Pt. I-Geographical relations of the dune floras. *Bot. Gaz.*, **27** (1): 95–117, 167–202, 281–308, 361–391.

CROIZAT, L. 1952. *Manual of phytogeography.* Junk, The Hague.

CROWTHER, J. G. 1969. von Humboldt: explorer of a new world. *New Scient.*, 11 Sept: 534–536.

DANSEREAU, P. 1957. *Biogeography: an ecological perspective.* Ronald, New. York.

DAVIS, W. M. 1899. The geographical cycle. *Geogrl J.*, **14** (5): 481–504.

DE CANDOLLE, A. 1885. *Géographie botanique raisonnée en exposition des faits principaux et des lois concernant la distribution des plantes d'époque actuelle.* Masson, Paris.

DE LAUBENFELS, D. J. 1964. *Plant geography and vegetation geography* (mimeo). Syracuse University, New York.

DE LAUBENFELS, D. J. 1968. The variation of vegetation from place to place. *Prof. Geogr*, **20** (2): 107–111.

DRUDE, O. 1897. *Manual ae géographie botanique.* Klincksiech, Paris.

EDWARDS, K. C. 1964. The importance of biogeography. *Geography*, **49** (2): 85–97.

EGLER, F. E. 1942. Vegetation as an object of study. *Philosophy Sci.*, **9**: 245–260.

ELTON, C. S. 1958. *The ecology of invasion by plants and animals.* Methuen, London.

ELTON, C. S. 1966. *The pattern of animal communities.* Methuen, London.

EYRE, S. R. 1968. *Vegetation and soils: a world picture*, 2nd ed. Edward Arnold, London.

GAUSSEN, H. 1954. *Géographie des plantes.* 2nd ed. Armand Colin, Paris.

GOOD, R. 1964. *The geography of the flowering plants*, 3rd ed. Longmans Green, London.

GLINKA, K. D. 1927. *The great soil groups of the world and their development.* Translated from the German by C. F. Marbut. Ann Arbor, Michigan.

GREIG-SMITH, P. 1964. *Quantitative plant ecology*. 2nd ed. Butterworths, London.

GRISEBACH, A. 1877–8. *La végétation du globe aprés la disposition suivant les climats*. Ballière, Paris.

HARDY, M. E. 1913. *A junior plant geography*. Clarendon Press, Oxford.

HARDY, M. E. 1920. *The geography of plants*. Clarendon Press, Oxford, 1920.

HERBERTSON, A. J. 1905. The major natural regions: an essay in systematic geography. *Geogr. J.*, **25** (3): 300–312.

HUMBOLDT, A VON and BOUPLAND, A. 1805. *Essai sur la géographie des plantes: accompangé d'un tableau physique des régions equinoxiales*. Paris.

JAMES, E. and JONES, F. 1954. *American geography: inventory and prospect.* Association of American Geographers, Syracuse University Press, N.Y.

KÖPPEN, W. 1918. Klassification der klimate nach temperatur, niederschlag und jahreslauf, *Petermanns Mitt.*, **64**: 193–203, 243–248.

LEMÉE, G. 1967. *Précis de biogéographie*. Masson et Cie, Paris.

MERRIAM, C. H. 1894. Laws of temperature control of the geographic distribution of terrestrial animals and plants. *Nat. geog. Mag.*, **6**: 229–238.

NEWBIGIN, MARION, 1948. *Plant and Animal Geography*. 2nd ed. Methuen,

OZENDA, P. 1964. *Biogéographie végetales*. Editions Dion, Paris.

PEARSALL, W. H. 1964. The development of ecology in Britain (The Jubilee address), in *British Ecological Society Jubilee Symposium*, 1963. Edited by A. Macfadyen and P. J. Newbould. Blackwell Scientific Publications, Oxford.

POLUNIN, N. 1960 *Introduction to plant geography and some related sciences.* Longmans, London.

SCHIMPER, A. W. F. 1903. *Plant geography on a physiological basis.* Translated by W. R. Fisher, revised and edited by Percy Groom and I. B. Balfour. Clarendon Press, Oxford.

SINKER, C. A. 1964. Vegetation and the teaching of geography. *Geography*, **49**: 105–110.

STODDART, D. R. 1965. Geography and the ecological approach: the ecosystem as a geographic principle and method. *Geography*, **50**: 242–251.

SORRE, MAX. 1948. *Les fondements de la géographie humaine*. Vol. 1, *Les fondements biologiques*. Armand Colin, Paris.

TANSLEY, A. G. 1920. The classification of vegetation and the concept of development. *Ecol.*, **8**: 118–149.

TANSLEY, A. G. 1937. *The British Islands and their vegetation.* Cambridge University Press,

WARMING, E. 1909. *Oecology of plants*. Clarendon Press, Oxford.
WEAVER, J. E. and CLEMENTS, F. E. 1938. *Plant ecology*. McGraw-Hill, New York.
WILHELM, E. J. 1968. Biogeography and environmental science. *Prof. Geogr.*, **20**: 123–125.
WYNNE-EDWARDS, V. C. 1962. *Animal dispersion in relation to social behaviour*. Edward Hafner, London.
WYNNE-EDWARDS, V. C. 1964. Population control in animals. *Scient. Am.*, August.

2
The Organic World

The organic world or 'biosphere' is that part of the earth which contains living organisms—the biologically inhabited soil, air, and water. In comparison with either the lithosphere or atmosphere it is a relatively shallow zone. It reaches its greatest depth in the oceans, extends no more than two or three metres below the land surface, and occupies mainly the lower hundred metres of the atmosphere. But for all its limited extent it is densely populated—teeming with myriad forms of life, bewildering in their variety and complexity. Its inhabitants range from plants and animals of microscopic size and simple one-celled structure to those of considerable proportion and complex form. That most readily visible part of the biosphere—composed of the larger plants and animals—is but a small percentage of the whole. A conservative estimate might put the total number of different species of plants at nearly half a million, of animals three to four times as many, while the number of individuals in any one part of the biosphere can attain staggering proportions. It has, for example, been estimated that in a gram of rich top-soil there may be as many as 100 000 algae (sixteen million moulds and other fungi, and perhaps several billion bacteria; the average number of animals living on an acre of forest floor in Panama has been found to be of the order of forty millions—and this does not include bacteria, fungi or other plants!

The biosphere is, however, more than just a collection of living organisms. No living creature exists in complete isolation and between all the inhabitants of the biosphere there is a complex interdependence and interaction. Animals cannot exist without plants nor plants without animals. Animals depend on plants or other animals for food. Any animal or plant will be affected directly or indirectly by the activities of the other animals and plants around it; in some cases such activities are mutually beneficial, in others antagonistic. Many plants rely on animals for pollenation or seed dispersal, while plants frequently provide shelter and protection for animals. Both plants and animals compete with each other, and among themselves, for space and food. Further, the biosphere could not exist nor can it be considered independently of the inanimate inorganic environment of land, air, and water within which it exists. It is wholly dependent on this environment for light, water, oxygen, carbon dioxide, nitrogen and other mineral nutrients necessary for the very existence of life. Conversely the nature of the inorganic environment is profoundly affected by the existence and interaction of the organisms which inhabit it. There is, in fact, a continual exchange of materials between the organic and inorganic parts of the biosphere. Plants take up inorganic elements, build them into organic substances which eventually are returned to the physical environment by the processes of growth and decay in both plants and animals.

In the complex web of interrelationships which exist between the various members of the biosphere and between them and their environment, plants, and more particularly *green* plants occupy a key position. They do so by reason of their ability, unique among living organisms, to use solar or light energy to manufacture from such simple inorganic elements as carbon, hydrogen and oxygen highly complex organic substances which, in turn, provide the food-energy necessary not only for their own growth but for the maintenance of all other forms of life on earth. This process of *photosynthesis* (sometimes referred to as carbon assimilation or primary biological production) can be summarised simply in the following manner:

$$\begin{bmatrix}\text{Carbon dioxide}\\ CO_2\\ +\text{ water}\\ H_2O\end{bmatrix} \rightarrow \begin{bmatrix}\text{light absorbed by}\\ \text{chlorophyll contained}\\ \text{in cells of green}\\ \text{parts of plants}\end{bmatrix} \begin{matrix}\text{carbo-}\\ \text{hydrate} + O_2\\ (CH_2O)\end{matrix}$$

The carbohydrates (for which CH_2O is a simplified formula) provide the basic 'building blocks' from which other food-stuffs such as proteins and fats can, in combination with such essential mineral nutrients as nitrogen, phosphorous, potassium etc., be produced. Of the light energy absorbed by the green plant about a sixth is used during the process of photosynthesis; the remainder is converted into potential chemical or food energy bound into the plant tissues themselves. This produces the energy necessary on the one hand for plant metabolism and for herbivorous animals on the other. The potential food energy is released in the reverse process of respiration, in both plants and animals, during which oxygen is used and carbon dioxide produced. Photosynthesis harnesses solar energy, respiration makes it available for all living functions. Photosynthesis moreover constitutes the basic and fundamental difference between most plants and all animals. The nutrition of green plants is, as a result of this process, *autotrophic* or wholly dependent on inorganic materials while that of animals is *heterotrophic* or wholly dependent on the organic food materials produced *initially* by plants. And until man can efficiently and economically duplicate the process of photosynthesis outside the green plant he is no less dependent on plant food than is the lowliest type of animal.

It is perhaps relevant at this stage to point out that the necessity for plants to carry out photosynthesis under a wide range of environmental conditions has played a major rôle in the evolution of the diverse forms of plant life that exist today. Plants obtain light and carbon dioxide by absorption through their surface and their efficiency in this respect will be dependent on the area of their absorbing surface in relation to their total bulk. Fossil records leave little doubt that plant life originated in the oceans; the remarkable similarity of chemical composition between sea water and that of the living cell has elicited the remark by one biologist that 'land and fresh water organisms are packets of sea water that have found ways of maintaining themselves in different environments' (Bates). Certainly the oceans provide a much less extreme or variable medium for photosynthesis and growth than does the land. In the sea the essentials of carbon dioxide, water, and mineral nutrients are combined in one enveloping medium, while on land they must be drawn from the two separate sources of soil and air. In the sea the plant is entirely surrounded by water which is much richer in dissolved carbon dioxide than the atmosphere. It is, therefore, not altogether surprising that the plant life of the oceans is the most prolific form of life on earth. It consists, however, almost entirely of the lower and more primitive (in the evolutionary sense) types of plants; the bulk is composed of microscopic uni-cellular organisms—the plant *plankton*—which float in the upper light-absorbent layers of the sea. Their minute size ensures a very large absorbing surface in relation to their volume. In the light of the efficiency of this simple form for photosynthesis and the remarkable uniformity of the environment, the absence of a greater variety of plant life in the sea is perhaps understandable.

On the other hand, the evolution of land plants required successful adaptation not only to a different but to a much more variable set of environmental conditions. The amount of carbon dioxide in the atmosphere remains relatively constant at 0·03 per cent.—much less than that dissolved in the sea. Plants growing on the land are subjected to much greater variations of temperature, from place to place and from season to season, than in the oceans. There is, in addition, a drastic and fundamental separation of the atmospheric environment, from which the plant obtains light and carbon dioxide, and the soil from which water and mineral nutrients must be drawn. It is necessary that the land plant be fixed in one spot and have the means of extending into and extracting water from the sunless area of the soil. It must also be tall enough to attain sufficient light in competition with other plants, and at the same time have as large a photosynthetic area in relation to volume as possible. The elaboration of the external form by branching and the sub-division of leaves has provided a means of maintaining maximum photosynthesising surface with increasing size. While in the sea large size would not be of any great advantage, on land, competition for light among immobile plants would obviously give those with greater size, and particularly height, an advantage. Height, however, creates problems associated with the passage of water and mineral nutrients from the soil to the photosynthesising organs, and, lacking the buoyancy of marine plants, with the mechanical support of even longer stems. Also, the necessity for the absorbing cells in the leaves to be protected from desiccation by an impermeable 'skin' or 'cuticle', while at the same time allowing carbon dioxide entry through minute openings called *stomata,* inevitably exposes the land plant to the loss of water by evaporation. The problems of attaining maximum photosynthesis on the land have therefore created those of maintaining an adequate water supply in the plant. The variety and complexities of form which have developed in the higher cone-bearing (*Gymnosperm*) and seed-bearing (*Angiosperm*) plants have enabled them to cope most successfully with the problems of attaining maximum efficiency of photosynthesis in competition with other plants under a diverse range of physical conditions; and have enabled them to become the dominant plants in the terrestrial vegetation cover today.

By the process of photosynthesis practically all but a very minute proportion of the energy necessary to sustain life enters the biosphere. The green plant is then the vital energy 'trapper and converter' and on its efficiency, and that of the earth's vegetation cover in this respect, depend the amount of primary food that the earth can produce and the amount of animal life it can support. And this has become a matter of increasing concern in a world where the pressure of a spiralling human population is making ever increasing demands on the biosphere. It has been estimated that of the solar radiation reaching the earth's surface most is expended in evaporation and in the heating of the atmosphere and the earth's surface. Only about one per cent. of the total light or about two per cent. of that of suitable wave-length (i.e. the visible white light which is absorbed by the chlorophyll) is involved in the process of photosynthesis. Of the total or gross amount of carbohydrate produced some is used in plant metabolism, the remainder which is built up into new tissues represents the *net photosynthetic production* over a period of time. The ratio between the energy value of the organic matter produced and that of either the light available (or that actually absorbed) is a measure of the plant's photosynthetic efficiency. The most recent and frequently quoted estimates suggest that the mean annual net photosynthetic efficiency for the earth's surface is somewhere in the order of 0·1 to 0·2 per cent, depending on whether calculations are based on total light available or that of suitable wave-length. While this would appear to be a very low order of efficiency it has, nevertheless, been suggested that it represents an annual conversion of over two hundred billion tonnes of carbon from carbon dioxide into sugar (equivalent to about a hundred times the combined weight of the world's chemical, metallurgical and mining industries!).

However any estimate of the mean annual global efficiency of photosynthesis must obviously obscure wide variations from one part of the earth's surface to another. There are deserts and ice caps where solar radiation cannot be utilised because either water or temperature conditions are insufficient for plant life; there are other areas where they limit or severely restrict photosynthesis for part of the year. On the other hand annual net efficiencies of two to five per cent. have been recorded for such crops as sugar beet, sugar cane and maize and of between 0·3 and 2·7 per cent. in British forestry plantations. It has further been suggested that a good, well-fertilised, well-watered field crop might be expected to have a photosynthetic efficiency as high as seven to ten per cent. at its period of maximum growth. Certain algal cultures grown under laboratory conditions have indicated that at low light intensities, efficiency of energy conversion of as high as twenty-five per cent. can be obtained, but the actual yield of organic matter in terms of quantity produced per unit area per unit time is low.

Expressions of photosynthetic efficiency vary according to the measurements on which the ratio between 'light' and 'food' energy are based. The former may be total incident light, that available to or that actually absorbed by plants; the latter may be expressed in terms of either *gross* or *net* plant production. Further, expressions of efficiency will depend on the unit of time during which the amount of photosynthesis is calculated, be it annual, growing season, monthly or daily. For these reasons most ecologists prefer to employ either gross or net primary production rather than efficiency when studying the conversion of solar radiation to plant material. The *productivity* of the photosynthesising plant or mass of vegetation is the *rate* at which energy is fixed. It can be measured either in terms of the rate of carbon dioxide absorption or oxygen production, or of the amount of organic matter 'manufactured'. The latter is the easiest to determine and the most frequently used. It can be expressed as: (1) *the dry weight* of; or (2) the amount of *carbon assimilated* by; or (3) the *energy equivalent* of the plant material produced per unit area per unit time. The amount of photosynthesis carried out by a plant depends on the rate of assimilation which is governed by such factors as light intensity, amount of carbon dioxide, availability of water and temperature conditions, and on the photosynthesising area (the combined leaf and green stem) capable of 'trapping' available light. Productivity will be greatest when optimum conditions for photosynthesis and maximum environmental-absorbing leaf area coincide. In this respect, and in terms of the sheer weight of organic matter produced, many crop plants lack the photosynthetic efficiency of natural vegetation; although their yield may be high it may be no higher and may even be less than that of say a woodland cover growing under similar environmental conditions. In the case of many crop plants their maximum photosynthesising area persists for a relatively short period of time. At the beginning of the growing season they cover only a small proportion of the ground area and hence do not make full use of all the available light. Later, as their foliage becomes denser, mutual shading reduces the efficiency of the total area. Further, the period when their photosynthetic area is greatest may not coincide with the period when climatic conditions are at a maximum for the process of assimilation: 'to obtain highest possible yields in terms of dry weights from annual crops, therefore, one should use large-seeded species which are able to produce leaves quickly following germination and sow them at such a time as to ensure that maximum leaf area is produced at the time of year most favourable to photosynthesis' (Fogg, 1963) is the obvious agricultural moral. In this respect it is worth noting that among the most efficient and productive crop plants are maize and sugar cane. They possess relatively broad, vertically or near-vertically set leaves which provide the maximum photosynthesising area to optimum light intensities with the minimum amount of mutual shading.

Variation in plant productivity with latitude and hence with growing season is illustrated by the following estimates of the general level of net *above ground* production of dry matter in metric tons per hectare per year, in certain areas:

Arctic	0·03
Sub-arctic	2·5– 4·0
Britain	4 –16
Tropical Rain Forest	20 –50
Bamboo Forest	50
Sugar Cane	70

A comparison of the production of organic matter in terms of the amount of carbon assimilated per year by the major types of world plant-cover reveal some interesting variations (Fig. 1). The productivity of the sea despite its prolific life and greater area is appreciably lower than that of the land. This has been attributed to the proportionally greater loss of material in respiration by the minute plant-plankton than by land plants, together with the fact that a higher proportion of the incident light is absorbed and reflected by water than by the atmosphere. In addition, it has been pointed out that plankton growth frequently depletes mineral nutrients in the superficial layers of the sea faster than they can be replenished and to amounts which limit production.

The greater photosynthetic productivity of forest in comparison even to cultivated land is now a well-known phenomenon; and under similar environmental conditions, that of evergreen conifers exceeds the production by deciduous trees. In the former the evergreen habit allows photosynthesis to proceed, albeit very slowly, when deciduous trees are leafless, and to take immediate advantage of the onset of favourable conditions when deciduous species have still to expend considerable energy producing new photosynthetic organs. Similarly 'warm' deserts are twice as productive as the cold 'tundra' deserts; in the latter both deficiency of light and temperature curtail the growing season in the winter, while the slow-growing perennials or rapidly growing annuals, adapted to the limited water supplies of the warm deserts, can make use of high light intensity at any time of the year. E. P. Odum recognises three orders of magnitude of productivity: *low*:

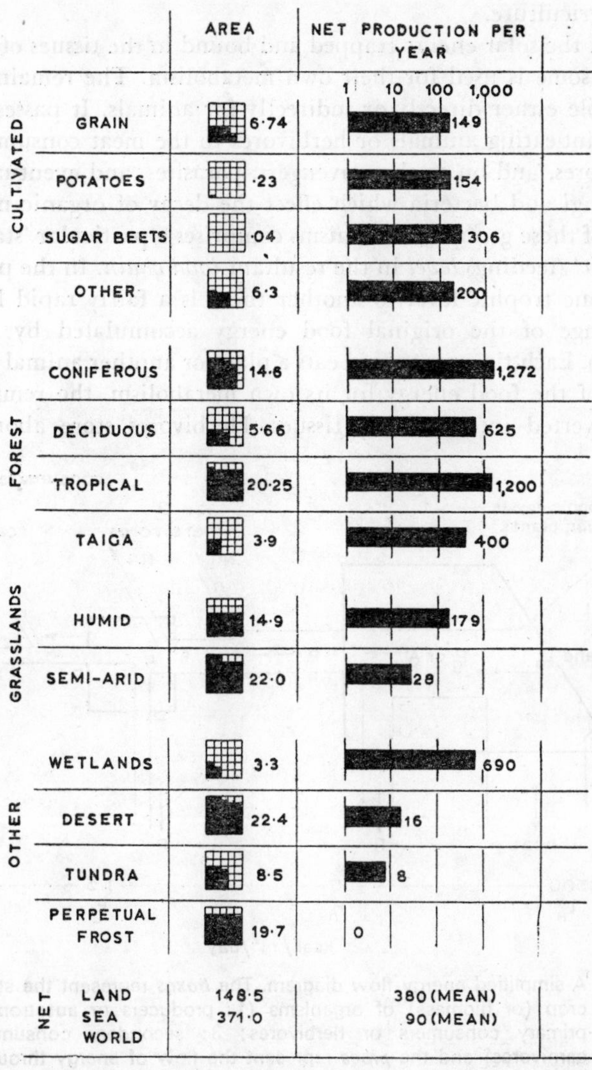

Fig. 1 Production of organic matter by the land vegetation of the world: *area* (in millions of square kilometres) occupied by given type of vegetation; *net production per year* expressed in grams of carbon per square metre incorporated in organic compounds; net production is the gross production of carbon compounds minus that consumed during the process of plant respiration. (Adapted from Deevy, Edward S, 1960)

less than one gram/m²/day characteristic of deserts, deep oceans and lakes; *medium*: one to 10 grams/m²/day—grasslands, coastal seas, shallow lakes and 'average agriculture; and *high*: 10-20 grams/m²/day—shallow water systems, moist forests and 'intensive' agriculture.

Of the solar energy trapped and bound in the tissues of green plants some is used for their own metabolism. The remainder is available either directly or indirectly for animals. It passes from the plant-eating animals or herbivores to the meat consumers or carnivores, and on to the scavengers, parasites, and eventually to the fungi and bacteria which effect the decay of organic matter. Each of these groups of organisms comprises a particular 'stage' or '*trophic*' (feeding) *level* in the resultant *food chain*. In the passage from one trophic level to another there is a fairly rapid loss or shrinkage of the original food energy accumulated by plants (Fig. 2). Each time an animal eats a plant or another animal it uses some of the food energy for its own metabolism, the remainder is converted into its body tissues. Herbivores store about ten

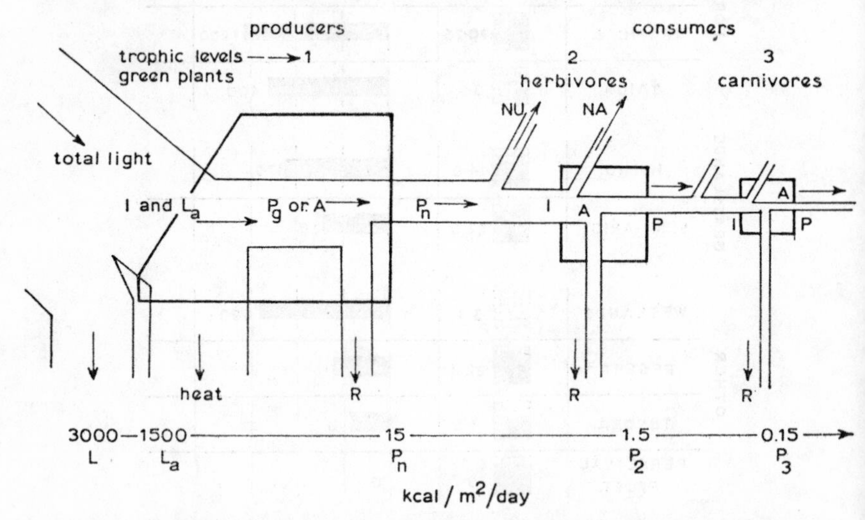

Fig. 2 A simplified energy-flow diagram. The *boxes* represent the standing crop (or biomass) of organisms (1: producers or autotrophs; 2: primary consumers or herbivores; 3: secondary consumers or carnivores) and the *pipes* represent the flow of energy through the biotic community. L = total light; L_A = light absorbed; P_g = gross production; P_n = net production; I = energy intake; A = assimilated energy; NA = non-assimilated energy; NU = used energy (stored or exported); R = respiratory energy loss. The chain of figures along the lower margin of the diagram indicates the order of magnitude expected at each successive transfer of energy starting with 3,000 kcal of incident light per m² per day. (After Odum, E. P. 1963)

per cent. of the energy provided by plants and carnivores about ten per cent. of that provided by their prey. Food chains can, as a result, rarely be composed of more than four or five links. Of the primary plant production energy, some passes directly via surface-living herbivores to carnivores and their prey and parasites: the remainder (together with the faeces and bodies of these animals) provides, in the form of dead and decaying organic matter, an important source of energy for saprophytic soil organisms. The relative importance of what have been designated the 'grazing' food chain, as in the first instance, and the 'detrital' chain, as in the later, varies from one part of the biosphere to the other. In the sea the former is more important than on the land where, in contrast, a relatively small portion of the living plant biomass is consumed by surface herbivores. While the stages by which energy passes from plant to animal forms a chain-like sequence, the feeding relationships are, in reality, rarely simple and linear. One type of plant may provide food for several different types of animals, both herbivorous and saprophytic; carnivorous animals usually have a variety of prey; and many animals, not least the omnivores, derive their food from more than one trophic level. The result is a complex *food web* (see Fig. 3).

In terms of human nutrition the implications of 'energy flow' through the biosphere and the resulting conversion losses are obvious and profound. More people can be 'kept alive' from one acre of wheat or rice than from the animal food that could be produced by an acre of grass or alfalfa. Hence the problem of feeding the world's growing population is aggravated by the need and the increasing demand for 'secondary' animal products rather than 'primary' plant foods. There is still room for increased agricultural production by improving the efficiency of existing crops, by irrigation in arid regions, and by the extension of cultivation into areas where crop production is possible though it may not be considered economically feasible at present. However, doubts have been expressed as to whether these combined methods can effect more than a ten-fold increase in food production or at a rate equivalent to the present rate of population increase. The time may come when, in order to support his increased numbers, man will have to shorten his food chains, even perhaps displace all other herbivores who compete with him for plant food, and utilise all available space for edible plants. And a growing concern with this vital problem has stimulated increasing interest in and studies of the *potential* primary productivity of the biosphere. And in 1964 an International Biological Programme (IBP)

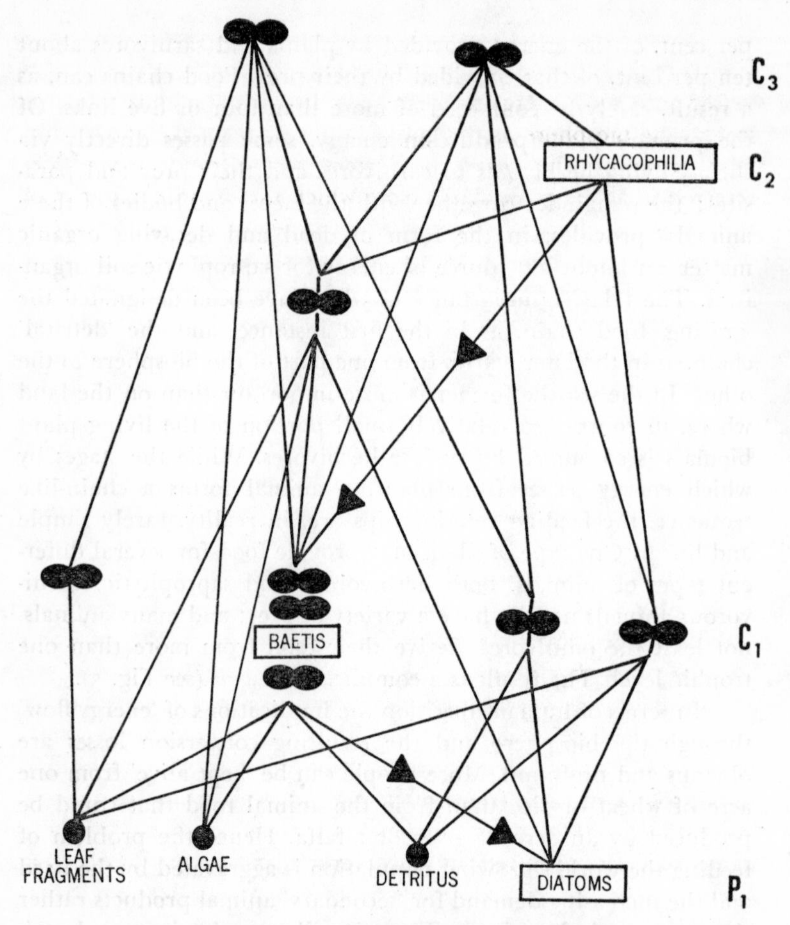

Fig. 3 Diagram showing relationship of food chain *Navicula viridula* (one species of diatom), *Baetis rhodani* (a mayfly) and *Rhycacophilia* sp, (ceaseless caddis) to part of food web in a fresh-water stream in Wales. P = primary producers, C_1, C_2, C_3. . . primary, secondary, tertiary consumers. (From: Phillipson, J. 1966, based on Jones. J. R. W. *J. Animal Ecol.* 1949)

was set up under the auspices of the International Council for Scientific Unions (ICSU). Entitled 'The Biological Basis of Productivity and Human Welfare', its object is 'to promote the world-wide study of (a) organic production on land, in fresh waters and in the seas, and the potentialities and use of new as well as existing resources; (b) human adaptability to changing conditions' in order to reconcile the need for increased food production with that for the rational use and conservation of organic resources.

The flow of energy into the biosphere is one-way, being constantly renewed by incoming solar radiation. The light energy converted into food energy is eventually dissipated as heat which is lost to the biosphere as out-going radiation from the earth's surface. However, the *chemical elements* or *nutrients* (sometimes referred to as 'biogenic salts') from which the organic material of the biosphere is built are continually being used and re-used. Their main point of entry into the biosphere is the green plant from which they move by shorter or longer routes finally to be released to the inorganic environment again by the action of decay-promoting bacteria and fungi. The more or less circular routes by which these elements move from the biosphere to the environment and back again are known as *nutrient cycles or biogeochemical cycles.* Through their operation the organic and inorganic worlds are inextricably bound together. Of the ninety-two known elements occurring in nature some thirty to forty have been identified as essential for living organisms. The macronutrients are those required in relatively large quantities and which play a key rôle in the composition of protoplasm. Of these carbon, oxygen, hydrogen and nitrogen, which together account for over ninety-five per cent. of the dry weight of protoplasm, are obtained either directly or indirectly in gaseous form from the atmosphere. The remainder—mineral nutrients such as phosphorus, potassium, calcium, magnesium and sulphur—have their initial source in rocks from which they are released by the process of weathering. The micro-nutrients or trace-elements are, as their name suggests, required in only very minute, but nonetheless essential amounts; they include iron, manganese, copper, zinc, boron, sodium, molybdenum, chloride, vanadium, cobalt, as well as others which are involved in the geochemical cycles but which, as yet, have no known biological function.

The rate and completeness with which these elements pass from the inorganic environment into plants, through various animals and back again vary. They depend on the element involved, the rate of organic growth and decay, and also on the activities of man. Some, like oxygen, carbon and hydrogen, returned by the processes of photosynthesis, respiration and transpiration, are circulated relatively rapidly and almost completely. In other cases the cycle may be slower or less perfect and some percentage of the particular element may be diverted for varying lengths of time into places or into forms inaccessible to organisms. The more complete and perfect cycles are those of the *gaseous type* whose reservoir is the atmosphere. They include oxygen, carbon, hydrogen

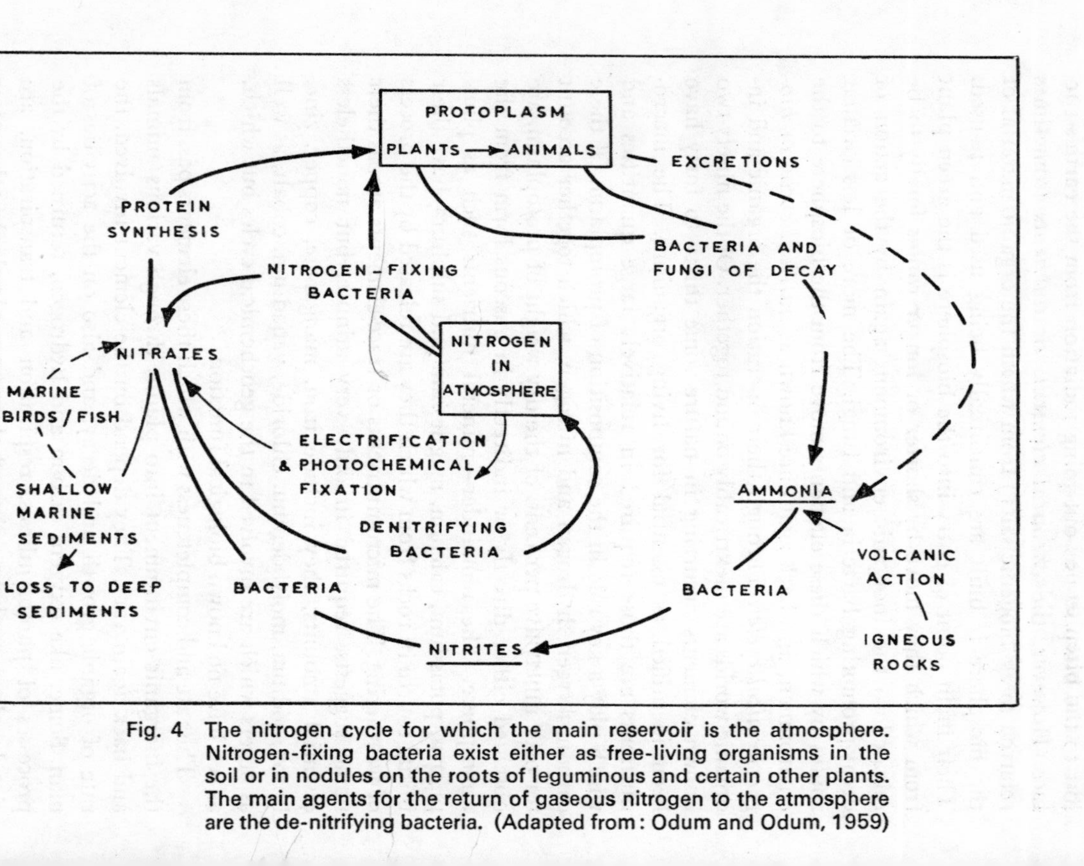

Fig. 4 The nitrogen cycle for which the main reservoir is the atmosphere. Nitrogen-fixing bacteria exist either as free-living organisms in the soil or in nodules on the roots of leguminous and certain other plants. The main agents for the return of gaseous nitrogen to the atmosphere are the de-nitrifying bacteria. (Adapted from: Odum and Odum, 1959)

and nitrogen (Fig. 4) which enter and leave the biosphere in gaseous form. What has been called the *sedimentary type* of cycle involves those mineral nutrients whose source is weathered rock; these tend often to be simpler, but less perfect cycles and as a result more susceptible to disruption particularly through the action of man. Under natural conditions the nutrients produced in rock weathering are absorbed in the soil solution by plants, and returned for re-use when these or the animals which eat them die and decay. The small proportion removed from the soil by leaching is, theoretically, made good by slow but continual rock weathering, supplemented in some instances by volcanic dust. Man, through his agricultural activities has in many instances removed, by way of his crops or animals, more nutrients than he has returned to the soil. In addition, he has in many areas greatly accelerated the processes of soil erosion and leaching. Nutrients so removed from the soil in drainage water may eventually be deposited, along with other products of erosion, on ocean floors. They are then lost to the nutrient cycle until such time as upwelling currents, sub-marine earthquakes, volcanism or major earth movements effect their re-emergence. As a result of man's activities mineral nutrients tend to be removed from or lost to the sedimentary cycles at rates faster than they can be replaced by the very slow process of rock weathering.

Of such an imperfect and precarious nutrient cycle, phosphorus is the classic example (Fig. 5). In comparison to the other elements and in relation to its biological importance, this element is relatively scarce in nature. As though in response to this, organisms have evolved means of accumulating phosphorus in their living tissues at a concentration much greater than occurs in the inorganic environment. The depletion of the phosphorus reservoir by erosion and sedimentation is renewed very slowly by rock weathering, airborne dust, volcanic gases, upwelling ocean currents and fish-eating guano birds. The accumulated droppings of the latter represent the most important and rapid return of phosphorus in the form of phosphate deposits. But this supply is extremely localised, being concentrated primarily on islands off the Peruvian coast. Moreover it accounts for less than three per cent. of that lost from the land. The increasing rate with which the supply of phosphorus is being lost to the nutrient cycles must be of serious consequence for the biosphere and for man. Deficiency of phosphorus is among one of the gravest agricultural problems today.

Not only has man become a recent and powerful agent in

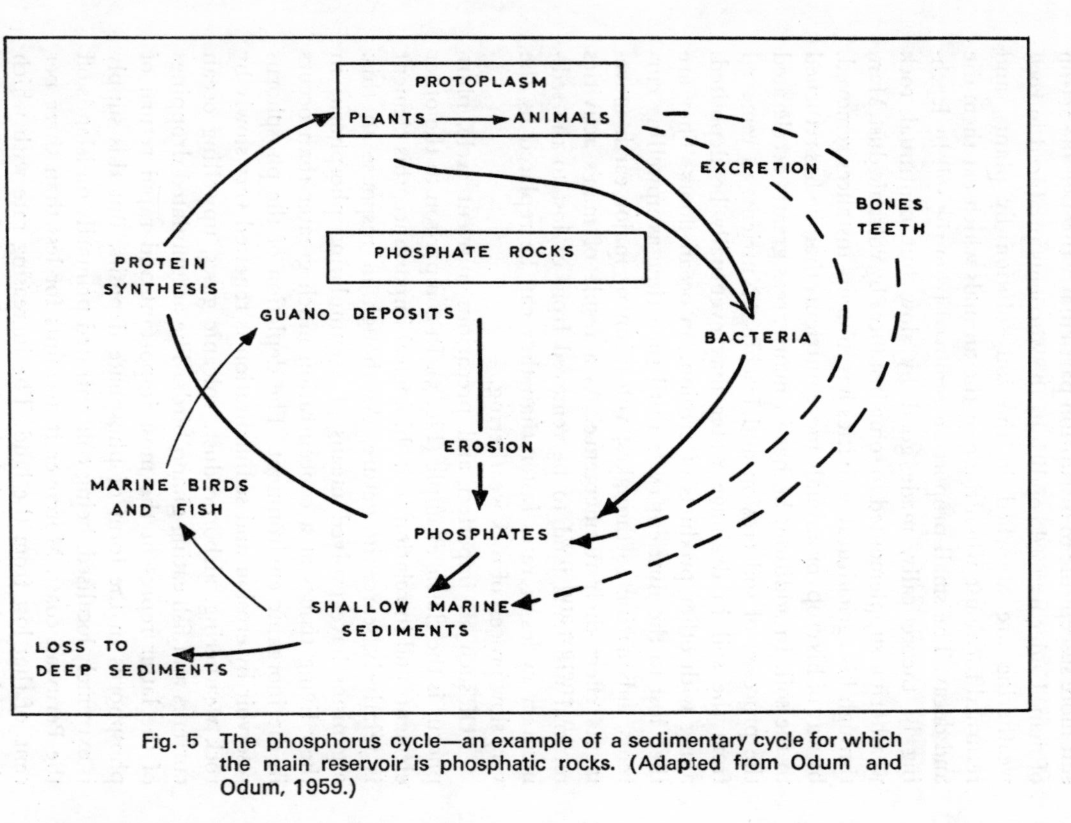

Fig. 5 The phosphorus cycle—an example of a sedimentary cycle for which the main reservoir is phosphatic rocks. (Adapted from Odum and Odum, 1959.)

accelerating the rate of biogeochemical cycles and in their disruption, he has also been responsible for the introduction of 'new' elements into the organic-inorganic cycle. Most important among these are chemical substances contained in herbicides, insecticides, and domestic and industrial effluents. Those which do not decompose rapidly and persist in the air, water, or soil can be absorbed either directly, or indirectly via plants, by animals including man. If they are not lost in the processes of excretion or respiration, they not only remain in the body tissues but increase in concentration as they pass from one trophic level to another. And, as will be discussed later, once incorporated into food chains, they may attain levels of concentration which can be toxic and lethal.

The two fundamental processes of *energy flow* and of *nutrient* or *geochemical circulation* are characteristic of and common to the whole biosphere. The former, as has already been illustrated, is that which forges the plant-animal food relationships; by reason of the latter, the biosphere and its inorganic environment are so inextricably interlocked and so mutually interdependent that it is difficult to separate them one from the other in nature. In both processes the green plant provides the master link—the point at which energy and nutrients enter the biosphere. The biomass is dependent on the amount and rate of energy which flows through it and on nutrient materials which circulate between it and the inorganic environment.

As a result of these two basic principles or, as they have been called by Odum, 'laws of nature', the biosphere together with its inorganic environment, forms what is referred to as a *biotic complex,* an *ecological system* or an *ecosystem.* Such a system or complex includes both organic and inorganic elements, each influencing the properties of the other, and both necessary for the maintenance of life as we know it. Its four basic components include (1) inorganic (or abiotic) substances; (2) producer organisms, the autotrophic or self-nourishing green plants; (3) consumers—the heterotrophic or 'other nourishing' animals; and (4) the reducers or decomposers, chiefly the micro-organisms, the bacteria and fungi which break down and effect the mineralisation of organic matter. Under natural conditions an ecosystem will be able to maintain and perpetuate itself when a balance is achieved between the supply and demand of nutrients and between energy input and energy loss. There will be a relative equilibrium between the numbers of different kinds of organisms and the energy or food supply available. Indeed, as it has been pointed out by Odum, 'ecosystem' is no more than a technical term to express what is meant by 'nature' and its operation as 'the balance of nature'.

Since all the component parts of an ecosystem are so intimately intermeshed it can be easily appreciated that any change in the amount, condition or rate of activity of one, will have repercussions on the whole. An increase or decrease in the amount of any of those substances such as oxygen, carbon dioxide, water or mineral nutrients, or a change in the light or temperature conditions necessary for efficient plant functioning, will directly affect plant growth and indirectly affect the animals dependent on them. For example a decrease in the amount of water below the minimum requirements for a particular type of plant may result in its disappearance from an ecosystem, and with it the animals which feed on it; or these animals may as a result be forced to feed on another plant and thus reduce the amount of food for, and consequently the number of animals who already depended on, the alternative supply, as well as of the herbivores who feed on them and so on. The removal of these plants may also affect the *other* plants in the ecosystem not only by reason of the adjustments in food chains that must be effected but by changing light and temperature conditions in the ecosystem itself and by altering soil conditions as a result of the part played by the original plant in the cycle of nutrients. In other words a change in any one component of the ecosystem necessitates a readjustment of the whole, the re-establishment of a new balance or equilibrium to cope with the new conditions.

The biosphere and its inorganic environment form the largest possible ecosystem—for which the term *ecosphere* has been suggested as meaning 'the sum total of life on earth together with the global environment and the earth's total resources' (Cole). In the past such major environmental upheavals as earth movements, ice ages and other climatic changes are known, from fossil evidence, to have upset, sometimes drastically, the 'balance of nature'; and one part or another of the ecosphere has had, from time to time, to readjust itself to a new set of environmental conditions. Not only have there been (and indeed there continue to be) minor and major fluctuations in the ecosphere but these have been accompanied by evolutionary changes as new forms of plants and animals—often better adapted to the new environmental conditions—have appeared while others, more important in the geological past, have disappeared or become less important. Also, within relatively recent times man has come 'crashing' into the ecosphere, and continues, with increasing intensity, to disrupt the 'balance of nature'. He has become an integral part of the ecosphere, but because of his superior skills he is now the dominant and most aggressive animal. He has in many instances taken more out of the ecosphere than he has returned to it, he has drastically reduced the numbers of certain plants and animals; many have been totally exterminated. As a result he has either changed the inorganic environment or the 'balance of plants and animals' in biotic communities. He has set off chain-reactions which he still cannot properly control because he does not yet know enough about the intricate mechanisms by which organisms are interlinked. He eliminates one pest or useless animals or plant only to find that he has caused a population explosion of another or even worse pest whose numbers in nature were controlled by the very organism he eliminated. Such is the dominance and increasing intensity of man's activities that it is impossible for the ecosphere to adjust itself to the new conditions he is creating. The adjustment must depend on man himself and his ability to use his organic resources in such a way that he will effect maximum production without depleting the capital resources available, and to realise that the intrinsic value of the ecosphere as a resource lies in its capacity to renew itself.

The ecosphere is the largest possible ecosystem: this latter term, however, can be applied to any interrelated group of plants and animals, any biotic community, irrespective of size, together with the environment in which its exists. The physical conditions of the atmosphere and of the surface of the earth vary considerably from one part of the globe to another, and varying combinations of environmental conditions are consequently reflected in differing combinations of plants and animals.

The ecosphere is, in fact, composed of a mesh of interlocking ecosystems. None, however, is isolated or self-contained, each one being linked to the other by an interchange of energy and nutrients effected by such processes as animal movements, and the transport of plant debris or soil nutrients. Nevertheless they can, for convenience, be delimited or described either in terms of the environmental conditions or *habitat* (i.e. 'home') with which they are associated, or the structure, composition and functions of their biotic components. The terms 'habitat' and 'environment' are often used interchangeably—the latter in a more general sense, the former more specifically with respect to a type of physical environment occupied by a group of organisms. The primary and basic distinction between marine and terrestrial habitats is reflected in two of the world's major and contrasting ecosystems.

Within the latter there is obviously an infinite variety of physical habitats dependent on variations of climate, rock and soil type, land form and so on, which can be defined on the basis of one or more attributes such as cold or warm, dry or wet, arctic, desert or tropical, sandy, lime-rich, dune, scree, flood-plain or mountain. The associated ecosystems may be extensive: of continental or larger proportions, or, in the case of localised *micro-habitats* be confined to a rock ledge, a patch of damp soil, a rotting tree stump.

In the ecological sense the concept of a habitat involves not only the physical conditions of a particular site but the organisms which occupy it; it implies the sum total of both physical and biological conditions. Without the latter one may have a potential habitat: an unoccupied site, a 'house' but not a 'home'! The presence of organisms modifies the physical conditions in which they exist, they create a particular environment. As a result each organism forms an integral part of the environment of every other one. Indeed for many animals the essential character of the habitat is dependent on the plants present from which they obtain food, shelter, support: their actual home may be on or in the tissues of a particular plant.

Plants, however, are more directly dependent upon and affected by, the nature of the physical site; and in turn they exert a greater environmental effect than do animals. They determine the type and volume of the latter. As a result plants play the major and fundamental rôle in the composition, structure, and function of most ecosystems; and because of their greater total mass or volume (biomass) they determine the 'appearance' of biotic communities. For these reasons terrestrial ecosystems, in particular, are frequently characterised (or classified) on the basis of their plant components, i.e. as forest, grassland, moorland, etc.

The distribution of different types of ecosystems is dependent on the spatial variation of a number of inter-acting, variable, ecological factors. The latter, in the words of A. G. Tansley, are 'any substance force or condition affecting either directly or indirectly' the behaviour and distribution of organisms. They can, for convenience, be classified according to whether they are a condition or function of the atmosphere (climatic), the soil (edaphic), plants and animals (biotic) or of man (anthropogenic). None, however, operates independently: each influences the other, and it is their combined inter-action which determines the character of the biosphere. The three succeeding chapters will attempt to analyse the effect of these major groups of ecological factors on plant distribution.

References

AUCHTER, E. C. 1939. The inter-relation of soils and plant, animal and human nutrition. *Science, N.Y.*, **89** (2315): 421–427.

BATES, M. 1964. *Man in nature*. Prentice Hall, New Jersey.

BILLINGS, W. D. 1964. *Plants and the ecosystem*. Macmillan, London.

COLE, LAMONT, C. 1958. The ecosphere. *Scient. Am.*, April.

CORNER, E. J. H. 1964. *The life of plants*. Weidenfeld and Nicholson, London.

DEEVY, E. S. 1960. The human population. *Scient. Am.*, September.

DICE, L. R. 1952. *Natural communities*. University of Michigan Press, Ann Arbor.

EVANS, F. C. 1956. The ecosystem as the basic unit in ecology. *Science, N.Y.*, **123** (3208): 1127–1128.

FOGG, G. E. 1957. Actual and potential yields in photosynthesis. *Advmt. Sci., Lond.*, **58**, 14(57): 395–400.

FOGG, G. E. 1963. *The growth of plants*. Penguin Books, Harmondsworth.

GALSTON, A. W. 1964, *The life of the green plant*. 2nd ed. Prentice Hall, New Jersey.

GATES, D. M. 1965. Energy, plants and ecology. *Ecology*, **46** (1 and 2): 1–13.

GATES, D. M. 1968. Toward understanding ecosystems. *Adv. ecol. Res.*, **5**: 1–34.

HUTCHINSON, G. E. 1949. On living in the biosphere. *Scientific Monthly*.

KORMONDY, E. J. 1965. *Readings in ecology*. Prentice Hall, New Jersey.

LINTON, D. L. 1965. The geography of energy. *Geography*. **50**: 197–228.

MACFADYEN, A. 1948. The meaning of productivity in biological systems. *I. Anim. Ecol.*, **17**: 75–80.

NEWBOULD, P. J. 1963. Production ecology. *Sci. Prog. Lond.*, **51** (201): 91–104.

NEWBOULD, P. J. 1964. Production ecology and the international biological programme. *Geography*, **49** (2): 98–104.

ODUM, E. P. 1963. *Ecology*. Holt, Reinhart and Winston, London.

ODUM, E. P. and ODUM, H. T. 1959. *Fundamentals of ecology*, 2nd ed. Saunders, Philadelphia and London.

OLSON, J. A. 1964: Gross and net production of terrestrial vegetation, in *British Ecological Society Jubilee Symposium*, 1963. Ed. by A. Macfadyen and P. J. Newbould, 99–118. Blackwell Scientific Publications, Oxford.

PEARSALL, W. H. 1959. Production ecology. *Sci. Prog., Lond.*, **47** (185): 106–111.

PHILLIPSON, J. 1967. *Ecological energetics*. Studies in biology, No. 1. Arnold, London.

Reid, L. 1962. *The sociology of nature*. Penguin Books, Harmondsworth.

Rowe, J. S. 1961. The level-of-integration concept and ecology. *Ecology*. **42** (2): 420–427.

Stoddart, D. R. 1965. Geography and the ecological approach: the ecosystem as a geographic principle and method. *Geography*, **50**: 242–251.

Tansley, Sir A. 1954. *An introduction to plant ecology*. Allen & Unwin, London.

Wassink, E. 1959. Efficiency of light energy conversion in plant growth. *Pl. Physiol., Lancaster*, **34** (3): 356–61.

Westlake, D. F. 1963. Comparisons of plant productivity. *Biol. Rev.* **38** (3): 385–425.

3
Atmospheric Factors

The atmosphere is essential to all life; the absence of a comparable combination of gases is the main factor precluding the possibility of life as we know it from other parts of the universe. Indeed, it is now thought that the present composition of the earth's atmosphere may well be a consequence of the very evolution of life itself. Not only is the atmosphere a renewable source of essential materials for plant growth but it acts as a 'filter' through which sunlight or radiant energy reaches the earth's surface and it provides, as a result, an insulating or thermostatic medium without which variations between day and night temperatures would be too extreme for the survival of any known form of plant or animal. Further it is the agency by which water is circulated through and distributed to the biosphere.

Plants depend directly on the atmosphere for certain fundamental materials, and conditions, necessary for their successful growth and reproduction. In its gaseous composition the *free* atmosphere is remarkably uniform and constant so that while oxygen and carbon dioxide are essentials, their proportions do not vary significantly enough to make marked differences in plant growth or distribution. On the other hand the actual or average physical state of the atmosphere—the weather or climatic conditions—varies considerably in time and place. Since different species of plants vary in their minimum requirements for, or in their tolerance of, particular climatic conditions, these conditions play a major rôle in determining where a particular plant can or cannot exist. The factors of greatest importance in this respect are those such as light, temperature and humidity, all of which are essential and all of which vary in amount or intensity from one part of the biosphere to another. All these factors, however, interact one with the other and operate in *combination* to produce those atmospheric conditions which will either permit the presence of certain plants in or exclude them from a particular habitat. The condition of any one of these factors will obviously have a direct effect on that of the others; light intensity and duration will influence temperature conditions, the temperature and humidity conditions of the atmosphere are interrelated, humidity conditions affect light intensity, and so on. As has already been indicated these climatic factors may be modified by the edaphic and biotic factors of any habitat. For this reason it is difficult, if not impossible and in many respects unrealistic, to try to isolate the independent effect of one particular climatic factor on plant growth or function. However, by means of controlled experiments much has been learned about the response of different kinds of plants to the effect of individual variable factors. Such efforts, and others, however limited and incomplete, are essential to a fuller understanding of the way in which these interacting variables operate under natural conditions, and to an assessment of the relative importance of the individual climatic factors for plant growth.

Light is necessary for photosynthesis, the process which provides the energy for all other plant functions, and, other requirements being satisfied, the amount of solar radiation that the green plant can utilise will set the limit to the maximum quantity of plant growth and production possible. Light, however, varies in quality, intensity and duration. That part of the solar spectrum absorbed by chlorophyll is the visible white light; infra-red and ultra-violet light are not utilised in photosynthesis, and their effect in other ways on plant life is still obscure. Ultra-violet light is known to be harmful to bacteria and is believed to exert a retarding effect on vegetation development. Much, however, is absorbed in the upper regions of the atmosphere and only a very small percentage reaches the earth's surface. Its effect is therefore greatest at high altitudes, where it is thought to be responsible for the shortened stems and, consequently, the flattened, rosette, leaf-form characteristic of many herbaceous plants which occupy such habitats. In general, however, it would appear that the quality of light, and particularly of the white light, is not so variable from one part of the biosphere to another as to be an important ecological factor.

Of much greater significance is variation in light intensity. This is, in the first instance, a function of the physical condition of the atmosphere, since water vapour, clouds, dust and certain gases all tend to decrease, by absorption or reflection, the amount of solar radiation reaching the earth's surface. This effect is further intensified in middle and high latitudes where the sun's rays must pass through a thicker layer of atmosphere than in tropical regions. The highest light intensities will be experienced in regions where the sun is most nearly overhead and the air is clear and dry. In high and mid latitudes there is, in addition, a more marked seasonal and diurnal variation in intensity, dependent on the changing angle at which the sun's rays strike the surface, than in tropical regions. Locally, the configuration of the land surface may result in pronounced differences in the light intensities received by north and south-facing slopes. Even within one particular habitat, the interception of light by the plants growing there gives rise to variations of light intensity within a collection of plants or even within one individual. Of the light falling on green leaves, it has been estimated that on average twenty per cent. is reflected, twenty per cent. is radiated as heat, forty-nine per cent. is used in evapo-transpiration and only about ten per cent. is transmitted through the leaf. The exact proportion is dependent on colour, thickness and anatomy. Hence plants growing in the shade of others have their light intensity very considerably reduced, while in any individual plant there will be mutual shading by the component leaves. In this latter respect it is interesting to note that the tapered 'conical' form of many coniferous trees is such as to permit the maximum exposure of all parts and cuts down the degree of leaf shading that occurs in trees with wider and denser crowns. The reflection and absorption of light by water is high: even when it is clear, only about fifty per cent. of light impinging on the surface penetrates to a depth greater than ten to twelve metres. A rapid decrease in light intensity therefore limits the possibility of photosynthesis to a maximum depth of about fifty metres in marine habitats.

For each species of plant there is a minimum light intensity essential for growth: this is the percentage of full daylight necessary for photosynthesis to produce new food material at a rate greater than it is being used up in the process of respiration. Similarly there is an optimum intensity beyond which the rate of photosynthesis decreases and increased light intensity may be deterimental or harmful to the plant. For most plants however the optimum is well below the maximum potential light intensity. Minimum and optimum requirements for light intensity vary from one species to another. There are those which can only grow and attain full development in light of low intensity, others which require bright conditions. The minimum requirements of the simple uni-cellular algae, such as plant plankton, is much lower than that for the higher land plants. They can survive where light conditions would be too weak for the latter; some deep-water algae, and also the algae and mosses which inhabit dark caves can grow in light intensities no greater than that of moonlight. Those plants which require high light intensity for optimum growth are the sun-loving plants (or *heliophytes*), in contrast to the shade-loving plants (or *sciophytes*). Most trees, all cereal crops, many grasses and herbaceous 'weeds' are heliophytic, while a great number of mosses and ferns, together with the herbaceous and shrubby plants of woodland habitats are sciophytic. There are, however, many plants which, though they attain maximum development in bright sunlight, can exist reasonably well under shade conditions and those which reach optimum development in shade but can tolerate bright light. Two common woodland plants which reveal this tendency are bracken (*Pteridium aquilinum*) and the bluebell (*Scilla nutans*); both can and do grow in deciduous woods and both exhibit greater vigour when clearing or felling exposes them to brighter sunlight.

The light requirements of a particular species, or conversely, its tolerance of shade, can vary during its life cycle. Trees are naturally heliophytic; their seedlings often have to or need to develop under lower light intensities than those necessary for the mature plant. Those of the beech, spruce, fir and yew in particular, and to a lesser extent those of oak, ash, and elm are tolerant of shade and may, in fact, thrive better under such conditions. In contrast, seedlings of willow, poplar, birch and Scots pine are heliophytic and under the shade of other plants may be unable to survive; since not only is their shoot but also their root development retarded, they are, consequently subjected to a greater danger from drought and mineral deficiencies. As well as influencing all the growth processes of a plant through its direct effect on the rate of photosynthesis, light intensity can affect both germination and reproduction. Insufficient light inhibits flowering in many species, while some seeds require exposure to light before they will germinate.

All plant life is inevitably subject to alternating periods of light and darkness. The length of daylight varies latitudinally from a constant twelve hours at the equator, to continuous sunlight for twenty-four hours during part of the year at very high latitudes. In mid and high latitudes there is, in addition, a seasonal variation as between the long summer and short winter daylight periods. For many plants this is a factor of little significance. There are others, however, for which a particular length of day (or night)—a given *photoperiod*—is an essential requirement for the production of flowers and seeds. Also leaf-fall from many deciduous trees of temperate latitudes in autumn is thought to be determined by day-length. The response of plants to duration of daylight is called *photoperiodism*. There are short-day species, such as chrysantheumums, certain species of tobacco and most of the spring and autumn flowering plants of temperate regions, which will not bloom until the day-length or photoperiod is less than a critical threshold of twelve to fourteen hours. Such should, perhaps, more accurately be called *long-night* plants, since the length of the dark period appears to be more important than that of daylight, and it has been demonstrated that successful flowering can be checked even if darkness is interrupted for only a brief interval. In contrast, the long-day plants, requiring over 12–14 hours daylight, include such plants as beet, lettuce, wheat, potato and many summer-flowering plants of temperate regions.

The significance of photoperiodism emerged as a result of a long period of experimentation, particularly with plants of economic value, under controlled laboratory conditions. It has had implications for the horticulturalist who can regulate the photoperiod artificially in his greenhouse to produce short- or long-day blooms all year round. Further it has been realised that certain crops deprived of optimum day-length for flower and seed production show increased vegetative development. For instance short-day onions will produce larger bulbs when grown under longer days, and long-day potatoes the biggest tubers under day-lengths somewhat shorter than the optimum for flowering. Obviously for those plants which exhibit photoperiodism the length of day may be the critical factor limiting their latitudinal distribution and their spread into areas where all other conditions might otherwise be quite suitable for their existence. In the low latitudes of the tropics long-day plants could not reproduce themselves. In high latitudes species have to be either long-day or indeterminate since the brief period when temperature conditions are suitable for growth occurs entirely when days are long. The length of summer days in these regions however compensates for the shortness of the period of suitable temperatures, and plants adapted to these conditions can complete their life cycles in a shorter time than at lower latitudes. In temperate latitudes a longer period of favourable temperatures permits the existence of both short-day autumn and spring, and long-day summer, flowering plants.

The vital processes activated by light can, however, only take place in the presence of water. Most living tissue is composed of a very high percentage of water, up to and even exceeding ninety per cent. of the fresh weight of many plants and animals. The protoplasm of very few cells can survive if their water content drops below ten per cent. and most are killed if it is less than thirty to fifty per cent. below saturation level (see Fig. 6). Water is the medium by which mineral nutrients enter and are translocated through plants. It is also necessary for the maintenance of leaf turgidity without which photosynthesis cannot operate efficiently.

Most land plants obtain water by way of their root systems from the soil. Some, particularly certain mosses and lichens can, however, absorb moisture direct from the atmosphere and are often capable of surviving extreme desiccation as well. Of the water taken up by a plant only a minute fraction is actually incorporated into the chemical compounds it synthesises. Something like ninety-eight per cent. passes through and is lost to the plant in the process of *transpiration.* Most of this loss takes place by evaporation from the minute pores or stomata situated on the surface of all green organs. The presence of these stomata, the primary function of which is to allow the entry of CO_2 and oxygen, means that all land plants are inevitably subjected to an almost continuous loss of water to the atmosphere; they must maintain a continuous stream of water through their tissues from the soil to the atmosphere.

The water requirements of plants varies; for every part of dry matter produced a given plant may transpire anything from 200–1 000 parts of water.

Wheat	507
Oats	614
Barley	539
Rye	724
Maize	369
Millets	275
Peas	800
Lucerne	1068
Buckwheat	578
Rape	441
Potatoes	448

Table 1 Transpiration ratios of selected crops, i.e., ratio of water transpired to dry matter produced. The ratio between grams of water transpired and grams of dry matter produced is a measure of the *transpiration efficiency* of a plant and for the same species it may vary according to environmental conditions. Plants such as lucerne (alfalfa) with a high efficiency are best suited to arid conditions. (From: Penman, H. L. 1963)

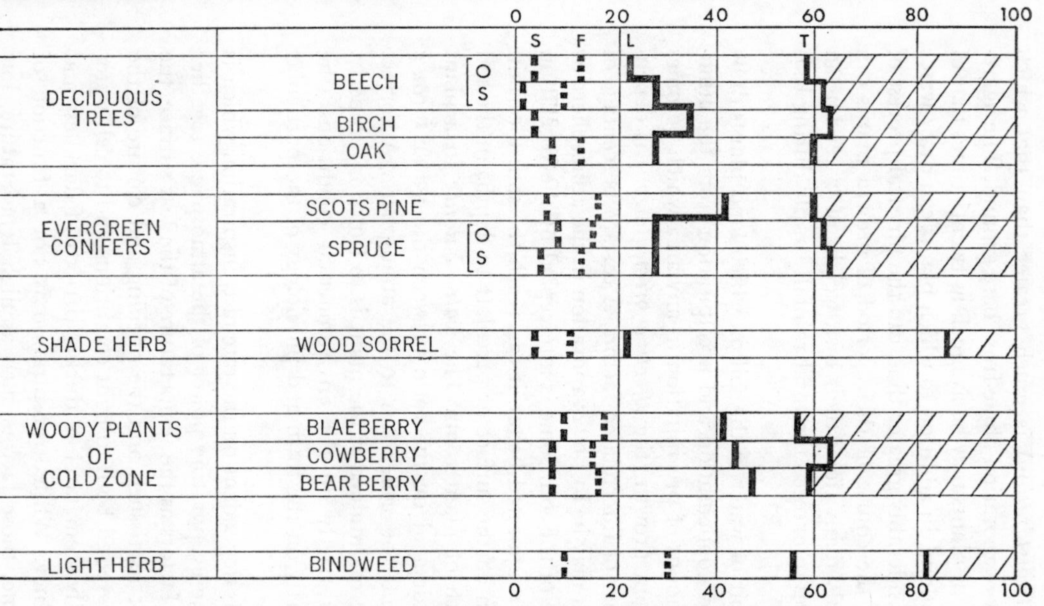

Fig. 6 Critical values for water content of leaves of certain types of vegetation in temperate and cold zones. S: water content at which stomatal closure begins: F: water content causing complete stomatal closure; L: lethal water content; T: Dry weight as a percentage of the total weight when tissues are saturated with water; (O = shade leaves; S = sun-leaves).
(From Birot, P. 1965)
Beech (*Fagus sylvatica*); birch (*Betula pendula*); Scots pine (*Pinus sylvestris*); spruce (*Picea excelsa*); wood sorrel (*Oxalis acetosella*); blaeberry (*Vaccinium myrtillus*); cowberry (*V. vitis-ideae*); bearberry (*Arctostaphylos uva-ursi*); bindweed (*Convolvulus arvensis*).

Obviously the larger the plant, the more foliage and hence the greater transpiring surface, the greater will be the amount of water it will be capable of transpiring: a maize plant may transpire 2.4 litres a day, an oak tree 675 litres. The loss of water from the soil by the combined processes of evaporation and transpiration (or *evapo-transpiration*) is much greater than that which would be lost by evaporation alone from bare ground. The aggregate leaf-surface of plants may be as much as twenty times that of the soil surface alone, in addition to which plants can draw water from depths considerably greater than can the physical process of evaporation. In the course of their period of growth plants can transpire not only many times their own weight of water but an amount which often exceeds the total rainfall supply during that period.

The reservoir of water in the soil on which land plants draw is replenished by condensation and precipitation from the atmosphere. That amount of precipitation in any area which becomes available for plant growth (the *effective precipitation*) is dependent on a number of interrelated factors. A certain percentage of the precipitation may be lost by evaporation in the atmosphere or from the surface of the soil; some is intercepted by the vegetation cover itself and evaporated before it can reach the soil; a certain amount may run off the surface or percolate to depths in the soil beyond the reach of plant roots. The *water balance* in a plant will, therefore, depend on the ratio between the *water income* from the soil and the *water loss* to the atmosphere. A correct balance must be maintained if the plant is to function properly. It is dependent partly on the environmental conditions (the habitat) and partly on the form and physiology of the particular plant.

Water loss is dependent on the rate of transpiration which is influenced by the evaporating power of the atmosphere together with such other factors as the temperature of the leaf surface, the water content of its tissues, and the opening and closure of the stomata which respond both to light conditions and to cell water content; generally stomata close during periods of darkness and of water deficiency. While transpiration is not a function of evaporation alone, those factors such as high temperature, low humidity, and vigorous air movement which promote evaporation have an important and major effect on the rate of water loss from a plant. The climatologist C. W. Thornthwaite has defined drought as a condition in which the amount of water needed for transpiration and direct evaporation (from the soil) exceeds the amount in the soil. The water income of the plant will be dependent on the amount of water available to it in the soil and the rate at which this is replenished. The rate at which plants absorb this water can however be affected by the temperature and the chemical composition of the soil solution. Low soil temperatures decrease the rate of absorption (and if the soil is frozen will inhibit it completely) as do extremely acid or alkaline conditions. Under such circumstances, although there may be no deficiency of soil water, the rate of absorption may not be great enough to make good the losses promoted by high evaporation. Such a condition is frequently referred to as one of *physiological* (rather than physical or absolute) drought.

If the amount of water, or rate at which it is absorbed by the plant, is not great enough to replace that lost by transpiration, the leaves will lose their turgidity, and wilting and eventually death will result. During the day, when stomata are open, many plants lose more water than they can absorb (or is available) and wilting may occur when light and temperature are at a maximum. However, during the night the balance is restored, since, with the closure of stomata in darkness, transpiration may be reduced to a negligible amount. For this reason it has been suggested that if night failed only once on earth many plants would perish. The amount of desiccation that can be tolerated and the ability to survive and recover from an unfavourable water-balance varies from species to species (see Fig. 6, p. 41), and is also dependent on the severity and duration of the drought conditions. Also, a particular plant's susceptibility to damage varies during its life cycle; seeds and mature plants are more capable of enduring drought than the seedlings or young, vigorously growing plants whose demands on water are proportionately greater than those of the mature plant.

Some plants, the *hydrophytes* and *helophytes,* can grow in water and permanently or seasonally water-logged soils respectively, while others, the *xerophytes,* are adapted to dry habitats. Between these two extremes are the *mesophytes* intolerant of either too much or too litle water. The means by which plants are adapted to conditions of periodic or permanent water deficiency are varied and complex, and by no means completely understood. Those capable of tolerating or 'enduring' a marked reduction in their cellular water-content for any length of time are relatively limited. None are as versatile in this respect as many species of algae, lichens, and mosses, some of which can survive a prolonged state of intense desiccation, resuming activity when water becomes available. Tolerance of desiccation is also characteristic of certain desert plants. But few higher plants can survive a reduction of over sixty per cent. of their water content for more than a few weeks. The majority of plants must either 'evade' or 'resist' desiccation. Short-lived, ephemeral annual plants, capable of completing their life cycles very rapidly when moisture is available can survive periods unfavourable to growth, as seeds. Similarly many perennial mesophytes can evade severe water stresses because their normal period of dormancy coincides with that when water is deficient; leaves die and, in the case of deciduous species, are shed and transpiration is, as a result, reduced to a minimum. Such plants which are alternately mesophytic and xerophytic, as in the case of many deciduous trees, are sometimes designated *trophophytic*.

On the other hand the 'true' xerophytes are those adapted to (capable of 'resisting' and remaining active in) conditions where water is permanently limited. They maintain the water content of their tissues at a constant and relatively high level by means of one or more physiological or anatomical features. These either reduce excessive water loss, facilitate absorption or permit water conservation in the plant's tissues. Among the so-called xeromorphic characteristics which are thought to aid in the reduction of transpiration are: (1) thick, wax-impregnated cuticles, possessed by the hard-leaved plants or *sclerophylls*, which minimise cuticular transpiration; (2) stomata protected from excessive exposure to evaporation by being sunk in 'pits' below the level of the leaf surface, by hairy leaves which help to maintain a layer of more humid air near their surfaces or by the rolling or folding of leaves which reduces the area exposed to transpiration; (3) the reduction in the size and *volume* of leaves (a feature of the small-leaved plants or *microphylls*) which in many desert plants may be reduced to mere vestigal spines or scales. Most important in this respect is the reduction in area of transpiring surface to volume. And in some small-leaved plants, size may be more than compensated by large numbers; although many evergreen conifers of cold and temperate regions reduce their transpiration to a very slow rate during winter, it is still several times that of a bare deciduous tree. The most effectives means of reducing transpiration, however, would appear to be the regulation of the opening and closure of stomata. Characteristic of many xerophytes (as will be noted again later) is the closure of stomata during that part of the day when evaporation potential is at a maximum, and their opening mainly at the beginning and end of the day when temperatures are lower. Other features characteristic of 'xeromorphic leaves' include a high density of 'veins' and a volume of water-conducting cells or vessels in relation to the dry weight of leaves much larger than in mesophytic forms; a reduction in inter-cellular spaces and in the size of cells, particularly those near the leaf surface.

Characteristic of xerophytes is the volume and rate of root development in comparison with that of leaves and stems. Vertical and lateral growth can be extremely rapid and extensive whenever soil moisture becomes available. In addition there are indications that some xerophytes—consequent on the somewhat higher osmotic presure of their cell solutions—are capable of absorbing soil water which would normally be unavailable to mesophytic plants. A feature of many perennial desert shrubs is the possession of exceptionally deeply penetrating roots which draw water from the capillary fringe above a permanent water-table situated at considerable depth below the surface. Although often classed as true xerophytes these *phraetophytes* (i.e. using *phraetic* or ground water) can obviously transpire freely and are often lavish in their use of water. They are, hence, capable of 'evading' rather than resisting drought.

Finally, there is a distinctive group of xerophytic plants whose adaption to drought is dependent on their ability to conserve water in their tissues. Some do so in underground organs—roots, tubers, etc.—some in their stems and trunks, others in their leaves. Of these the most distinctive are the true succulents which include all the Cacti and some of the Euphorbiaceae. They are characterised by the combination of a very thick cuticle, a low surface: volume ratio of the surface shoots, and stomata which are not only sparser but which remain closed throughout the hours of daylight.

There has been a tendency to assume that all these features commonly associated with plants of dry habitats are the means whereby they have become adapted to conditions where moisture is deficient—that they have a direct and positive survival value. In biogeographical literature in the past, it has too often been assumed that the form of a plant (of its leaves or other organs) must have some ecological significance and that there is a direct causal relationship between form and environment. Form or morphology of an organ is not, as is now realised, always a reliable guide to its physiology or to the way in which it functions. Much doubt has, in fact, been expressed among botanists as to the exact significance of many of the so-called xeromorphic features, particularly of leaves.

As has already been noted, some plants can withstand drought of varying intensity and duration without any obvious morphological adaptations. Many species possessing some or all of the accepted xeromorphic features can transpire freely and as much, if not more, per unit area as can mesophytic plants when water is available and their stomata are wide open. It is well known that, in such cases, although the stomata may be smaller, they are more numerous, so that their total surface-area is no less than that of other plants. Also, certain xeromorphic characteristics are shared by plants which grow in salt marshes (such as the glassworts, *Salicornia spp*) or on heath-land (such small and hard-leaved species as heather, *Calluna vulgaris* and crowberry, *Empetrum nigrum*) where soil moisture is by no means deficient. In these cases the characteristics of succulence, hard-leaves and so on, have commonly been explained as an adaptation to physiological drought consequent upon the decreased rate of water absorption under these particular conditions. But this explanation is not unanimously accepted by botanists and ecologists; the antagonists point out that the succulent plants of salt marshes transpire quite as freely as others. It has been noted that all the xeromorphic features noted tend to develop in many non-xerophytic plants subject to water stress or in some cases to nutrient deficiency. As a result, it has not been established conclusively the extent to which many of the anatomical and morphological features are a *result* of water deficiency or a *means* of reducing water loss.

The grouping of plants as xerophytes, hydrophytes, and mesophytes is fundamentally on the basis of the environmental conditions or habitats in which they normally exist; the plants themselves are not necessarily distinguished by similarity of form nor is there any clear-cut distinction between these groups. Some plants are more or less xerophytic, some more or less hydrophytic than others. Nevertheless the form or structure of plants appears to reflect more obviously the water-balance conditions under which they grow than any other single factor of the climatic environment. The number of species which can tolerate xerophytic conditions are fewer in number and more specialised in form than those where deficiency of water does not pose an environmental problem.

There is, however, another climatic factor, which though not *essential* for plant growth, can exert a considerable influence on the form of a plant. This is wind. Increased wind force, particularly when combined with low atmospheric humidity can, through its effect on evaporation, greatly increase the rate of transpiration. Even under conditions of plentiful soil moisture it may put a strain on the plant's water balance as severe as that in completely arid regions. In areas exposed to high wind force, as along coasts and at high altitudes, the height to which plants grow may be limited by their ability to absorb and transport water upwards rapidly enough to replace that lost by transpiration. The 'elfin' timber on the upper limits of the tree-line on high mountains rarely exceeds a few inches in diameter and often is not more than a foot in height. Plants in exposed habitats are subjected to wind-shearing which imparts a recumbent or asymmetrical form, since those parts exposed to high wind-force may have their growth retarded or inhibited by excessive transpiration. In addition persistent wind of high velocity can cause the curvature and malformation of branches and trunks of trees, as well as physical damage.

The temperature condition of the atmosphere through its effect on relative humidity and hence on the evaporating power of the air influences both the effectiveness of precipitation and the rate of transpiration. Moreover, as well as being an indirect factor in plant growth, temperature has a direct effect on practically every plant function. It provides the 'working condition' for all organisms and controls the rate of their biological activity and processes. Van't Hoff's Law that the speed of chemical reactions doubles with every temperature increase of 10°C is applicable within limits, and up to a certain optimum, to the metabolic functions of plants and animals. Although there are few places in the world where temperatures are continuously too cold or too hot for life, most of the biosphere operates within a range of between 0°C and 50°C. There are some simple algal plants that can grow and reproduce at temperatures below zero and others which can tolerate temperatures as high as 70–80°C in hot springs.

For each species of plant there are minimum, optimum and maximum temperatures necessary for its various metabolic activities. These are called the *cardinal* temperatures. The minimum (the 'threshold' or 'base') temperature is that below which a function cannot operate; plants native to warm regions (melons, sorghums and date palms for example) require minimum cardinal temperatures of between 15–18°C for growth, temperate cereals of between −2° to 5°C, while some evergreen conifers have minimum threshold temperatures for photosynthesis of −3°C. An increase of temperature above the necessary minimum will be

Crop	Cardinal points °C			Number of days required for germination (the breaking through of roots) at indicated temperatures °C			
	Min.	Opt.	Max.	4·38	10·25	15·75	19·00
Wheat	3–4·5	25	30–32	6	3·0	2·0	1·75
Rye	1–2	25	30	4	2·5	1·0	1·0
Barley	3–4·5	20	28–30	6	3·0	2·0	1·75
Oats	4–5	25	30	7	3·75	2·75	2·0
Maize	8–10	32–35	40–44	—	11·25	3·25	3·0
Sorghum	8–10	32–35	40	—	11·5	4·75	4·0
Rice	10–12	30–32	36–38	—	—	—	—
Timothy	3–4	26	30	—	6·5	3·25	3·0
Flax	2–3	25	30	8	4·5	2·0	2·0
Tobacco	13–14	28	35	—	—	9·0	6·25
Hemp	1–2	35	45	3	2·0	1·0	1·0
Sugar Beet	4–5	25	28–30	22	3·75	3·75	3·75
Red Clover	1	30	37	7·5	3·0	1·75	1·0
Alfalfa	1	30	37	6	3·75	2·75	2·0
Peas	1–2	30	35	5	3·0	1·75	1·75
Lentils	4–5	30	36	6	4·0	2·0	1·75
Vetch	1–2	30	35	2	5·0	2·0	2·0

Table 2 The cardinal points for germination of some important crops. (From: Klages, K. H. W. 1942)

accompanied by an increased rate of activity up to an *optimum*, thereafter the rate declines until a *maximum* temperature is reached above which activity ceases. Whether a plant will be able to grow in a certain area or not will depend on whether the requisite amount of heat is available for a period long enough to allow the completion of its life cycle. While the degree and duration of heat may be sufficient to allow vegetative growth, unless these are also sufficient for the production of seeds a plant obviously cannot maintain itself. An early and broad classification of plants according to the temperature conditions distinguished between (1) megatherms: plants of tropical and sub-tropical regions with at least four months averaging over 20°C; (2) mesotherms: temperatures between 10–20°C; (3) microtherms: in regions with eleven to eight months averaging less than 10°C, and (4) hekisotherms: of the polar belts where all months had average temperatures of less than 10°C.

With a decrease of temperature below the minimum necessary for growth, the plant's metabolic activities slow down, and may cease completely. Unless the plant is prepared for this contingency its tissues are liable to cold damage and to desiccation;

the rate of water absorption may be retarded to such an extent that it cannot adequately compensate for even a low rate of transpiration. The ability of plants to endure low temperatures varies considerably between species. Tropical plants with high cardinal temperatures may be injured or even killed by conditions well above freezing. For those native to mid-latitudes temperatures must drop below zero to become critical. In other parts of the world species, such as the conifers of the Boreal forest, can survive temperatures of –60°C and periods when their tissues may be frozen solid. Some plants can grow and reproduce under a cover of snow. Tolerance of temperatures below the cardinal minimum however depends also on the stage of development of the particular plant, as well as on the intensity and duration of such conditions. Plants native to cool and cold climates, with marked seasonal variation of winter and summer temperatures, must be capable of completing their cycle of growth within the period of favourable temperatures available and of surviving a period of greater or shorter duration when temperatures fall below their minimum requirements. Successful survival of winter cold is dependent basically on the accumulation and storage of food and on protection from desiccation. Annual plants do so in the form of seeds, herbaceous perennials as tubers, corms, bulbs and rhizomes—food storage organs in which the resting buds are incorporated. Trees and shrubs form buds, resistant to cold and drought, from which growth will be renewed; they also reduce their rate of transpiration and in this latter respect leaf-fall is the most effective method. While evergreen and particularly coniferous trees reduce their transpiration in winter they cannot do this so effectively as deciduous trees.

Plants of temperate regions, then, pass the period of low winter temperatures in a state of 'suspended activity', in a dormant or quiescent state. It would appear, however, that a period of rest or dormancy is a natural function of many such plants but is not necessarily *enforced* nor is it a direct response to environmental conditions. After completing growth many plants become dormant although temperature and other climatic conditions may seem, to all intents and purposes, still favourable for biological activity. It has been noted that a beech tree completes its growth in the first two or three weeks of spring; many spring flowers complete their whole life cycle before the period of maximum summer temperatures. Even in the perpetually warm and moist tropics there are plants which exhibit a regular rhythm of growth and dormancy which is unrelated to any climatic rhythm. Nor do plants of the southern hemisphere always change their period of growth and rest when introduced to northern latitudes. Dormancy then would appear to be an inherent characteristic of certain plants which permits their adaptation to regions with a marked seasonal variation of temperature, and indeed also drought. Furthermore, many plants of temperate regions will not recommence growth, (the dormancy of their seeds, tubers or buds, cannot be broken) unless they have been subjected to a period of low winter temperatures. This *winter stimulation*, as it is called, is as necessary for renewed growth as is the onset of favourable temperature and light conditions. Trees and shrubs native to Britain require a winter temperature of usually less than 9°C for periods varying from 200–3000 hours. In addition, certain plants, particularly the winter varieties of many cereals, flower sooner if exposed to low temperatures (around freezing point) after germination; winter wheat, for example, if planted in spring will not flower in the same year. Biennial plants such as celery and various kinds of beet form leaves and tubers in the first year of growth and flower only in the second year, after subjection to winter chilling. The artificial cold stimulation of wheat, particularly, has been called *vernalisation* or *yarovisation* (or bringing into 'spring' condition) by the biologist, Lysenko, who was responsible for the application of this technique in Russia.

Not only is winter stimulation a necessity for many cool and cold region plants but so also is an alternation of low night and higher day temperatures. The response of plants to such rhythmic seasonal and diurnal fluctuations of temperature is referred to as *thermoperiodism*. Certain plants will flower only when night temperatures are below a required mimimum. The common daisy only grows when days are cool and night temperatures are between 8°–13°C: it cannot reproduce under continuously warm conditions and will die if kept in a heated greenhouse. The seeds of the well-known agricultural grasses, meadow grass (*Poa pratense*) and cocksfoot (*Dactylis glomerata*), germinate most successfully when subjected to conditions where there is a fluctuation of day and night temperatures. Laboratory experiments have proved that tomato plants produce optimum growth with day temperatures at 20°C and nights at 10°C; while the potato, a close relative, will form tubers only if night temperatures fall within a fairly narrow range of between 10°C–21°C. It will now be obvious that the need either for low winter or for low night temperatures must limit the latitudinal and altitudinal distribution of certain plants; many species of cool and cold regions could not exist within the tropics any more successfully than warmth-demanding species could at high latitudes.

In order to be able to exist in a given area all plants need a particular length of time during which their requirements of temperature, moisture and light for successful growth and reproduction can be satisfied. This is usually referred to as the *growing season* or *period*. Various efforts have been made to determine and define quantitatively the length and 'quality' of this period for particular plants or in the different climatic regions of the world. One approach to the problem has been by means of *phenological* studies, that is to say the study of the relationship between climatic factors and the periodic or recurring events in the life cycles of plants (or animals, for that matter). Phenological observations involve the recording of the dates at which such plant functions as germination, bud-bursting, flowering, leaf-fall, fruit-ripening and seed-dispersal occur in particular species. From such data the length of time it takes a plant to mature in a given habitat can be calculated. On the basis of records collected over a period of years by the Meteorological Office the average date at which twelve common wild spring flowers come to bloom has been calculated for a number of meteorological stations in the British Isles. Lines joining places with the same average flowering date—(e.g. *floral isophenes*) reveal a time-lag in this event with latitude and altitude (Fig. 7). From similar studies conducted in the U.S.A. it was estimated that phenological events are, on average, about four days later for every degree of latitude northward and for every 120 metres of increase in altitude. Assuming a decrease of 0·5°C for each degree of latitude, every 120 metres of altitude and every 5° of longitude, attempts were made to establish *bioclimatic zones* which would reflect regional variations in the climatic potential for plant, and particularly crop growth.

However, attempts to correlate phenological events with the actual weather conditions at the time of their occurrence (or to correlate the average date of the event with average climatic conditions) have not been very successful. The onset of flowering or leaf-fall in temperate regions may be as much a function of an increase or decrease in day-length as of concurrent temperature changes. The actual time of the commencement of growth after dormancy or of flowering may be dependent not only on the weather conditions at the time of the event but also on those of the preceding summer or winter.

In those regions of the world where temperature would appear to be the most important climatic factor limiting growth

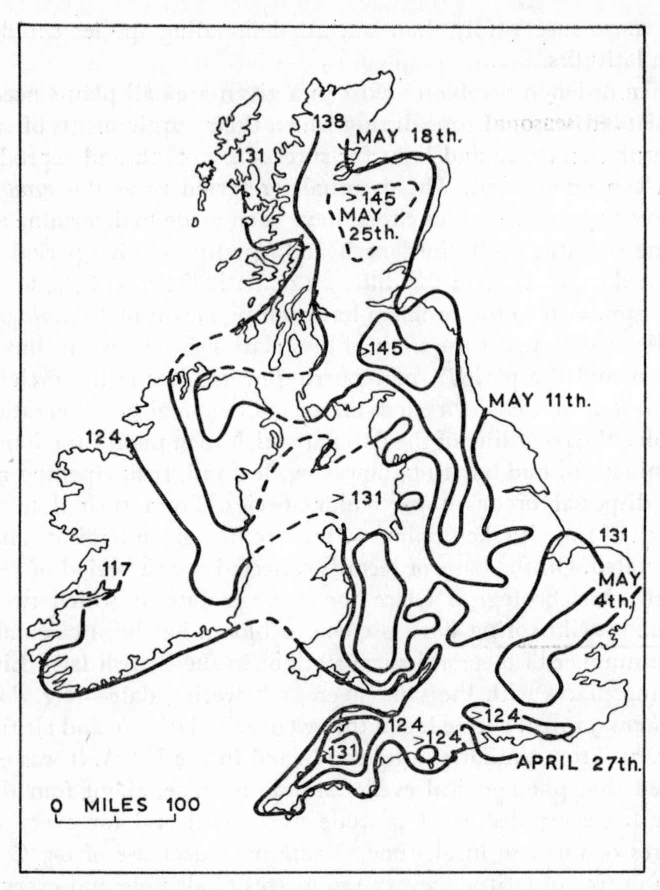

Fig. 7 Average floral isophenes (1891-1925) or lines of equal flowering dates (each line is given a date *and* number of days since beginning of year) for twelve selected plants in the British Isles. (Map based on Phenological Report (1934) in *Quart. Journ. Roy. Met. Soc.*, Vol. 61, No. 260, Phenological Number, 1935)

during part of the year, the growing season is generally defined as that period when temperatures (expressed as daily or monthly means) are above a selected *minimum* or threshold value for growth. The threshold temperature chosen to define the beginning and end of the growing season can vary from 0°C to 10°C; though that most commonly employed value is 42° or 43°F (6°C) which is regarded as representing the minimum temperature necessary for the commencement of growth in temperate cereals. The growing season so defined is really only valid when applied to mid and high latitudes. In fact, to postulate on this basis that tropical regions have a year-long growing season means very little. Many of the plants adapted to warm climates require much higher threshold values while, as has already been noted, the lack of marked seasonal or diurnal variations of temperature in such regions can preclude cool temperate plants from them. In the U.S.A. on the other hand, the growing season is more commonly defined as the 'length of time between the (average date of) last killing frost of spring and the (average date of) first killing frost in autumn', as being a more realistic measure of the period required for the satisfactory and economic growth of frost-sensitive crops. The frost-free season is then taken as the number of consecutive days during which temperatures are continuously above 0°C.

One of the obvious limitations of these definitions is that, whatever threshold value is selected, they give only a measure of the *length* of the 'temperature-favourable' period. They fail to reveal differences in the quantity of heat that can be received by two stations with a similar average length of growing season (see Table 3 and Fig. 8).

		Alt.	J	F	M	A	M	J	J	A	S	O	N	D
Aberdeen	57°N 2°W	8	3	3	4	6	9	12	14	13	11	8	5	4
Chicago	42°N 88°W	823	−3	−2	2	8	14	20	23	23	19	12	5	0

Table 3 Mean monthly temperatures for Aberdeen and Chicago. For both stations the length of the growing season (above the threshold value of 6°C) is seven months. During this period however Aberdeen has 1095 *day degrees. Chicago* 2328 *day degrees.*

Attempts, therefore, have been made to express the quality of heat; what may be called the efficiency of the growing season, in terms of *accumulated temperatures*. These are obtained by summing the mean daily temperatures (expressed as *day-degrees* or *heat units*) above the chosen threshold. This value has also been referred to as 'the temperature summation' or the 'remainder index'. Other workers have proposed the use of *photo-thermal units* (day-degrees × average hours of daylight at a particular latitude) since a decrease in the length and intensity of the growing season with latitude is compensated for by longer day-length. Further efforts have also been made to refine the accumulated temperature index by taking into account the increased rate of

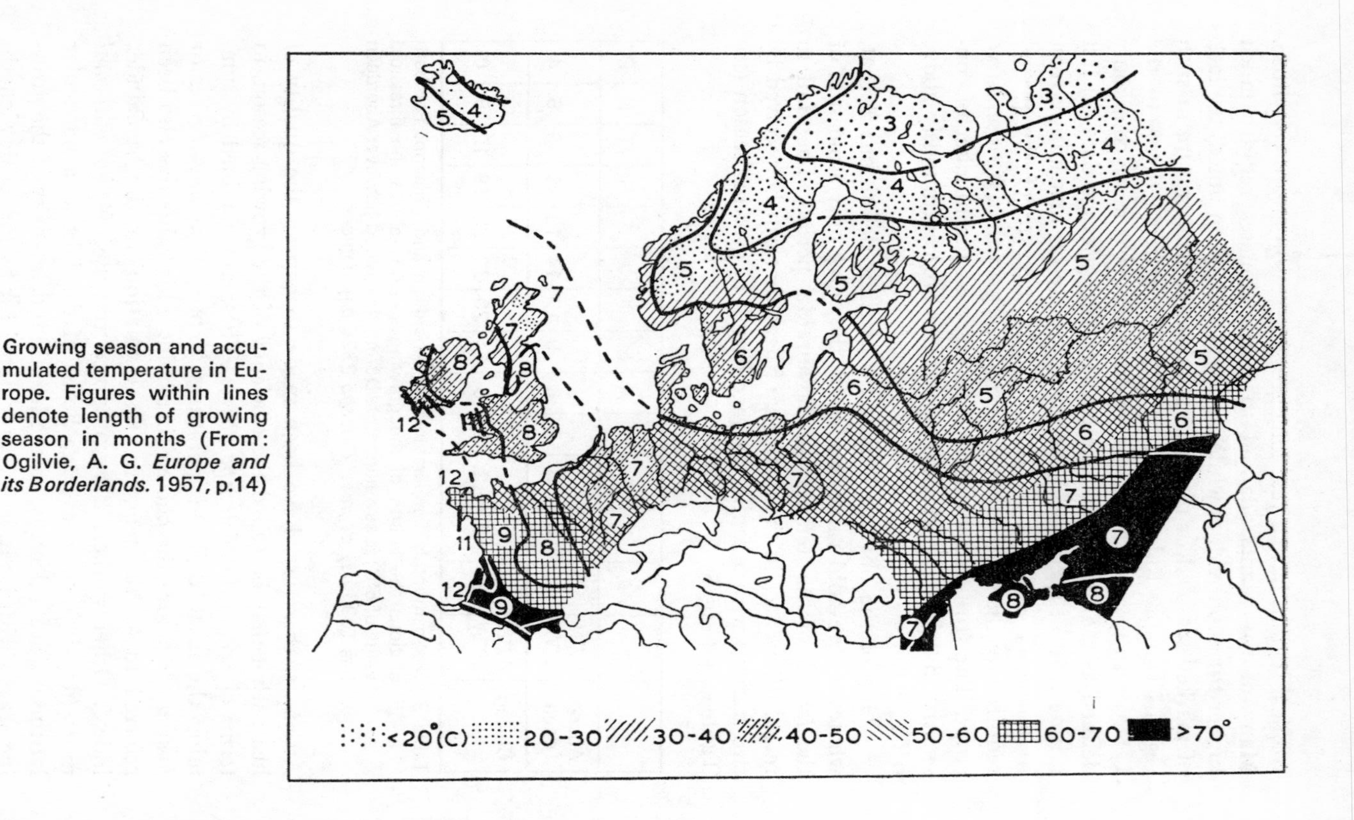

Fig. 8 Growing season and accumulated temperature in Europe. Figures within lines denote length of growing season in months (From: Ogilvie, A. G. *Europe and its Borderlands.* 1957, p.14)

growth with increased temperature and the existence of optimum cardinal temperatures for the rate of plant growth.

Such methods of defining the growing season are based on the use of mean daily or mean monthly air temperatures. They suffer from obvious limitations. They obscure the variations in the temperature requirements of different plant species within a region so defined, and they also fail to recognise that night or winter temperatures may be as critical for the growth of plants as the day-time or summer values. It is also well to remember that plants react to *actual* temperatures and that mean values may obscure maximum and minimum values whose specific effect is quite different from that represented by the mean. In addition, since the calculations involved are derived from air temperatures they do not reveal the modifying effect of such factors as soil, aspect, or the vegetation cover itself or the duration or amount of heat available for plant growth in a particular habitat. Despite these limitations, however, the growing season so defined provides a useful basis for the broad and general comparison of actual or potential conditions for growth between one region or habitat and another, and it provides a guide to possible variations in the length of this season with latitude, altitude and distance from the sea. It gives, however, a general measure only of the *thermal* growing season.

There are many parts of the world where on the basis of this most commonly accepted and used definition of the growing season, temperatures are favourable throughout the year but where a deficiency of moisture is the critical factor limiting plant growth. Even in temperate regions with a marked thermal season, the length and efficiency of this period can be curtailed by an insufficiency of moisture. The problems of defining the growing season in terms of the moisture available for plant growth are exceedingly difficult and have not been, as yet, satisfactorily solved. Plants vary in their minimum water requirements and also in their ability to tap soil- and ground-water supplies. Moisture becomes critical for growth when the amount available in the soil drops below that necessary to make good the loss by transpiration. This value is dependent not just on the amount but on the effectiveness of the precipitation a place receives, and the latter, as we have already noted, is a function of a variety of inter-related factors among the most important of which are those which influence evaporation and transpiration. Attempts have been made to express the effectiveness of precipitation for plant growth, and to define water deficiency or 'drought' in terms of

c

the relationship between precipitation and evaporation in a given area. The assumption is often made that the limit between humid and arid conditions occurs when evaporation (E) is equal to precipitation (P). In Australia it has been suggested that the growing season commences when precipitation is more than one-third of evaporation from an open-water tank. But there are relatively few (compared with either temperature or rainfall) reliable records of evaporation from which such calculations can be made. In their absence use has been made of saturation deficit (which is easier to derive from existing humidity records) as a measure of the evaporation-potential of the atmosphere; and the N/S or P/SD quotient or ratio is calculated by dividing annual precipitation in millimetres by the absolute saturation deficit of the air expressed in millimetres of mercury.

Many geographers and biologists have favoured the use of precipitation: temperature ratios as an expression of the effectiveness of rainfall and the degree of aridity in a given area: the obvious limitation of such ratios lies in the assumption that evaporation is a function of temperature alone. More recently the climatologist C. W. Thornthwaite introduced the concept of potential *evapo-transpiration*: that amount of moisture which would be evaporated from the soil and transpired from vegetation if it were available. Drought is that condition when the amount of water needed for potential evapo-transpiration exceeds the amount available in the soil, and assuming that the soil has a capacity to store approximately 10 cm of rainfall. But there are even fewer measurements available of either actual or potential evapo-transpiration than there are of evaporation from open-water tanks. By empirical methods C. W. Thornthwaite constructed a mathematical formula whereby potential evapo-transpiration values can be calculated from known temperature data and light conditions. The relationship between the precipitation and the calculated potential evapo-transpiration is then expressed in terms of a surplus or deficit of water *in the soil*. The graphs showing the march of average monthly rainfall and potential evapo-transpiration illustrate the occurrence of these events throughout the year under a variety of different climates (Fig. 9). However it has been shown that when C. W. Thornthwaite's formula (which was derived from known facts about temperate climatic conditions) was applied to tropical regions, calculations of soil water deficit or surplus did not always give results which accorded with the actual conditions.

The concept of the growing season (which is generally

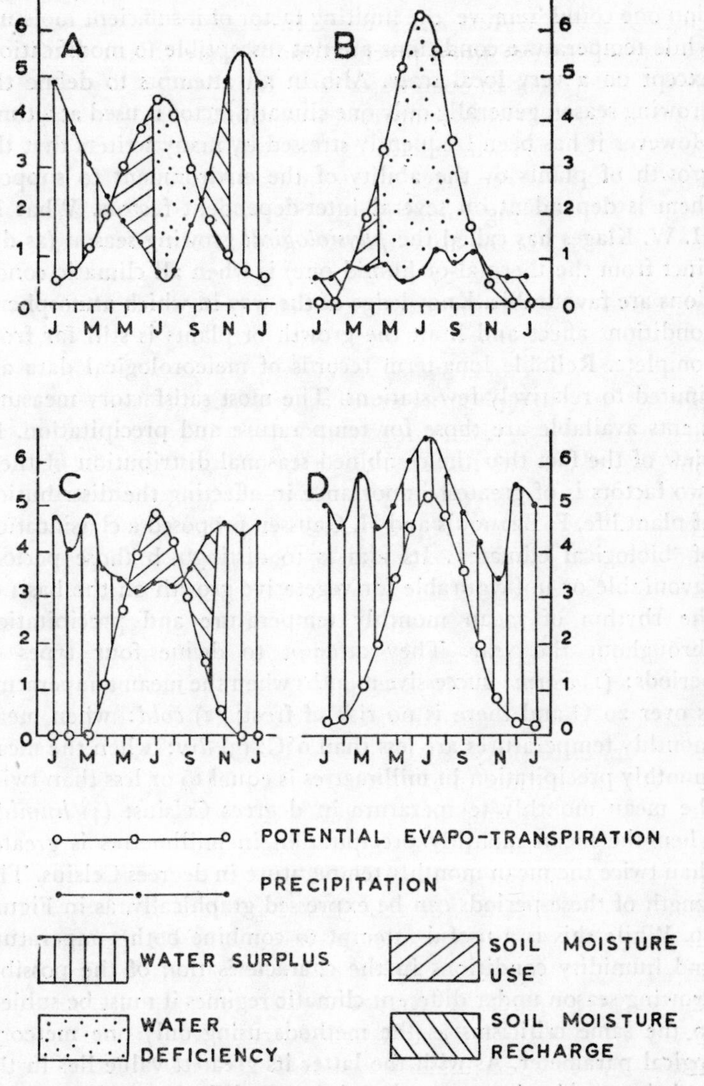

Fig. 9 Annual march of precipitation and potential evapo-transpiration at selected stations. A. Seattle (Washington); B. Grand Junction (Colorado); C. Bar Harbour (Maine); D. Brevard (N.C.). (From: Thornthwaite, C. W. 1948)

synonymous with the *thermal* growing season) was based originally on the assumption that, of the two main factors limiting plant growth, temperature was the more fundamental: through irrigation one could 'remove' the limiting factor of insufficient moisture while temperature conditions are not susceptible to modification, except on a very local scale. Also in all attempts to define the growing season generally only one climatic factor is used at a time. However it has been frequently stressed by many writers that the growth of plants or the ability of the environment to support them is dependent on several inter-dependent factors. What K. H. W. Klages has called the *physiological* growing season (as distinct from the thermal or humid one) is when all climatic conditions are favourable. Knowledge of the way in which atmospheric conditions affect and limit the growth of plants is still far from complete. Reliable long-term records of meteorological data are limited to relatively few stations. The most satisfactory measurements available are those for temperature and precipitation. In view of the fact that the combined seasonal distribution of these two factors is of greatest importance in affecting the distribution of plant life, F. Bagnouls and H. Gaussen proposed a classification of 'biological climates'. Its aim is to distinguish those periods favourable or unfavourable for vegetative growth on the basis of the rhythm of mean monthly temperature and precipitation throughout the year. They attempt to define four types of periods: (1) *warm*: successive months when the mean temperature is over 20°C and there is no risk of frost; (2) *cold*: when mean monthly temperatures are less than 0°C; (3) *dry*: when the mean monthly precipitation in millimetres is equal to or less than twice the mean monthly temperature in degrees Celsius; (4) *humid*: when the mean monthly precipitation in millimetres is greater than twice the mean monthly temperature in degrees Celsius. The length of these periods can be expressed graphically, as in Figure 10. While this is a useful attempt to combine both temperature and humidity conditions in the characterisation of the possible growing season under different climatic regimes it must be subject to the same criticisms as the methods using only one meteorological parameter. As with the latter its greatest value lies in the basis it provides for comparison between different environmental conditions.

The relationships between climatic factors and plant growth are complex and are still imperfectly understood. Attempts to discover the optimum climatic conditions for the growth of a particular species, and the limits within which it can exist are

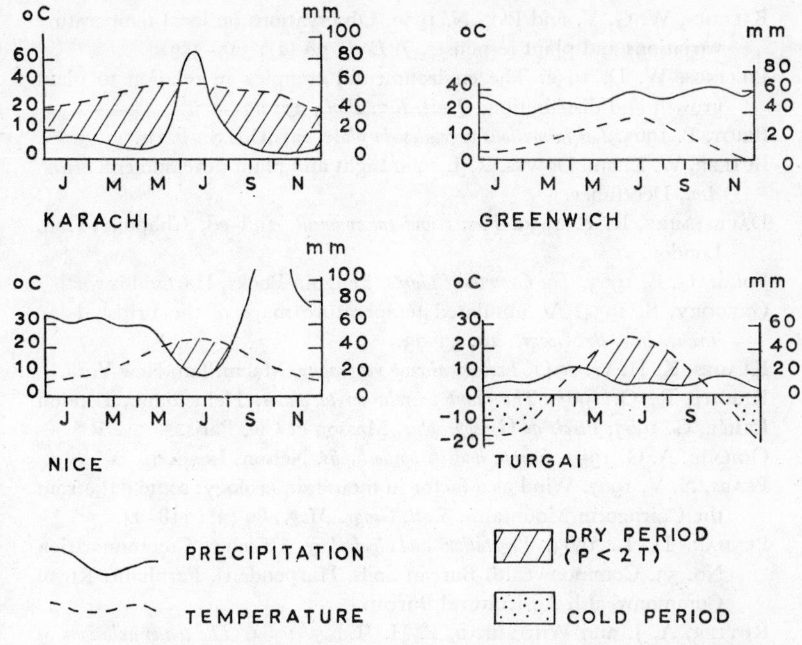

Fig. 10 Graphical representation of seasons unfavourable or favourable for growth at selected stations. Mean monthly precipitation in millimetres: mean monthly temperature in degrees Celsius. Temperature represented by a scale double that for rainfall. P = 2 T°C is limit between humidity and aridity; therefore when curve for rainfall is below that for temperature (P < 2 T° C) the period is 'biologically' dry. (From: B. Bagnouls et Henri Gaussen 1957)

exceedingly difficult. Whether a particular plant or group of plants can occupy a given habitat does not depend entirely on the presence of favourable climatic conditions. These are essential. But soil and other biotic conditions must also be suitable. Climate sets the scene, climate provides certain opportunities—a certain potential—for plant growth; the extent to which these are 'exploited' by a particular species may depend on soil conditions.

References

Ashby, M. 1961. *Introduction to plant ecology*. Macmillan, London.

Bagnouls, F. and Gaussen, H. 1957. Les climats biologiques et leur classification. *Annls Géogr.*, **66** (355): 193–220.

Bainbridge, R. *et al.* (Eds). 1966. *Light as an ecological factor*. A Symposium of the British Ecological Society. Blackwell Scientific Publications.

Baron, W. M. M. 1967. *Physiological aspects of water and plant life*. Heinemann, London.

Balchin, W. G. V. and Pye, N. 1950. Observations on local temperature variations and plant responses. *J. Ecol.*, **38** (2): 345–353.
Billings, W. D. 1952. The environmental complex in relation to plant growth and distribution. *Quart. Rev. Biol.*, **27**: 251–65.
Birot, P. 1965. *Les formations végétales du globe.* S.E.D.E.S., Paris.
Butler, W. L. and Downs, R. J. 1960 Light and plant development. *Scient. Am.*, December.
Daubenmire, R. F. 1959. *Plants and environment*, 2nd ed. Chapman-Hall, London.
Fogg., G. E. 1963. *The Growth of plants.* Penguin Books, Harmondsworth.
Gregory, S. 1954. Accumulated temperature maps of the British Isles. *Trans. Inst. Br. Geogr.*, **20**: 59–73.
Klages, K. H. W. 1942. *Ecological crop geography.* Macmillan, New York.
Knight, R. O. 1965. *The plant in relation to water.* Heinemann, London
Lemée, G. 1967. *Précis de biogéographie.* Masson et Cie, Paris.
Ogilvie, A. G. 1957. *Europe and its borderlands.* Nelson, London.
Pears, N. V. 1967. Wind as a factor in mountain ecology: some data from the Cairngorm Mountains. *Scott. Geogr. Mag.*, **83** (2): 118–24.
Penman, H. L. 1963. *Vegetation and hydrology.* (Techn. Communication No. 53. Commonwealth Bureau Soils, Harpenden). Farnham: Royal Commonwealth Agricultural Bureau.
Rutter, A. J. and Whitehead, F. H. (Eds.). 1963. *The water relations of plants.* A Symposium of the British Ecological Society. Blackwell Scientific Publications, Oxford.
Salisbury, E. J. 1926. The geographical distribution of plants in relation to climatic factors. *Geogr. J.*, **67** (4): 312–342.
Synopsis of Symposium held on March 19th, 1958, on the subject of 'The Growing Season' (Mimeo): Department of Geography, University College of Wales, Aberystwyth.
Taylor, J. A. (Ed.). 1967. *Weather and agriculture.* Pergamon Press, London.
Thornthwaite, C. W. 1948. An approach towards a rational classification of climate *Geogr. Rev.*, **38** (i): 55–94.
Went, F. W. 1957. Climate and Agriculture *Scient. Am.*, June.
Whitefead, F. H. 1957. Wind as a factor in plant growth. *Symposium on Control of Plant Environment*, edited by J. P. Hudson. Butterworth, London.
Wilsie, C. P. 1962. *Crop adaptation and distribution.* Freeman, London

4 Edaphic Factors

Edaphic factors are those soil properties which affect plant growth and distribution. For the majority of land plants the soil is that medium, that part of their habitat, in which they are anchored and from which they obtain, by way of their root systems, the water and most of the mineral nutrients necessary for their existence. But the soil is far more than just an anchorage and nutrient reservoir for plants, it is essentially a *product* of the plants which inhabit it and of the interaction between them and the underlying weathered rock mantle. The soil and the plants which contribute to its formation are inseparably linked by a continual interchange of materials. The living plants draw from the soil nutrients which are eventually returned and released for re-use when their tissues die and decay. Stated very simply, soil is what plants grow in, and strictly speaking soils cannot form on those parts of the earth's surface where plants are unable to exist. Also, just as the green shoots of plants provide the essential source of food for all land animals, so the organic tissues which plants return to the soil support a large population of other soil organisms. The latter include countless millions of microscopic bacteria and fungi, as well as the larger, visible, soil-inhabiting animals such as insects, spiders, mites, worms, moles etc. All these, and many more besides, obtain food energy either directly or indirectly from plant tissues; in doing so they effect the disintegration and decomposition of organic matter, during which process the nutrients originally obtained by plants either from the atmosphere or from the rock mantle are made available again. The soil is, then, the biologically inhabited and active layer of the earth's crust. It has been compared to a sort of 'factory' or 'laboratory' in which a whole series of complex biological and chemical processes, instigated by the activities of soil organisms, maintain the essential 'turnover' of nutrients necessary for plant growth.

The 'air of the soil' differs from that of the atmosphere in many significant respects. The former is devoid of light and, in the absence of photosynthesis, the organic processes of respiration or oxidation predominate. As a result soil-air contains less oxygen and a much higher percentage of carbon dioxide than does the free atmosphere. Also the soil climate is somewhat more equable than that above ground; temperature fluctuations, either daily or seasonal, are less marked and decrease with depth. The main soil components or ingredients include the inorganic or mineral material produced by rock weathering, and the organic-matter content most of which is supplied by plant tissues. This structural framework influences the amount and availability, as well as the composition of the soil-water and air without which plant roots could not survive.

Examination of any soil sample, except peat, would show that over two-thirds of its volume was comprised of the mineral or inorganic component. Its chemical and physical properties exert a direct and indirect effect on both plant growth and soil formation. It is produced by the shattering, disintegration, and 'rotting' of rocks in the process of weathering. The amount, the size and the chemical composition of the mineral fragments present in any particular soil will obviously be dependent on the type of rock and on the type and intensity of weathering to which it has been subjected. The latter is a function of both climate and of biological activities. In arid or very cold regions, particularly where rock is directly exposed to the effects of alternate heating and cooling or freezing and thawing, the mechanical shattering of rock into smaller and smaller fragments is common. The mineral nutrients contained in these rock fragments will, however, not become available for plant growth until they have been 'released' by chemical rotting or weathering. This results from the attack on rocks by water, and its effectiveness is considerably heightened by being 'acidified' and by increasing temperature. Organisms contribute to both types of weathering. Plant roots can enlarge and widen rock cracks, while carbon dioxide produced by roots and decaying organic matter, together with other organic acids produced by dead or living organisms, make the soil-water a more effective agent of chemical weathering. Such plants as lichens, for instance, are capable of existing on bare rock surfaces; they obtain their mineral requirements by secreting acidic substances which bring about a chemical corrosion of the rock; they are, as a result, important agents of weathering.

The material produced by rock weathering is composed of

mineral particles which may vary in size from large, coarse grains to those so fine that they cannot be distinguished individually by the naked eye. They are usually differentiated and described on the basis of size-classes (*soil separates* or *fractions*). The diameter limits used (somewhat arbitrarily) to define them vary slightly from one country to another. The size limits established by the International Society of Soil Science in 1926 differ from those used by the U.S. Department of Agriculture.

Soil Separate	International System	U.S. Dept. Agric.
Very coarse sand (fine gravel)	— — —	2.00—1.00 mm
Coarse sand	2.0—0.2 mm	1.00—0.50 mm
Medium sand	— — —	0.50—0.25 mm
Fine sand	0.2—0.02 mm (0.2—0.05)	0.25—0.10 mm
Coarse silt (or very fine sand)	— — —(0.05—0.02)	0.10—0.05 mm
Silt	0.02—0.002 mm	0.05—0.002 mm
Clay (or colloidal minerals)	less than 0.002 mm	less than 0.002 mm

Table 4 Classification of soil particles: more recent subdivision based on diameter limit of 0.05 mm which corresponds to a significant change in the properties of the particles.

In most soils, stones (over two mm diameter) will normally be pieces of rock. The sand and silt fractions on the other hand are usually composed of small fragments of the more resistant rock-forming minerals; of the latter the most impervious to both chemical weathering and mechanical pulverisation is quartz (SiO_2) and it is not, therefore, surprising that the sand and silt fraction is commonly composed of ninety to ninety-five per cent. quartz grains of varying size. The gravel, sand and coarse silt particles contribute little, if anything at all, to plant nutrition. In the case of rock-fragments, the minerals have not been 'weathered out', while the quartz grains represent the end-product, the inert residue, of the weathered material from which all the essential mineral nutrients have long since been removed.

The 'clay fraction' on the other hand is distinguished by the fact that it is composed of secondary minerals. In a given soil these may be the direct products of rock weathering or of soil forming processes. They are the 'alteration products' which together form what is often referred to as the 'weathering complex'. The clay fraction is composed mainly, though not necessarily exclusively of the *clay minerals*. The chemical composition of the latter is exceedingly complex and is dependent not only on the type of rock from which they were produced but also on the type and duration of weathering. Most contain varying proportions of oxygen, hydrogen, silica, aluminium and iron. Various groups of clay minerals have been identified, two of the most widespread being the 'silicate clays' characteristic though not exclusive of temperate regions and the 'iron and aluminium hydrous oxide' clays of tropical regions. The importance of the clay fraction is, however, as much related to its physical as to its chemical composition—though the former tends to be dependent on the latter. Two of the most important attributes of the clay particles are, first, their minute size and, second, the colloidal states of those less than 0.001 mm in diameter. By reason of this latter property some clay mineral particles can absorb two to three times their own volume of water, swelling when wetted, shrinking when dried. Related to this property is their ability to adhere to each other so that, when moist, clay minerals become plastic and when dry, the particles cohere into a hard impenetrable mass. Such characteristics are much more marked in some types of clay minerals (i.e. the silicate clays in particular) than in others, and the cultivation of soils in which they form a significant fraction is particularly difficult. In addition, clay particles have, to a much greater degree than any other of the soil fractions, the ability to 'attract' and 'hold' water and minerals on their surfaces. This is a property related partly to their colloidal nature but also to their size; the combined surface-area of clay particles is very much greater than that in a comparable volume of larger grains; it has been estimated that the external surface-area of one gram of colloidal clay is at least 1 000 times that of a gram of coarse sand. Because of these properties clay is, biologically and chemically, the most important of the soil fractions. It is the 'active' fraction, and its water- and nutrient-holding properties make it, together with organic matter, the major determinant of soil fertility.

In any soil the proportion of the various soil fractions present determine that physical property known as its *texture*. As can be seen from Table 5 and Figure 11 a great variety of textural classes are possible dependent on the varying proportions of the fractions present. The texture of a soil is one of its most important and basic properties: from the farmer's point of view probably *the* most important. It determines the relative ease with which roots can penetrate into the soil. It affects the nutrient-supplying ability of the soil. It influences, too, the water content, aeration and temperature, all of which are essential for the living

Textural Class	Per cent soil fractions		
	Sand	Silt	Clay
Loamy sand	85	10	5
Sandy loam	65	25	10
Loam	45	40	15
Silt loam	20	60	20
Silty clay loam	15	55	30
Clay loam	28	37	35
Clay	25	30	45

Table 5 Percentage composition of a number of selected textural classes. Size limits of soil fractions as in Figure 11 (From: Buckman and Brady. 1960)

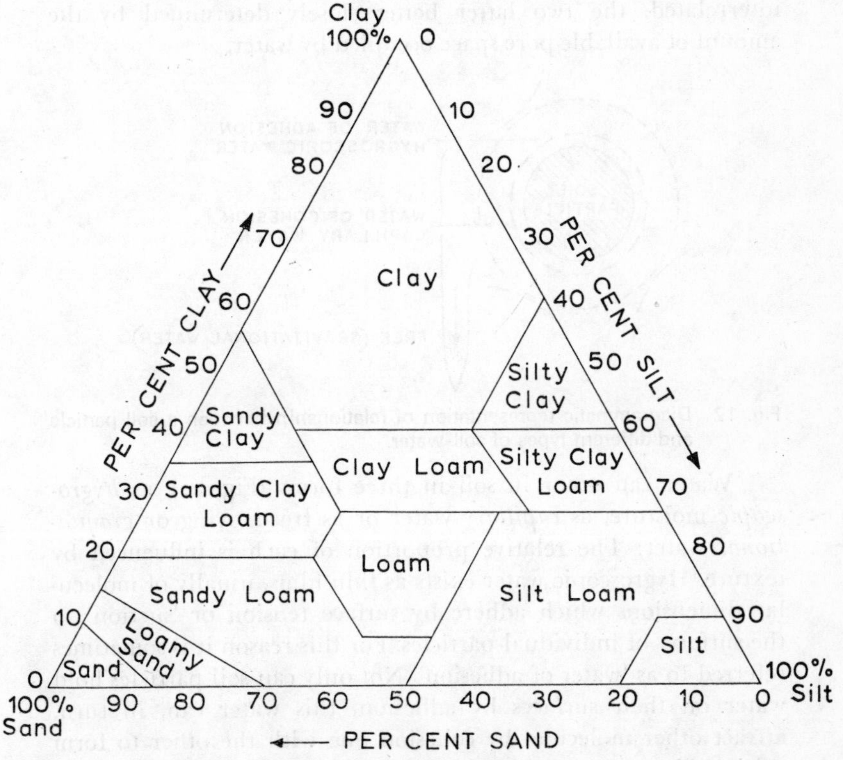

Fig. 11 Texture triangle: graphical method by which textural classification of soils may be determined. The thirteen types are based on American diameter limits of clay: less than .002 mm; silt 0.002—0.05 mm; sand 0.05—2.0 mm. To use the diagram points corresponding to percentage of silt and clay in the soil are located on the relevant side of the triangle; the compartment in which lines projected inward (parallel to the clay and sand sides of the triangle respectively) intersect, indicates the class name of the soil in question.

inhabitants of the soil. These latter three conditions are to no small extent influenced by the number and size of the spaces, the *pore spaces,* which occur between the individual particles. Porosity is a function then of texture and is directly dependent on the size and arrangement of the mineral particles. Coarse and loosely packed sand will have large pore spaces though these may only occupy about twenty to forty per cent. of the total volume. Conversely in fine-textured soils the great number of minute pores may account for forty to sixty per cent. of its bulk. More significant, however, than the volume occupied by pore space is their size. The larger spaces or so-called *macro-pores* allow much freer movement of air and water than the smaller *micro-pores.* The water, air and temperature conditions in a given soil are intimately interrelated, the two latter being closely determined by the amount of available pore space occupied by water.

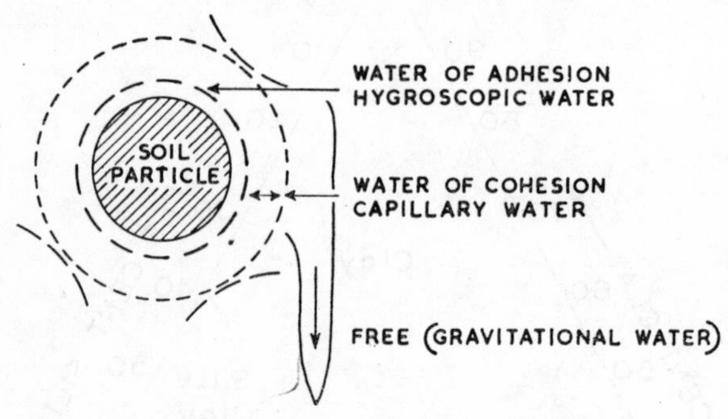

Fig. 12 Diagrammatic representation of relationship between a soil particle and different types of soil-water.

Water can occur in soil in three forms (Fig. 12), as *hygroscopic* moisture, as *capillary* water or as free-moving or *gravitational* water. The relative proportion of each is influenced by texture. Hygroscopic water exists as thin films, usually of molecular dimensions which adhere by surface tension or 'suction' to the surfaces of individual particles. For this reason it is sometimes referred to as 'water of adhesion'. Not only can soil particles hold water on their surfaces by adhesion, this water can, in turn, attract other molecules by cohesion one with the other to form thicker films of *capillary* water. The amount and tenacity with which water can be held or retained by these forces of adhesion and cohesion depends on the soil texture: it is greatest in fine clay and least in large sand particles. This capillary water is, in addition, capable of some, though very slow, movement comparable to that of water through a porous material dipped in water. It tends to move from wetter to drier areas of the soil by the process known as *capillarity*. This movement only really becomes significant when there is a source of saturated soil adjacent to dry unsaturated material, as exists at the junction of a water-table. Even under these circumstances the distance water will move upwards (or laterally) from a pond or stream by capillary action is extremely limited. It will be greatest the finer the texture and hence the smaller the capillary spaces. Experiments have shown that in a clay-loam soil capillary movement upwards from a water table does not exceed seventy-five centimetres, while in coarse sand it is no more than thirty centimetres (see Fig. 13).

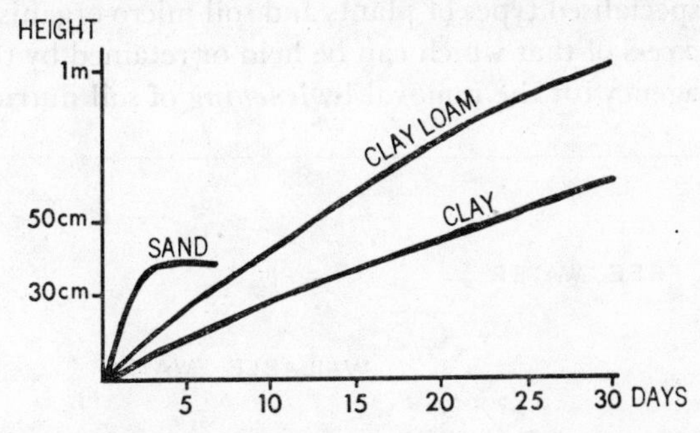

Fig. 13 Amount and rate of rise of capillary water above a water-table in soils of varying texture. (From: Duchaufour, Philippe, 1965)

Water in the soil in excess of that which can be held by adhesion and cohesion will come under the forces of gravity and will percolate and move down through the soil pores as free-draining or gravitational water. The amount not retained and its ease of percolation will be dependent on the number of macro-pores in a particular soil.

The significance of these three types of water for plant growth vary (Fig. 14). Hygroscopic and the inner thinner films of capillary

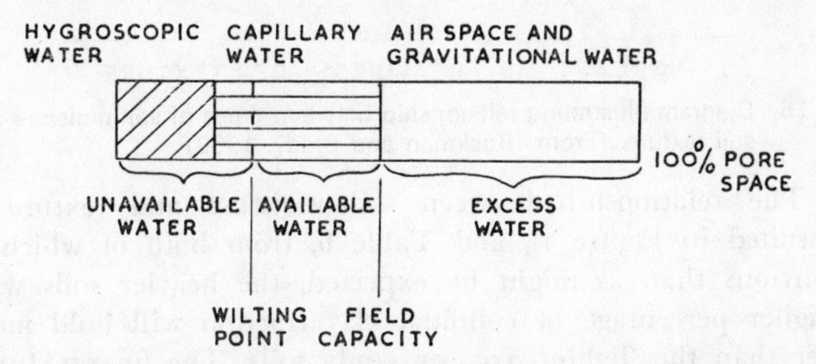

Fig. 14 Diagram illustrating relationship between types of soil-water and that available to plants. (From: Buckman and Brady, 1960)

water are held with such force as to be virtually unavailable for plant growth. They are released too slowly to prevent wilting. Most of the capillary water is readily available for absorption by plant roots, and this, in fact, is the main source of soil water used by plants. A soil is said to be at *field capacity* (Fig. 14) when it is holding the maximum amount of capillary water possible; under these circumstances the micro-pores will be filled with water but the macro-pores will be occupied by air. Moisture in excess of field capacity is of limited use and may even be detrimental. If all pores are filled with water, the soil is saturated and under conditions of slow drainage the resultant exclusion of air will inhibit all but the more specialised types of plants and soil micro-organisms. Also water in excess of that which can be held or retained by the soil is the main agency for the removal by *leaching* of soil nutrients.

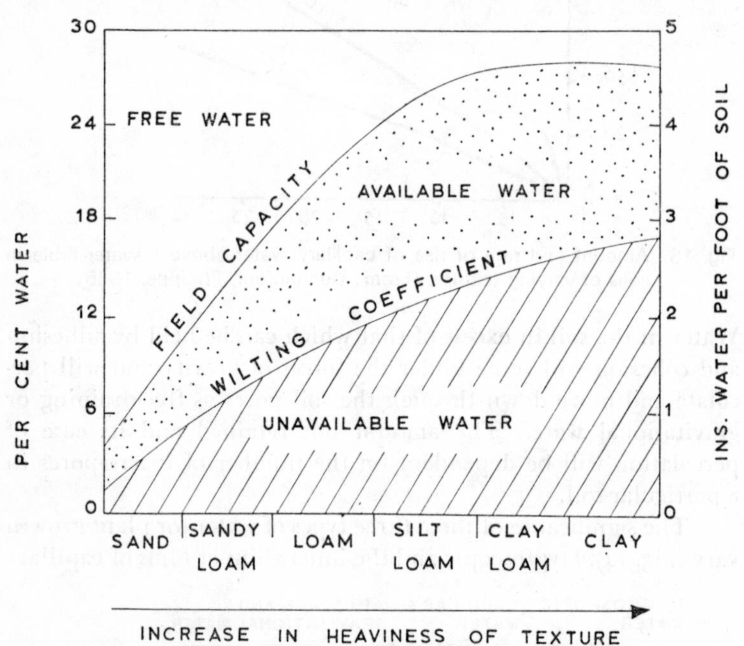

Fig. 15 Diagram illustrating relationship between types of soil moisture and soil texture. (From: Buckman and Brady, 1960)

The relationship between soil moisture and texture is illustrated in Figure 15 and Table 6, from both of which it is obvious that, as might be expected, the heavier soils with a higher percentage of colloidal clay fraction will hold more water than the 'lighter' coarser sandy soils. The finer-textured soils will also impede water movement, will be slow to drain and will be more susceptible to water-logging and, consequently, oxygen deficiency. On the other hand, the coarser soils are characteristically well-drained but as a result are more highly leached and suffer more readily from drought. They will however warm up and lose heat more rapidly than the heavier soils;

Soil	1 Organic Matter	2 Hygroscopic Coefficient	3 Field Capacity	4 Capillary Water (2 + 3)	5 Max-retentive Capacity
Fine Sand	2·13	3·4	7·6	4·2	44.5 per cent
Sandy Loam	3·01	6·9	15·5	8·6	58·0 ,, ,,
Silt Loam	3.58	10·4	24·0	13·6	76·5 ,, ,,
Silty Clay	5·91	16·1	30·4	14·3	87·0 ,, ,,

Table 6 Percentage hygroscopic and capillary water capacities of various soils. (From: Buckman and Brady. 1960)

the latter are 'cold' because of their higher moisture content as a result of which they warm up more slowly but retain their heat longer. Within an area of relatively uniform atmospheric climatic conditions the soil texture of such contrasted habitats as a sand dune and clay flat may modify the soil 'micro-climatic' conditions sufficiently to make the one too dry, the other too wet, for certain plant species. Soil temperatures on the colder clays may also delay the commencement of growth for several weeks after that on the drier warmer sandy habitat.

In many soils there is a tendency for the individual soil particles to cohere together in 'clumps' or aggregate groups to give what have been described as 'compound particles'. These give to the soil a particular *structure* or what the farmer would describe as its 'tilth'. The size and shape of such aggregates vary; they may be large and blocky, small and nodular or thin and platy. That most favourable to organic life in the soil and to plant growth, and which the farmer endeavours to produce by his methods of cultivation, is a *crumb* or granular structure, i.e. small 'lumps' of about 3mm–6mm diameter. The exact mechanism of crumb formation is not fully understood but among the contributory factors are organic matter which along with colloidal substances and organic secretions help to bind mineral particles together. The presence of lime promotes the flocculation of fine clay particles. Also, it has long been known (though not fully understood why) that grass roots appear to be instrumental in producing crumb structure. The crumbs provide a very favourable balance between soil, air and moisture; they are porous but also contain fine water-retaining particles; and a soil with such a structure therefore combines the advantages, and mitigates the disadvantages, of the excessively coarse or excessively fine-textured soil.

The physical and chemical properties of the mineral fraction in any soil are, however, profoundly influenced by the presence of organic matter. Between these two components there exists such a close inter-relationship that an independent analysis of their characteristics and of their effects on plant growth must, at best, be a text-book expediency. The organic matter in a soil consists of an accumulation of plant and animal tissues in various stages of decomposition. The primary and most abundant source is supplied by green land plants whose contribution far exceeds that from all other organisms combined. In fact they return to the soil much more material than they ever extract from it and in doing so provide food for a vast population of other living organisms within the soil, which, after 'eating' and 'being eaten', themselves become part of the soil organic matter. The presence of these organisms is (as has already been mentioned) essential to the soil and to the plants which it supports. Without them organic matter would be of little use, and if they were to fail other forms of life could not be maintained. All assist directly or indirectly in the decomposition of organic tissues. The majority are microscopic; those of plant nature, particularly the bacteria and fungi, are most numerous and of greatest importance since they effect the greatest amount of organic decomposition. The soil animals, however, also contribute to this process; some prey on living or dead plants in the soil, some on other animals. Many of the larger members such as rodents, insects, millepedes and centipedes, slugs and snails, spiders and earthworms through their eating and burrowing habits not only help to break down organic matter but to distribute it more evenly through the mineral material. In this respect, particularly in temperate regions of the world, none is so important or effective as the various species of earthworm. In a good, well-aerated soil, there may be as many as 250 000–1 000 000 per hectare; it has been noted that the weight of worms in pasture land may be greater than that of the livestock grazing on it. They are the natural tillers of the soil. They eat their way through the soil. As it passes through their digestive tracts organic matter is decomposed and thoroughly mixed with the mineral particles ejected at the same time. It has been estimated that the weight of worm casts produced per hectare can range from one to twenty-five tonnes and that the weight of soil passing through their alimentary tracts can be anything from four to thirty-six tonnes per hectare per year. They progressively bring soil material from below and re-deposit it near or on the surface: evidence from archaeological sites suggests that this may be, in some instances, at a rate of fifteen to twenty centimetres a century. In tropical regions, termites or white ants fulfil much the same rôle as earthworms.

The breakdown and decomposition of organic matter in the soil takes place in a series of stages with, in the latter phases, the various species of bacteria and fungi being the predominant agents. This process is accompanied by the liberation of such substances as water, carbon dioxide, complex organic compounds including organic acids and eventually simple end-products—the nitrates, sulphates and phosphates, soluble in the soil water, and capable of being absorbed by plant roots. The final product of decomposition is a brown to dark-brown, amorphous, chemically complex material called *humus* (though, in many instances, the term humus is used to describe all the organic matter, at whatever stage, in the soil). Humus is the greatly altered, highly modified or resynthesised remnants of the most resistant materials in plant tissue. Just as clay is the end-product of the weathering of complex mineral matter, so humus is the final stage in the process of organic decomposition called *humification*. Its properties, and its effect on the soil, as will be explained later, are in many ways similar to those of the clay fraction. The two often occur together in very close association in the soil to give what is usually referred to as the clay-humus complex or the 'colloidal complex'.

The amount of organic matter and humus in any soil will depend on the balance between the amount and rate of supply on the one hand, and on the rate of decomposition of plant tissues on the other. Both are affected by the climatic conditions, the type of vegetation and the physical and chemical properties of the soil in a given habitat. The most complete and efficient type of decomposition is that which takes place in the presence of oxygen through the activities of aerobic bacteria and associated fungi. This process of oxidative 'combustion' eventually results in the complete 'mineralisation' of organic matter—the final breakdown into what has sometimes been graphically described as 'a drop of water, a puff of gas (CO^2) and a pinch of mineral salts', i.e. the simple elements from which organic matter was originally synthesised! That even the humus finally decomposes is testified by the fact that under naturally stable conditions its amount remains relatively constant although fresh organic matter is continually being added to the soil. Active decomposition is promoted by the same factors which promote plant growth; and this convenient synchronisation of bacterial and plant activity ensures that the nutrients released in decomposition become available when required and are protected from unnecessary wastage by leaching. For efficient functioning aerobic bacteria require moisture, oxygen, mineral nutrients (for which they actually compete with the growing plants) and 'mild' non-acid soil conditions.

The rate of decomposition is considerably increased by high temperatures; ordinary soil temperatures seldom are so high or so low as to kill bacteria which attain optimum activity between 21–38°C. Rapid organic decomposition (and not only in the soil) is a characteristic feature and problem of daily life in the humid tropics. Also the rate of bacterial activity can be quickened by increased aeration such as occurs when a protective vegetation cover is removed or by the opening-up of soils by cultivation. Organic decomposition will also be more rapid in the coarse-textured soil than in those where a high proportion of the clay fraction may engender oxygen deficiency.

Under conditions of continuously high temperature and humidity, where plant growth is rapid but where the rate of organic decomposition is correspondingly intense, the organic-matter content, as in many tropical soils, may be less than one per cent. In cold and cool temperate regions, where temperatures are lower, bacterial activity is consequently slower and may, in addition, be inhibited during periods of winter cold or summer aridity; hence the amount of organic matter and humus in the soil is greater. However in most mineral soils it rarely exceeds fifteen per cent. of the volume. There is a marked and characteristic difference between the amount and distribution of organic matter in temperate grassland and forest soils. In the latter the major contribution is from leaf-fall on the surface and this becomes incorporated gradually with the mineral particles below through the activities, particularly, of earthworms. In grassland, on the other hand, most of the organic matter is produced beneath the surface by the dense ramified network of roots and rootlets. Each year these add a very much greater weight of organic matter than do the sub-aerial leaves. Under comparable site conditions the humus content of forest soils is usually less than fifty tonnes per hectare, whereas in grassland soils it may be as much as 600. It has also been noted that the densely matted turf of grassland soils may check aeration and, in retarding the rate of bacterial activity, contribute to the greater amount of humus which accumulates.

	Ash	K_2O	CaO	M_5O	P_2O_5	S_1O_2
Deciduous forest	194	13	89	15	12	56
Evergreen coniferous forest	104	10	54	8	7	20

Table 7 Pounds of minerals taken up annually from one acre. (From: Daubenmire, R. F. 1959)

The chemical composition of the plant tissues, however, also has a considerable effect on the rate of decomposition and, as a result, on the type of organic matter and humus that will be produced. Some plants are 'more demanding' in their mineral requirements than others (see Tables 7 and 8) and, provided these are available in the soil, they will contribute organic matter relatively rich in such nutrients as nitrogen, phosphorus, potassium and calcium. In a well-aerated soil this mineral-rich organic matter will promote rapid and efficient bacterial activity, a relatively rapid turn-over of nutrients and a mild well-decomposed, crumbly, '*mull*' type of humus. With these conditions will also be associated a large number of soil animals, particularly earthworms, as a result of which the incorporation and dissemination of the humus through the soil layers is fairly rapid and uniform. Soils initially deficient in, or later depleted of nutrients, will however only be able to support those plants which are less exacting in their requirements. These plants return to the soil organic matter low in

Beech	2·46 per cent.
Oak	1·07 ,, ,,
Pine	0·99 ,, ,,
Bracken	0·83 ,, ,,
Heather	0·44 ,, ,,

Table 8 Calcium content of different leaf litters (percentage dry weight) (From: Salisbury, E. J. 1947)

nutrients, with a higher proportion of carbon than nitrogen and insufficient calcium to counteract the effect of organic acids produced during decomposition. The resulting acidity inhibits the action of aerobic bacteria and that of the acid-tolerant fungi is somewhat less efficient and complete: humification takes place slowly, nutrients are released slowly and a raw, peaty '*mor*' humus results. Under conditions of excessive acidity there may be a progressive accumulation of partially decayed organic matter. Not only is this type of organic matter deficient in nitrogen and other nutrients, but the acid conditions also inhibit the action of nitrifying bacteria and those which 'fix' gaseous nitrogen. Its relative poverty is further reflected in a sparsity of soil animals and, in the absence of earthworms, mor humus tends to accumulate on the surface and does not become readily incorporated into the underlying mineral soil.

The progressive and pronounced accumulation of organic matter is a characteristic of those soils where drainage is impeded or imperfect. When soils become fully saturated with water the result is a deficiency of oxygen which may so retard decomposition that partially or incompletely decayed organic matter forms a progressively thickening layer of peat. If the water is mineral-rich (or *eutrophic*) it will support a fairly demanding type of vegetation and the resulting peat will be rich, slightly alkaline 'fen' peat. When drained and exposed to oxidation this decomposes easily and provides a very fertile medium for plant growth. When accumulation takes place under acid, nutrient-deficient (or *oligotrophic*) conditions, the resulting organic matter will form a much poorer, acid bog-peat whose nutrient status is comparable with mor humus and into which it often grades.

In most soils however the amount of organic matter in relation to the mineral component is relatively small. The nature of this organic matter is extremely complex and is still only imperfectly understood. It is not at all easy to isolate or distinguish between decomposing organic matter and its end-product, humus: between the two there exists a continuous gradation. As important as its composition, however, is its function in the soil: and its significence in this respect is far greater than its modest quantity might suggest. It is the major source of plant nutrients; the bulk of the minerals necessary for plant life are 'produced' or made available in the decomposition of organic matter. Much of the phosphorus, potassium, sulphur, calcium etc., originally derived from weathered rock is held and kept in circulation in organic compounds in the soil. This is the only form in which nitrogen, that nutrient required in the largest amounts, is available for plant growth. Nitrogen does not exist in inorganic compounds; the only source is atmospheric nitrogen, an inert gas which higher plants cannot utilise directly. It is known that small quantities of nitric acid produced by the electrical discharges during thunderstorms, are carried into the soil by rain water, but there is much disagreement as to the significance of the contribution this makes to the soil. It may however have been an important initial source and may well be of relatively greater importance as a source of nitrogen in tropical regions than elsewhere. More important are the nitrogen-fixing bacteria of the soil. Some are independent free-living organisms which utilise atmospheric nitrogen to build up their tissues which eventually are added to the organic matter. Of greater significance, however, are those nitrogen-fixing bacteria which form nodules on the roots of certain plants with which they maintain a symbiotic relationship. They are more particularly associated with the leguminous plants—clovers, lupins, peas, beans and other members of this large and varied family which includes many trees and shrubs, particularly in tropical regions; there are, in addition, several non-leguminous species—such as the common alder—which are nodule-bearing. From the plant the bacteria derive carbohydrates while they in turn build up nitrogenous compounds on which plants can draw. The nitrogen-fixing bacteria form an essential link in the nitrogen-cycle; they replenish the only source in the soil, the organic matter.

Finally organic matter has, like the clay fraction, colloidal properties which are of great importance. Its capacity to imbibe water (up to eight or nine times its original volume) is even greater than that of clay. As has already been noted, it helps to bind mineral particles together and hence increase their stability. Since it exists in a very finely divided state its 'surface attraction' is even greater than that of clay and it thus greatly reinforces the water and nutrient retaining properties of the soil. Not only are the clay and organic matter comparable in many ways but they are so closely associated that the properties of the clay-humus or 'colloidal-complex' can be considered together. The chemical and physical properties of a soil are largely controlled by this complex and it plays a major part in the supply of nutrients that become available for plant growth.

All the essential plant nutrients except carbon, hydrogen, and oxygen are obtained from the soil. Some of the macro-nutrients are required in relatively large amounts, while only a 'trace' may be needed of the micro-nutrients (see page 25). Plant growth can be retarded, or the presence of a particular species excluded from a given habitat because: (1) these are lacking in the soil; (2) they become available for use by plants too slowly; or (3) they exist in incorrect proportions and are therefore not adequately balanced in the soil. The main sources of nutrients are either rocks or organic matter. In these they exist as complex insoluble compounds and it is not until they are released, by weathering in the former, and decomposition in the latter, as simple *soluble* forms that they become available for plant use; plant roots can only absorb these nutrients from the *soil solution*. Two examples given by H. O. Buckman and N. C. Brady of this transfer of plant nutrients from complex insoluble unavailable to simple, soluble available forms are:

(1) *Nitrogen*

Organic Matter → Ammonium Salts → Nitrate Salts → Available soluble nitrate salts

NH_4^+ → NO_2^- → NO_3^-

Decomposition and Nitrification

(2) *Potassium*

Rock Mineral e.g. Microline Feldspar + Carbonic acid + water → Hydrated Silicate + Available soluble Potassium Carbonate and insoluble Silica

$(2KAlSi_3O_8) + (H_2CO_3) + (H_2O) \rightarrow$

Chemical Weathering

$(H_4AlSi_2O_9) + (K_2CO_3) + (4SiO_2)$

Clay Mineral

The nutrients, when released, may be washed out of the soil completely. Some may be absorbed immediately by the plant roots and other micro-organisms, while some may be retained temporarily by the soil clay-humus fraction. The latter, in fact, acts as a sort of store-house of available plant nutrients and thereby checks their loss in drainage water.

In the soil solution inorganic mineral salts frequently exist as electrically-charged units called *ions*; they may be positively charged *cations* or negatively-charged *anions*. For example, in solution, a compound like common salt (sodium chloride or NaCl) will tend to break-up or dissociate into Na^+ cations and Cl^- anions; similarly water (H_2O) will dissociate into H^+ and OH^-, lime ($CaCO_3$) into $CO_3^{=}$ and Ca^{++}. The more important ions present in the soil solution may be those of:

Nitrogen	NH_4^+, NO_2^-, NO_3^-
Phosphorus	$HPO_4^{=}$ $H_2PO^-{}_4$
Photassium	K^+
Carbon	$CO_3^{=}$ HCO_3^-
Calcium	Ca^{++}
Magnesium	Mg^{++}
Sulphur	$SO_3^{=}$ $SO_4^{=}$
Water	H^+ OH^-

The anions are present mainly in the soil solution, while the positively-charged cations can be attracted to and held on the surfaces of the minute, colloidal clay/humus particles which behave like negatively-charged anions. In humid regions of the world, the most abundant cations are those of calcium and hydrogen with smaller amounts of magnesium, potassium and sodium. A soil's capacity to 'hold' or 'adsorb' minerals in this way depends on its texture, particularly on the proportion of clay particles and organic matter it contains, in which this property is vested. Coarse sandy soils low in organic matter are frequently low in mineral nutrients and those that become available are very susceptible to leaching; from the agricultural view point they are 'hungry' soils because a high proportion of the nutrients supplied in artificial fertilisers may be lost in drainage water.

The cations held by the clay/humus particles can however be replaced by, or exchanged for other ions of equivalent 'chemical value' (or valency); for this reason they are called *exchangeable ions.* The tenacity with which ions are held and the ease with which they can be replaced vary. The order of strength or retention or adsorption is $H > Ca > Mg > K > NH_4 > Na$, i.e. sodium is the most, hydrogen the least easily displaced or exchanged. Hence it is not surprising that hydrogen and calcium are the most abundant ions in humid soils, and that of the basic minerals, calcium is the most abundant. In humid regions water is constantly moving downwards through the soil and dissolving carbon dioxide from the atmosphere and the organic matter and hence forming a weak carbonic acid. The latter, plus the organic acids produced during decomposition, are continually adding hydrogen ions to the soil solution. These can displace the exchangeable calcium ions retained by the colloids and this base, being readily soluble, is easily leached out of the soil in drainage water. Unless there is an abundant source of calcium in the parent mineral-material, or in the organic matter, to counter-balance the hydrogen ions added, the latter may replace most of the calcium ions. With the progressive loss of calcium the soil tends, as a result, to become *acid* in reaction. The acidity or alkalinity of a soil is expressed in terms of the hydrogen-ion concentration of the soil solution, or its pH. This value is expressed as a negative logarithm; that is to say if the concentration is one-millionth of a gram per litre it is $1 \div 10^6$, or 10^{-6} which is, by convention, recorded as pH 6. High pH values then indicate a smaller hydrogen concentration and therefore greater alkalinity, while the lower the values the greater the degree of acidity (see Table 9)

	0 lower limit of scale
	1
	2
Extremely acid	3
Acid	4·0–5·0
Moderately acid	5·0–6·0
Slightly aicd	6·0–7·0
Neutral point	6·5
Slightly alkaline	7·0–7·5
Moderately alkaline	7·5–8·0
Alkaline	8·0–9·0
Extremely alkaline	over 9·0
	10.
	11.
	12.
	13.
	14. upper limit of scale

Table 9 pH scale: soil solutions are unlikely to be less than 3 or greater than 10.

Various interrelated factors tend to promote soil acidity. An acid parent-material derived from rocks low in bases, such as sandstone, quartzite, or granite, will support only those plants tolerant of nutrient-deficient conditions, which produce on decomposition a poor humus and abundant organic acids. Carbon dioxide produced during decay and respiration in the soil combines with rainwater to form carbonic acid. The pollution of air and water with industrial wastes—particularly sulphuric acid—may be an additional source. Where these conditions are combined with high rainfall and coarse texture the removal or *leaching* not only of the easily exchangeable and soluble calcium, but of other minerals normally little soluble in pure water will be the more severe. Under conditions of very high acidity, particularly when promoted by the presence of organic acids, the colloidal clay and humus fraction of the soil becomes unstable. In the absence of calcium the fine clay particles tend to deflocculate and can be leached downwards. Also the clay mineral itself tends to undergo decomposition liberating compounds of iron, aluminium and manganese which also become soluble and can be carried down through the soil. Part of the humic matter goes into solution and gives the brown coloration so characteristic of water draining from peaty soils.

The pH of a soil also affects the relative availability for plant growth of those mineral nutrients present. At approximately pH 6·5—when the reaction of the soil solution is just about neutral—all minerals are sufficiently available to satisfy plant requirements and there is enough calcium to counteract acidity, maintain the stability of the clay/humus complex, and promote crumb structure. Under conditions of increasing alkalinity, however, certain minerals may be rendered insoluble or their

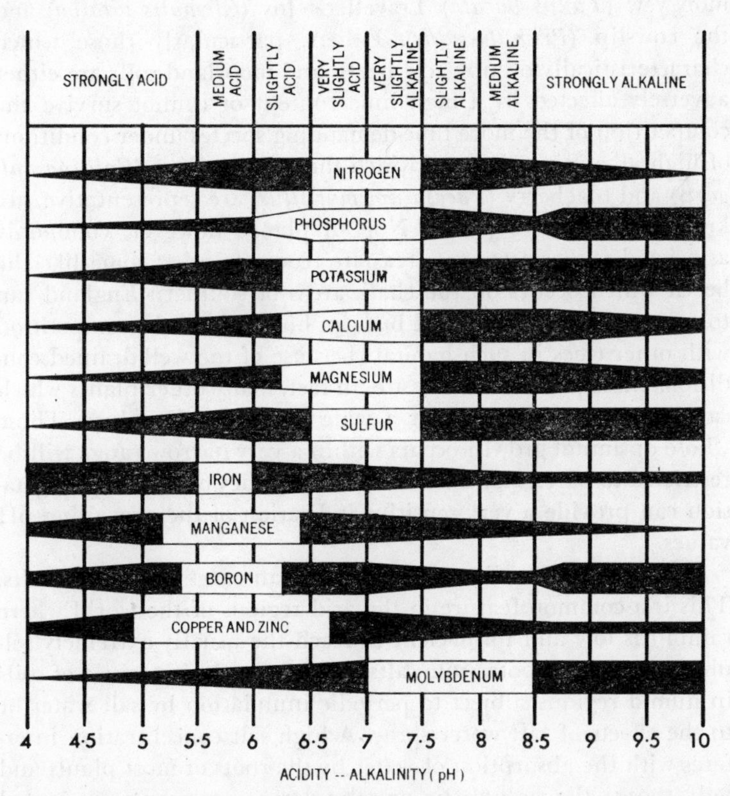

Fig. 16 Diagram illustrating availability of plant nutrients in relation to soil pH. (From: Pratt, Christopher J. 1965)

absorption by the roots of certain plant species inhibited in some way (see Fig. 166). The uptake of phosphorus, iron boron, copper and zinc may be checked despite their presence in the soil; and this type of deficiency may inhibit the growth or preclude the presence of certain demanding species. On the other hand extremely acid conditions can promote the solubility of iron and aluminium to such an extent as to be toxic for some plants; and phosphorus may be rendered unavailable since it tends to combine with these minerals to form insoluble compounds.

Plants and also the various soil micro-organisms vary in their minimum and optimum requirements for soil nutrients and also in their ability to tolerate certain degrees of soil acidity or alkalinity. Most agricultural crops in Britain thrive best in neutral to very slightly acid soils (pH6–6·5). Some plants however will grow only on soils, like chalk and limestone, which have a high calcium content—these are the so-called *calcicole* species such as the common yew (*Taxus bacata*) Traveller's Joy (*Clematis vitalba*) and the cowslip (*Primula veris*). Others, particularly those which characteristically occupy acid heath and moorland soils are either adversely affected by a high lime content or cannot survive the competition of the more lime-demanding species under conditions of high alkalinity; these, of which the common ling (*Calluna vulgaris*) and blaeberry (*Vaccinium myrtillus*) are representative, are known as *calcifuge* plants. Not all plants, however, commonly associated with calcareous areas are true calcicoles; some like the beech which occurs on the chalk areas of southern England can tolerate a wide range of pH but can hold its own in competition with other trees in such habitats because of the well-drained conditions these provide. There are, in fact, many other plants which can grow successfully under a wide range of pH values. Those whose optimum growth occurs within a very narrow range will be restricted to particular habitats and when they occur in the vegetation can provide a very sensitive indication of the prevailing pH values.

Finally, some soils may suffer from an excess of sodium salts. This is a common feature in the arid regions of the world where rainfall is low and insufficient to leach the usually extremely soluble sodium and potassium salts. It is also a characteristic of soils in humid regions subject to periodic inundation by salt water or to the effects of salt water spray. A high salt concentration interferes with the absorption of water by the roots of most plants and only those—the *halophytic* or salt-tolerant species—are adapted to grow under these rather extreme conditions. Among the halophytes there are those which can tolerate a high degree, others only a low degree, of salinity. A distinction is usually made between saline and alkaline soils. In the former there exists an excess of free salt sufficient to inhibit the growth of non-halophytic plants. The amount of exchangeable sodium is less than fifteen per cent. and the pH is usually below 8·6. Where the water table comes near the surface or irrigation is practised intense evaporation of water from the soil gives rise to a concentration of salts, on or just beneath the surface which produces a whitish 'bloom' or 'crust'. In alkali soils the amount of exchangeable sodium normally exceeds fifteen per cent. and the pH is between 8·5 and 10. Under these conditions both clay and humus become dispersed. The resulting lack of aeration and poor drainage, together with the high sodium, result in extremely infertile soils and inhospitable habitats.

References

ALBRECHT, W. A. 1957. Soil fertility and biotic geography. *Geogr. Rev.*, **47** (1): 86–105.

ARNOLD, P. W. 1964. Soil plant relationships. University of Newcastle upon Tyne, *Inaugural Lectures*.

BLACK, C. A. 1968. *Soil plant relationships*, 2nd ed. Wiley, London.

BRADE-BIRKS, G. S. 1944. *Good soil*. English Universities Press, London.

BUCKMAN, H. O. and BRADY, N. C. 1960. *The nature and properties of soils*, 6th ed. The Macmillan Co.

BURGES, A. 1958. *Micro-organisms in the soil*. Hutchinson University Library, London.

DAUBENMIRE, R. F. 1959. *Plants and environment*. Wiley, London.

DUCHAUFOUR, P. 1965. *Précis de pédologie*, 2nd ed. Masson et Cie, Paris.

DONAHUE, R. L. 1965. *Soils: an introduction to soils and plant growth*. 2nd ed. Prentice-Hall, London.

GARRETT, S. D. 1963. *Soil fungi and soil fertility*. Pergamon Press, London.

HALL, Sir A. D. 1945. *The soil*. 5th ed. (New and revised edition by G. W. Robinson). John Murray, London.

HANDLEY, W. R. C. 1954. *Mull and mor formation in relation to forest soils. Bull. For. Comm., Lond.*, **23**: H.M.S.O., London.

JACKS, G. V. 1954. *Soil*. Nelson, London.

JENNY, H. 1941. *Factors of soil formation*. McGraw-Hill, New York.

JENNY, H. 1958. Role of the plant factor in the pedogenic functions. *Ecology*, **39** (1): 5–16.

MUIR, J. W. 1955. The effect of soil-forming factors over an area in the South of Scotland. *J. Soil Sci.*, **6** (1): 84–93.

PEARSALL, W. H. 1950. *Mountains and moorlands*. Collins, London.

PRATT, C. J. 1965. Chemical Fertilisers. *Scient. Am.*, June.
ROBINSON, G. W. 1951. *Soils, their origin, constitution and classification.* Longmans, London.
ROBINSON, G. W. 1937. *Mother earth: being letters on soil addressed to Prof. R. G. Stapledon.* Murby, London.
RUSSELL, E. W. 1961. *Soil conditions and plant growth*, 9th ed. Longmans, London.
RUSSELL, Sir J. E. 1959. *The world of the soil*, 2nd ed. Collins, London.
SALISBURY, E. J. 1947. *Downs and dunes*, Bell, London.
SATCHELL, J. E. 1958. Earthworm biology and soil fertility. *Soils Fert.*, **21** (4): 209–219.
TAYLOR, A. J. 1960. Methods of soil study. *Geography*, **45** (1/2): 52–67.

5
Biotic and Anthropogenic Factors

In order that a particular type of plant may exist in a given area the physical environment of soil and climate must satisfy at least its minimum requirements for growth and reproduction. But whether or not it will be able to occupy a potentially favourable physical habitat will depend on the effect of other plants and animals which are capable of living under similar conditions. No plant exists in isolation: it grows in company with other plants of the same or different species and with the numerous animals dependent on them for food and shelter. Most physical habitats are suitable for occupation by many different types of plants: indeed the more favourable the physical environment the greater the variety of plants, and hence of animals, that could theoretically avail themselves of its possibilities. The actual presence, abundance and vigour of a particular species in a given area will therefore be dependent upon its ability to obtain its essential requirements and to maintain itself successfully *in company with* other organisms.

All the organisms that occupy any part of the biosphere form an integral part of the environment of every other organism in the same place. Between all the plants and animals that occur in a particular habitat there exists a complexly interwoven web of mutual interdependence and interaction. Some organisms are beneficial to or even essential for the existence of others; some are detrimental and may inhibit the co-existence of others. The very presence of plants and animals modifies the atmosphere and soil in which they live. The environment of any plant is, as a result, partly physical and partly biological, and the biological or *biotic* factors, which influence plant growth and distribution, are those which result from the action of living organisms. Biotic factors are, in many ways, more diverse and complex than physical or inorganic factors since they are dependent upon the activities of such a wide variety of organisms. Also, the interaction between the physical and biological environment is such that it is not easy, nor indeed even possible, to analyse independently the relative effects of any one biotic factor. Some may have a direct effect as a result of the actual physical contact between organisms; others operate indirectly through their influence on the physical environment. By and large, however, it can be stated that, in relation to plant growth and distribution, the most important effect of other plants is that exerted indirectly through their modification of the physical environment; while that of animals is through their direct physical contact with the plants on, or among, which they live.

Green land plants modify almost all the physical attributes of a given habitat. By their presence, and particularly their shade and shelter, they alter the light intensity, the temperature and humidity conditions and the movement of the lower layer of the atmosphere which they occupy. They create their own particular micro-climate. Also, through their use of water, mineral nutrients and their contribution of organic matter they influence the condition of the soil to a considerable extent. As a result one type of plant may make a particular physical habitat more or less suitable for other plants. The establishment of a given species in an accessible and physically favourable site will depend on its ability to compete for space, light, water, and soil nutrients with other potential occupants. *Competition,* or 'the struggle for existence' between plants of the same or different species arises because the resources of a habitat are insufficient to meet the demands of all the plants available and capable of growing there. The relative success of individual plants or species in this competitive struggle will be dependent upon their ecological requirements, their life-forms, their vigour and density of growth and their seasonal development. Plants vary in their optimum requirements for successful growth. The 'more demanding' or 'more aggressive' may, because of their greater reproductive capacity, vigour of growth and size, so modify climatic and edaphic conditions as either to depress or to exclude the less demanding and aggressive from a particular habitat.

One of the main forms of competition between plants is for light; and the effect of plants through shading and the consequent reduction of light intensity is by far the most important way by which one type of plant may suppress or exclude another. If the shade cast by one species reduces the light intensity below that required for the optimum growth of another, the rate of growth and vigour of the latter will be depressed and, consequently, its ability to compete for soil nutrients and water will also be reduced. The reduction of light below the minimum for photosynthesis will inhibit its presence completely. It is, however, as well to remember that the effects of such competition are selective; while the reduction of light intensity can depress or prevent the growth of light-demanding species, it can also stimulate and provide the very conditions necessary for the successful growth of shade-plants.

In the struggle for light the competitive ability of plants is closely related to their life-forms and habit of growth. Obviously the tallest plants capable of growing in an area will be the most 'aggressive'; trees will have a competitive advantage over shrubs, which in turn will have an advantage over low-growing herbaceous plants. Among plants of similar height-range, however, ability to compete for light and degree of aggressiveness will be dependent upon the density and seasonal duration of their vegetative cover. Deciduous trees such as beech, oak and ash all have a comparable potential height-growth. Because of the arrangement of its leaves and branches, however, the beech casts a much denser shade, both in winter and summer, than the oak; the ash is even more translucent and rarely reduces the light intensity beneath its canopy to less than seventeen per cent. of full sunlight. Not only has the beech a greater competitive ability but the number of species capable of tolerating its heavy shade are much fewer than in an oak or ash wood of similar age growing on comparable sites. The effect of a stand of trees whose crowns form a continuous canopy will be much greater than that of a widely dispersed open wood or parkland. The paucity or even absence of undergrowth in a dense plantation of evergreen conifers contrasts markedly with the relatively rich variety of shrubs and herbs that can exist under the less continuously shaded conditions in deciduous woodland. The struggle for light is probably nowhere greater than in the tropical rain forest; trees of varying light requirements and with no marked seasonal rhythm of growth form a dense evergreen canopy some thirty to forty metres in depth which reduces light on the forest floor to less than one per cent. of that outside the forest.

Above ground, competition for light dominates the struggle for existence. This influences, and indeed its final outcome may be determined by, competition for soil nutrients and water. Plants vary in their optimum requirements for both; and their ability to obtain these in completion with others will depend on the vertical and horizontal extension of their root system, and the rapidity with which this develops. These rooting characteristics are dependent partly on the species of plant and partly on soil conditions. Root competition will be most severe when plants are drawing from the same level in the soil. It is, however, not at all easy to distinguish clearly between the relative effects of root and shoot competition. Competition between roots of newly germinating plants tends to commence sooner; and the more vigorous and demanding species or individuals may so deplete the supply of nutrients and water that the rate of growth of the less demanding plants will be severely checked. Obviously competition for light, nutrients and water are closely linked. An inability to compete for nutrients depresses vigour of growth and increases the liability of shading-out by more vigorous and aggressive plants. Similarly shading will reduce the vigour of both shoot and root growth and hence the ability to compete for nutrient and water supplies successfully. A crucial stage in the re-afforestation of moorland areas in Scotland is that immediately succeeding the transplantation of young conifers from the forest nursery; competition from a well-established and vigorous growth of heather or bracken may check the young tree's growth completely. An initial application of super-phosphates to the tree roots is often necessary to ensure that the shoot grows rapidly above the level of the competing 'natural' vegetation.

Generally speaking competition is most intense between individuals of the same or different species which make similar demands on the same supply of light, nutrients or water at the same time. It attains its greatest intensity among individuals of the same species, particularly at the seedling stage. Most plants have a very high reproductive capacity, but relatively few progeny (one in every 2 000 beech seedlings, for instance) reach maturity. High 'seedling mortality' is a result of the severe competition between a greater number of plants than can be supported by the habitat. The thinning of crops and forest plantations of smaller, less vigorous plants cuts down competition and ensures the optimum growth of the remainder. Basically, competitive ability depends on the greater growth and vigour, of both roots and shoots, of some individuals at the expense of others. A plant will have its greatest competitive capacity when growing under optimum physical conditions. Under sub-optimum conditions, near the limits of its climatic range or on nutrient deficient soils it may be crowded out by plants better adapted to the particular physical environment. Many plants can be cultivated under a much wider range of climate and soil than they would tolerate in the wild. Protected from competition many plants can grow successfully and withstand conditions under which they could not otherwise survive. Few crops deprived of the protection of man would long maintain themselves in face of the more aggresive undomesticated plants.

On the other hand many plants are excluded, by the competition of even more demanding and vigorous species, from those areas in which they could grow with greater vigour. As a result they may be restricted to habitats which do not permit optimum growth; this is seen by the greater size and vigour certain plants achieve when they colonise, or are cultivated in, a more favourable physical habitat than they normally occupy. The rampant growth of annual weeds on cultivated land is a case in point. It is certainly true of many annual, light-demanding species which are found where harsh conditions of soil and climate provide an open, well-illuminated habitat where they alone can maintain themselves. The opinion has also been expressed that the preponderance of annuals in deserts may be due as much to the absence of competition as to any special adaptation to climatic conditions. It is also known that some halophytic plants attain greater vigour when grown in non-saline soils free of competition. However, a salt tolerance greater than that of other species allows them to survive competition under those conditions of high salinity which would either depress or exclude plants not similarly adapted. The preference of many plants for soils of a certain mineral composition or pH status can be explained not so much by the direct effect of the chemical condition on the plants themselves but by the relative competitive ability of the plants that can grow under such conditions. E. J. Salisbury notes that while sorrel (*Rumex acetosa*) shows a decided preference for acid soil conditions, deprived of competition it can grow with much greater vigour on limed than on non-limed soils. A. G. Tansley quotes the classic example of the heath bedstraw (*Galium saxatile*) which avoids calcareous soils not because it *cannot* grow on them but because, under such conditions, its seedlings develop only very slowly and are, as a result, severely handicapped in competition with other plants which can grow more vigorously. Another form of 'competition', but one about which less is known, is that related to what has been called the anti-biotic effects of organisms. There is evidence—though limited as yet to relatively few species—that some plants produce on decay, or as exudates from living roots, harmful chemicals or toxins. These can apparently kill seedlings of the plant itself or others. Certain desert shrubs, prairie forbs and kauri pine trees are known to 'poison' the soil in this way. It has been suggested that, in some instances, the wide and often exceedingly regular spacings of desert shrubs may be explained by this factor rather than competition for water; but as a result competition for water is avoided.

Competition is the principal biotic factor which, under natural conditions, determines those plants which will be able to co-exist in a given habitat. The physical factors of soil and climate are *permissive,* that of competition is *selective.* The latter operates indirectly through the combined effects of green plants on the physical environment. In addition, however, direct contact and interaction between plants may also be mutually beneficial or detrimental and may, under certain circumstances, become significant in determining the presence or absence of a particular type of plant in a given habitat. Climbing plants, such as vines and other lianas, though rooted in the soil, are dependent on others for physical support. This particular form gives them a competitive advantage in warm and moist environments where the struggle for light may be intense. Others including certain algae, fungi, mosses and liverworts, and orchids, are *epiphytic;* they must rely on larger plants completely for support, though not for nutrition. In some instances their growth and abundance may be such as to depress or even damage the plant to which they are attached. In this respect, however, *parasites* which derive the whole or part of their nutrition from their host plant are more powerful and destructive biotic factors; the majority are micro-organisms—bacteria and fungi—but include some higher green plants such as mistletoe. Under natural conditions a balance is normally achieved between host and parasite, and the distribution of the latter will be dependent on that of the former. The extent to which parasites may limit the distribution of plants is unknown. However they can depress the growth and vigour of their host plants to the extent that their competitive ability is reduced. Bacterial or fungal attacks virulent enough to deplete or destroy a particular type of plant do occur. This however tends to happen more often than not in modified or disturbed habitats. A classic and oft-quoted example is that of the Chestnut blight which has all but eliminated the Sweet Chestnut tree from the eastern United States; this was the result of the 'accidental' introduction of a parasite into a new environment in which the host trees did not possess a natural resistance to attack. Also, in choosing and helping to create the most favourable conditions for the growth of crop plants man provides the optimum condition for the concentration and proliferation of its parasites. The ravages of the cotton boll-weevil in the U.S.A., the potato blight in Ireland, and the phylloxera infestation of vines in France all testify to the potency of the parasite as a biotic factor.

There are, however, many plants whose relationships and interactions are *symbiotic*: they live together to their mutual advantage. In many such cases, though not necessarily all, the presence of one plant may be dependent on that of the other. Probably the most important symbiotic relationships exist between bacteria and fungi on the one hand and green plants on the other. The symbiotic nitrogen-fixing bacteria which exist in nodules on the roots of leguminous and other plants have already been referred to; such nodules also form on the leaves of some tropical plants. The presence of nodulated plant on certain poor, nitrogen-deficient soils is undoubtedly aided by this symbiotic relationship. Also common to many herbaceous plants and to most trees and shrubs is the close association between fungi and roots: the former, penetrating and existing within the living cells of the root, form what is called a *mycorrhiza*. This condition has been most closely studied in relation to the pines and heath plants usually associated with acid soils. In many such cases it has been demonstrated that a mycorrhiza must be present for the successful growth of the particular species. Attempts to introduce pines into areas where soils are poor and the necessary fungal symbiont is lacking have failed. It would also appear that in an area long deforested the mycorrhizal fungi may die out and hence make natural or artificial re-afforestation difficult. Several members of the heath family—including azaleas, rhododendrons and the common heather or ling (*Calluna vulgaris*) are dependent on the presence of mycorrhiza that require a relatively high degree of soil acidity. Under alkaline conditions the disappearance of the symbiotic fungi results in the death of the plant; consequently the soil *must* be acid to allow the successful growth and reproduction of such species.

Biotic factors, however, also include those related to the activity of the animals that live among and are dependent upon the plants growing in a particular area. In contrast to that of plants the effects of animals are primarily direct: in some cases they may be necessary for, in others antagonistic to, the successful growth of certain plants. Many plants are dependent on animals for seed dispersal and for cross-pollination of their flowers. In the latter instance animals—and more particularly insects—may be absolutely essential for the survival of a plant species in an area; it will not be able to exist if the pollenating agent is scarce or absent. Murice Ashby notes the example of the greater bindweed (*Calystegia sepium*), which seldom sets fertile seed in Britain because the convolvulus hawk-moth upon which it depends for pollenation has become rare. This form of interdependence can be so specialised as to create considerable problems for the introduction of plants from native to exotic habitats. For instance it proved difficult to grow good Smyrna figs in Calfornia until the necessary pollenating wasp was also introduced; while the production of red clover in Australia was possible only after the importation of the bumble bee.

The most outstanding and conspicuous effects of animals however tend to be antagonistic rather than beneficial. Plants are the basic food producers and are therefore naturally and inevitably subject to the depredations of the animals which feed on them. In any habitat a close relationship exists between the types and numbers of plants and those of the animals that can be supported directly or indirectly by them. Any individual plant may, in fact, sustain a great number of different types of animals. For instance, Turrill notes that 'the oak provides food and environment for more species of insects and other invertebrates than does any other British plant'. Obviously the growth and reproduction of plants will be influenced by their animal 'dependants'. Under natural conditions a balance tends to be maintained between the numbers, growth and reproduction of plants and those of the animals that feed on them. Under natural conditions a particular type of plant will rarely be completely excluded from a habitat solely because of the destructive effects of animals, since the presence and survival of the latter will be dependent on that of the former. However, the depredation of animals may be a contributory factor in areas where unfavourable physical conditions or excessive competition weaken the vigour and lessen the numbers of certain plants and therefore make them more susceptible to elimination. In Britain the beech grows near its climatic limit; succesful natural regeneration is therefore limited to years of full mast which may occur only once every three or four years. In other years, not only is a smaller number of seeds produced but these and the seedlings which may germinate are often completely destroyed by various kinds of animals. Fluctuations of animal populations do occur, however, and the increase in the numbers of a particular type—a plague of locusts for instance—may be such as to reduce the number of a certain type of plant, or even destroy it. Under natural conditions, Malthusian checks usually operate long before food runs out, which will tend to reduce, through increased mortality, the excessive numbers of the inflated animal population.

The dominant, and now probably the most powerful and destructive animal in the environment of any plant to-day is *man*. There are relatively few areas of the earth's land surface where, to a greater or lesser degree, his activities have not had repercussions on plant growth and distribution. There are relatively few areas left where the plant cover is completely natural, in the sense of being the result solely of the interaction of physical and non-human biotic factors. Man's actions have been both direct and indirect. On the one hand he has modified the physical environment in such a way as to encourage or discourage the presence of certain plants in particular areas. He has created new habitats, be they quarries, coal bings, railway cuttings or bomb-sites. He has altered or modified others through the addition or extraction of minerals and water to the soil, and through his pollution of soil, air and water he has been responsible for making certain physical habitats more, or less, suitable for particular plants. On the other hand the direct effects of man on plant life are manifold. He has accidentally or deliberately introduced plants to areas where formerly they did not exist. He has, by clearing and cultivation, given preference and protection to some plants at the expense, or to the exclusion, of others. Further, he has been directly instrumental in intensifying the destructive, antagonistic effect of animals on plants. He has drastically increased the numbers of certain types of animals at the expense of others, and there are many animals whose numbers man has reduced almost to the point of extinction. He has increased the number of his domesticated grazing animals, sheep, cows, goats, horses, etc., and reduced the number of animals that would normally compete with them for food. He has also through the provision of an increased plant-food supply or through the elimination of competitors or predators effected an increase in the numbers of certain wild herbivorous animals such as deer, rabbits, voles, mice and squirrels. As a result man has been directly responsible for the *intensification of grazing* and the emergence of this animal activity as an important ecological factor. Closely allied to his promotion of grazing has been his use, both accidental and deliberate, of *fire* as a means either of modifying or destroying the natural plant cover.

Grazing becomes an important destructive or selective habitat factor when its intensity is such as to injure or prevent the successful growth and reproduction of plants. Susceptibility or resistance to various types and intensity of grazing depends on the type of animal involved and also on the palatibility and growth-form of the particular species available. Animals differ in their method of eating and in their preference for particular types of plants. Horses, sheep and cattle graze mainly on herbaceous species, while goats and deer more frequently 'browse' on trees and shrubs. Sheep, rabbits and certain other rodents are notoriously close croppers, while cattle, because they crop vegetation by twisting it round their tongues then pulling and tearing rather than biting it off, prefer taller and ranker plant growth. But the preference of animals for particular types of plants is also closely determined by their nutritive requirements and by the relative palatibility or desirability of the species available. Many plants are avoided by grazing animals because they are distasteful, hairy, prickly, or even poisonous. The persistence and continued spread of bracken (*Pteridium aquilinum*) on heavily grazed rough pasture in Scotland is undoubtedly aided by the fact that it is slightly poisonous, particularly to young stock. And the survival and prevalence of shrubs such as elder (*Sambucus nigra*), gorse (*Ulex spp*), broom (*Sarothamus scoparius*) and the common weeds rag-wort (*Senecio jacobaea*) and creeping thistle (*Cirsium arvense*), in face of grazing can be attributed to their lack of palatibility for one reason or another. Their freedom from attack, however, will be relative to the availability of other more desirable forms of herbage, and the degree of selectivity of the animals present. In this respect sheep are notoriously selective of the more palatable and nutritious plants; cows and in particular goats are much less fastidious.

Among the most susceptible to the effects of grazing are often, therefore, the taller, more conspicuous or 'tastier' species. Annual plants and tree seedlings are particularly vulnerable. They may be destroyed completely, their numbers thereby drastically reduced and their regeneration limited or, under extreme circumstances, completely prevented. More resistant to grazing are perennial herbaceous plants with underground food-storage organs, such as bulbs, corms, rhizomes etc., whose buds are well-protected from injury or removal. Such plants are capable of making new growth despite the removal of some or all of their vegetative parts. However a balance must be maintained between the rate of removal of their green photosynthesising parts and the storage of food for continued growth. For instance it has been demonstrated that if bracken is cut regularly during its growing season for at least three consecutive years, storage of food in underground rhizomes will be so reduced as to weaken or completely depress its growth for several years, if not destroy it completely. Intensity of grazing may be such that a particular plant is 'eaten out'; in many moorland areas of Britain the common heather or ling has disappeared from areas where it has been unable to maintain growth in face of heavy grazing pressure.

No group of plants is better equipped to withstand grazing and trampling than the grass family (and also the closely related sedges). Not only are many species of grass particularly resistant to grazing but their growth may, in fact, be stimulated by it. The natural environment of many large herbivorous animals is the grassland areas of the world. The evolution of the one was probably closely related to that of the other. The capacity of perennial grasses especially, to endure grazing is a function primarily of three important characteristics of their growth-form: their mode of growth, their method of branching and the nature of their root system. Unlike many plants whose growing points are located at the apex of leaves and shoots, the reproduction of fresh tissue occurs chiefly at the *base of the grass leaves*. This part is least likely to be damaged by grazing and allows regrowth to continue at the same time as material is being removed. This rapid growth-renewal is also combined with a rapid production of new shoots. Characteristic of many species of grass is the ability to produce numerous fresh lateral shoots by *tillering*—that is to say from new axial buds which form at the base of older shoots. Further, this process is stimulated by grazing or cutting. Not only does it provide the plant with a most effective means of recovery but it allows growth and photosynthesis to proceed at a high rate. Finally grasses are characterised by the production of numerous *fibrous, adventitious roots,* the rapid growth of which allows them to survive damage by either trampling or burrowing.

The general effect of grazing then is to reduce the proportion of those plant species less able to withstand certain intensities of grazing than others. It tends to result in a reduction of plant variety in any habitat—a sparser vegetation and fewer species than would otherwise exist given the prevailing physical conditions. Some species are eliminated. The destruction of tree and shrub seedlings may effectively prevent the regeneration of woody growth. The depredations of sheep, goats and, in some instances, pigs, have not only contributed to deforestation but have directly prevented the regeneration of natural woodland in the long-settled areas of Europe, not least in Mediterranean countries.

Sheep have played a major rôle in maintaining the treeless condition of many upland areas in Britain; the rapid colonisation by birch seedlings of well-drained heath and moorland cleared of grazing animals testifies to their limiting effect. The necessity to protect new plantations from attack by sheep, deer and other grazers of tree seedlings considerably increases the expense of re-afforestation in such areas.

Many plants, though not necessarily eliminated, are reduced in number; the vitality and vigour of the survivors is weakened, and competition from other less-affected plants is thereby favoured. Relieved of competition from tree growth, bracken grows with greatly increased vigour. With the disappearance of heather, grasses and other plants can become more abundant. The greater vigour and competitive ability of many species more resistant to or less affected by grazing is further stimulated by concurrent soil changes. The continued grazing, particularly of the more nutritious and demanding plants, extracts from the soil mineral nutrients and potential organic matter which can never be fully replaced by animal excreta alone. On unimproved grazings the gradual decline in soil fertility which inevitably results contributes to the greater competitive capacity and aggressiveness of poorer, less demanding species. The spread of the tough, wiry, moor mat grass (*Nardus stricta*) and heath rush (*Juncus squarrosus*) over considerable areas of rough grazing in Scotland are indicative of a concomitant decline in grazing quality and soil fertility. Such changes are the outward and visible signs of *over-grazing*, a condition under which intensity of grazing (either in terms of numbers of animals or duration) causes a depletion either of a particular species, or of the vegetation cover as a whole, at a rate greater than they can be renewed and maintained by re-growth. Under extreme circumstances over-grazing can result not only in the elimination of certain species and decline of soil fertility but also in the weakening or removal of part or most of the vegetation cover. The exposed, unprotected soil becomes, particularly where slopes are steep or drought occurs, susceptible to accelerated soil-erosion.

Comparable to grazing in its effect on the growth and distribution of plant life is *burning*. Fire, that 'first great force employed by man', is an environmental factor as important as grazing and probably of even more widespread occurrence. Like grazing, fire is not a completely 'artificial' man-initiated process, though its emergence as a major ecological factor is undoubtedly due to man's accidental and deliberate promotion. Vegetation fires can be started naturally by lightning or volcanic activity. In those parts of the world with a prolonged warm, dry season during which desiccated and very combustible plant debris accumulates, climatic conditions increase the inherent risk of fire. For such reasons, combined with the fact that fire is not in itself an 'organic activity', some would regard it as a physical rather than a biotic or anthropogenic factor. The extent, however, to which fire was an important ecological force before man evolved is difficult to assess. He has undoubtedly been responsible for making it such a widespread and frequent occurrence that the present vegetation of vast areas of the world's vegetation cover can only be fully understood in the light of its effects.

Archaeological and botanical evidence have made it increasingly obvious that the burning of vegetation is as old as man himself. As Carl Sauer remarks, 'even to Palaeolithic man, occupant of the earth for all but the last one or two per cent. of human time, must be conceded the gradual deformation of vegetation by fire.' Charcoal and herbaceous pollen preserved in peat deposits indicates that forest burning in Europe began to increase markedly in Neolithic times. That the vegetation of North America had long been subjected to fire in the pre-Columban period is now widely acknowledged. Stumps of ancient redwood trees (*Sequoia sempervirens*) in California reveal a frequency of at least four fires a century during the last 1 100 years. In prehistoric times many vegetation fires must have been started—as they are today—accidentally. Since man first set his 'camp-fires' the risk and frequency of accidental fires have increased enormously. Deliberate firing of vegetation was, and indeed still is, used by primitive man to open up forest and woodland areas in order to facilitate hunting and collecting and to drive game into the open. The use of fire to eliminate useless or dead vegetation in order to stimulate the growth of new, fresh, green shoots of grasses and other forage plants is a time-honoured practice in pastoral areas; it is exemplified today in such widely separated and contrasted areas as the moorlands of Britain and the savannas of inter-tropical Africa.

Fire is similar to grazing in being both a limiting and selective habitat factor. Its effects on vegetation are dependent on the type, frequency and intensity of burning and on the relative resistance of different species of plants to injury or destruction. Ecologists frequently distinguish between three types of fires. *Surface fires,* as the term suggests, sweep over the ground surface rapidly burning off litter and the above-surface parts of low-growing plants. The intensity of such fires can be increased by high wind-force, and exceptionally dry climatic and soil conditions combined with an excessive accumulation of surface litter. When the soil contains a thick organic layer this may, during a particularly severe burn, be ignited and continue to smoulder for a long period after a surface fire has been extinguished. Such *ground-fires* may result in the destruction of roots and underground plant organs, as well as part or all of the soil organic matter. The most damaging type of fire is the *crown-fire* which in forest, wood or shrub land spreads from the crown of one plant to another and, unless the ground surface is particularly moist, may completely destroy all surface vegetation and soil organic content. Such are the holocausts which in a few hours can destroy several hundred years of forest growth.

Plants vary in their susceptibility to different intensities of burning. Those which can survive and which, in some cases, are stimulated by fire are the so-called *pyrophytes* (the fire or fire-resistant plants). Most liable to damage or destruction are annuals, tree seedlings, and woody growth in general. Among the most resistant are perennial plants with underground food-storage organs from which renewed growth can take place after surface burning. These, such as for example heather, bracken, and many perennial grasses, often have new growth stimulated by the removal, particularly, of old or dead surface parts. For this reason burning has long been recognised and used as a means of promoting the vigorous growth of those plants which are of the maximum use as forage for grazing animals. Above all, fire favours grass and other perennial herbs, at the expense of trees and shrubs, since the former are so much better adapted to withstand its effects. From grasses fire removes only one year's growth much of which may already be dead; from trees it may be tens or even hundreds of years' growth. Perennial grasses not only recover quickly, often growing with increased vigour, but they can produce abundant seed in one or two years after germination. Most trees and shrubs, however, require several years before producing seeds and fire may recur so frequently as to prevent them completing their life cycle and hence regenerating.

Some trees and shrubs are, however, more resistant than others. The woody monocotyledons (the palms and palmettoes of inter-tropical regions) are particularly well adapted to withstand injury. They do not possess the vulnerable growth tissues (the cambium) in their trunks as do other types of trees. Among the latter, greater resistance to fire may be because of particularly thick or tough barks as in the cork-oak (*Quercus suber*) or the Japanese larch (*Larix leptolepis*), commonly used as a fire-break round new plantations in Britain. The foliage and wood of many deciduous and broad-leaved evergreen trees burn less readily than those of the highly resinous conifers. The ability, characteristic of many trees except conifers, to produce new shoots or 'suckers' from the base or lower parts of their trunks also aids recovery from burning. Certain trees, particularly some species of pine, can produce seed at a very early age, as little as two or three years after germination. This allows the plant to complete its life-cycle before enough debris has collected to initiate the next fire. Others have particularly hard-coated seeds which can withstand very high temperatures. In many of these respects, conifers—and particularly the pines—are outstanding in their resistance to and recovery from fire. A number of pines, such as the North American jack-pine (*Pinus banksiana*), the Mediterranean (*Pinus pinaster*) and the Lodgepole (*Pinus contorta*) of western North America, have serotinous fire-cones. The ripe, but unopened cones remain attached to the tree for many years, only opening and releasing their seeds when, as the tree gets older, they begin to dry up. The opening of such cones is in fact promoted by the heat and dryness during and following forest fires—thus allowing these trees to regenerate rapidly in an area in which other types of plants and their seeds have been destroyed. The long-leaf pine, one of the common species of the 'pineries' of south-eastern U.S.A. is particularly well-adapted to recurring fires. After germination the seedling grows a few centimetres in the first few weeks developing a dense ring of long needles which surround and protect the stem and terminal bud. Above-ground growth then stops and for a period of three to seven years the plant concentrates its activities into forming a deep and extensive root-system and in building up food reserves. During this period the seedling can withstand burning more successfully than can competing hardwood seedlings. This stage is then followed by rapid stem-growth of as much as one to two metres a year which carries the sensitive growth-points above the level of surface fires. Equipped with a thick, corky bark it is then resistant to all but the most intense fires.

While man has modified and intensified the effects of fire by his deliberate use of it he has also, in more recent times, attempted to control and reduce its frequency. In many instances his methods of fire protection have reduced the frequency of fires but have, unwittingly, greatly increased the intensity and destructiveness of those that continue to occur. Protection from fire may allow the accumulation of plant debris to such an extent that when fire occurs its severity is even greater than on previously more frequent occasions. Recent studies have demonstrated that the effects of fire-exclusion have been much more drastic than grazing on the composition and form of the Ponderosa pine forests of south-western U.S.A. Here trees originally grew in scattered groups in an open, grassy parkland formation. This pattern had been established and maintained by lightning and fires set by Indians over a long period of time. Indians are known to have occupied the area for at least ten thousand years. Such fires, which probably occurred at regular intervals of three to ten years acted as a natural thinning agent and kept down the amount of surplus combustible vegetation. The introduction of domestic livestock (which reduces the inflammable grass-cover) together with an efficient fire-protection programme has within the last fifty to sixty years eliminated regular fires. The result has been an increase in pine cover and the establishment of often impenetrable pine thickets. The absence of thinning combined with high tree-stocking rates is leading to stagnation of growth, while the gradual build-up of excess fuel has increased the potential fire-hazard to dangerous levels.

In general, as with grazing, the effect of fire on plant growth is selective. The less fire resistant species may be destroyed completely or their number and vigour drastically reduced. Relieved of competition the growth of the survivors is greatly stimulated. Regular burning tends to reduce the number of types of plants that can grow in a given area and to result in the predominance of one or two fire-resistant species. The dominance of coniferous over broad-leaved trees in the warm, humid areas of south-eastern U.S.A. has been favoured by the frequent recurrence of fire. Burning, combined with grazing, has contributed to the greater extent of treeless grassland in the prairie, steppe and savanna areas of the world than can be explained by climatic conditions alone. Indeed Carl Sauer goes so far as to suggest that the origin and preservation of grassland in both temperate and tropical regions is due mainly to burning.

The effect of burning on soil conditions can also be considerable, in some cases increasing (if only temporarily) in others decreasing fertility. The combustion of organic matter releases mineral nutrients more rapidly than would otherwise take place by bacterial activity alone. The initial fertility of soils cultivated after 'slash-burn' clearing or the shifting cultivation practised in tropical areas is due to the 'fertilisation' by ashes from burnt vegetation. It has been suggested that the wild game of Africa and the deer of the chaparral areas of California are dependent on fire to promote a flush of new palatable grass or other foliage. Vigorous regrowth after fire is partly a result of lack of competition from other plants excluded by fire and partly due to soil-enrichment by plant ash. Indirectly fire may also help, through such soil enrichment, to accelerate bacterial activity and, hence, a more rapid turnover of nutrients than might otherwise take place. It appears that in hot, dry regions it may in fact act as the principal 'decomposer' particularly under circumstances where organic litter has become so dry that the action of the fungi and bacteria of decomposition is retarded.

The detrimental effects of over-burning due either to increased intensity or frequency of fire can, however, be comparable to and quite as disastrous as over-grazing. In areas where surface vegetation is completely destroyed or its recovery is inhibited by too frequent burning, soil is laid bare. Exposure to insolation and the direct impact of rain may lead to a greatly accelerated breakdown and loss of organic matter. The resulting loss of water-absorbing and water-retaining properties and soil stability can then lead to severe soil erosion by water or wind.

References

Alhgren, I. F. and Alhgren, C. E. 1960. Ecological effects of forest fires. *Bot. Rev.*, **26** (4): 483–532.

Ashby, M. 1961. *Introduction to plant ecology*. Macmillan, London.

Beadle, N. C. W. 1940. Soil temperatures during forest fires and their effect upon the survival of vegetation. *J. Ecol.*, **28** (1): 180–192.

Bonner, J. 1950. The role of toxic substances in the interaction of higher plants. *Bot. Rev.* **116** (1): 51–65.

Braun-Blanquet, J. 1932. *Plant sociology*. McGraw Hill, New York.

Cooper, C. F. 1961. The ecology of fire. *Scient. Am.*, April, **204** (4): 150–160.

Cooper, C. F. 1960. Changes in vegetation, structure and growth of South-western pine forests since white settlement. *Ecol. Monogr.*, **30** (2): 129–164.

Crisp, D. J. (Ed.). 1964. *Grazing in terrestrial and marine environments.* A Symposium of the British Ecological Society. Blackwell Scientific Publications, Oxford.

Daubenmire, R. F. 1959. *Plants and environment.* Wiley, London.

Daubenmire, R. F. 1968. Ecology of fire in grasslands. *Adv. Ecol. Res.*, **5**: 209–266.

Eiseley, L. C. 1954. Man the fire-maker. *Scient. Am.*, September, **191** (3): 52–57.

Fenton, W. E. 1953. *The influence of man and animals on the vegetation of certain hill grazings in south-east Scotland.* I: Technical Bull. No. 4, Sept. 1951; II. Technical Bull. No. 5, Sept. 1952. *The influence of man and animals on the vegetation of certain hill grazings in mid-East Scotland.* Technical Bull. No. 7, Sept. 1953. The Edinburgh School of Agriculture. The Edinburgh and East of Scotland College of Agriculture.

Gimmingham, C. H. and Whittaker E. 1962. The effects of fire on regeneration of *Calluna vulgaris* from seed. *J. Ecol.* **50** (3): 815–822.

Grant, S. A. 1968. Heather regeneration following burning. *J. Br. Grassld Soc.* **23** (1): 26–33.

Hanson, H. C. 1939. Fire in land use and management. *Am. Midl. Nat.*, **21** (1/2): 415–431.

Harper, J. L. 1961. Approaches to the study of plant competition. In Symposium of the Society for Experimental Biology. No. 15. *Mechanisms in Biological Competition.*

Milthorpe, F. L. 1961. The nature and analysis of competition between plants of different species. *Ibid.*

Olsen, C. 1961. Competition between trees and herbs for nutrient elements in calcareous soils. *Ibid.*

Salisbury, E. J. 1929. The biological equipment of species in relation to competition. *J. Ecol.*, **17** (2): 197–222.

Sauer, C. O. 1947. Early relation of man to plants. *Geogr. Rev.*, **37** (1): 1–25.

Sauer, C. O. 1963. Fire and early man. Chap. 14 in *Land and life* Edited by John Leighly. Univ. of California Press.

Sauer, C. O. 1950. Grassland climax, fire and man. *J. Range mgmt*, **3**: 16.

Stewart, L. 1951. Burning and natural vegetation in the United States *Geogr. Rev.*, **41** (2): 317–320.

Tansley, Sir A. 1954. *Introduction to plant ecology* (3rd.ed.) Allen and Unwin, London.

Thomas, W. L. (Ed.) 1956. *Man's role in changing the face of the earth.* Chicago Univ. Press, Chicago.

Turrill, W. B. 1948. *British plant life.* New Naturalist Series. Collins, London.

Woods, F. 1960. Biological antagonisms due to phytotoxic root exudates. *Bot. Rev.*, **26**: (4): 546–569.

6 Plant Evolution and Distribution

We must now turn from the environmental conditions which influence plant growth and distribution to a consideration of some of the characteristics of plants themselves. Plants vary in appearance and physiological constitution, in requirements for growth, and in ability to tolerate environmental conditions—characteristics which, as has already been explained, are often closely interrelated. The smallest and most easily recognisable group of plants that can be differentiated on the basis of form or morphology is the *species*—the basic unit of organic classification. The problem, however, of what exactly constitutes a species is one with which biologists have long been wrestling without as yet reaching any solution. Basically a species is a 'population' of individual organisms (plant or animal) which can be distinguished from the individuals of another population in terms of three criteria: (1) form or external appearance—the members of one species resembling each other more than they do those of another; (2) the ability of the members of a species to interbreed, and their inability to interbreed, at least freely and easily, with those of another species: the basic reason why species retain their identity; (3) ecological requirements and tolerance of external environmental conditions: morphologically quite distinct species can live together because of their ability to make different uses of (to occupy different niches in) the same habitat and, hence, to avoid direct competition. On the other hand, very closely related species rarely occur together, either because their requirements are so comparable as to give rise to excessive inter-competition or because of their ease of cross-fertilisation and the resulting inability to retain their identity.

The principal character most commonly used by botanists to identify, describe and classify plants is that of morphology. In this respect, however, plants do not possess quite the same degree of diversity or complexity as animals. The former are simpler in 'body-structure' and organisation and have not developed anything like the same number of different kinds of organs as the latter. Although plants differ in size, degree of branching, form and arrangement of leaves etc., many different kinds share the same basic organisation into roots, stems and leaves and a general

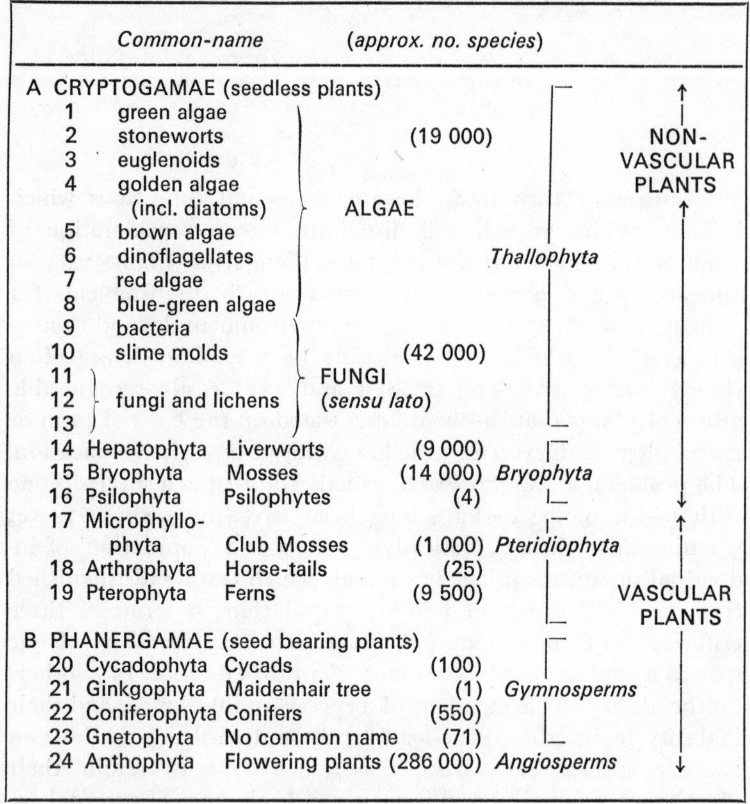

No.	Division	*Common-name*	Group	*(approx. no. species)*	Grouping	
A	CRYPTOGAMAE (seedless plants)					
1		green algae	ALGAE	(19 000)	*Thallophyta*	NON-VASCULAR PLANTS
2		stoneworts	ALGAE		*Thallophyta*	NON-VASCULAR PLANTS
3		euglenoids	ALGAE		*Thallophyta*	NON-VASCULAR PLANTS
4		golden algae (incl. diatoms)	ALGAE		*Thallophyta*	NON-VASCULAR PLANTS
5		brown algae	ALGAE		*Thallophyta*	NON-VASCULAR PLANTS
6		dinoflagellates	ALGAE		*Thallophyta*	NON-VASCULAR PLANTS
7		red algae	ALGAE		*Thallophyta*	NON-VASCULAR PLANTS
8		blue-green algae	ALGAE		*Thallophyta*	NON-VASCULAR PLANTS
9		bacteria	FUNGI (*sensu lato*)	(42 000)	*Thallophyta*	NON-VASCULAR PLANTS
10		slime molds	FUNGI (*sensu lato*)		*Thallophyta*	NON-VASCULAR PLANTS
11–13		fungi and lichens	FUNGI (*sensu lato*)		*Thallophyta*	NON-VASCULAR PLANTS
14	Hepatophyta	Liverworts		(9 000)	*Bryophyta*	NON-VASCULAR PLANTS
15	Bryophyta	Mosses		(14 000)	*Bryophyta*	NON-VASCULAR PLANTS
16	Psilophyta	Psilophytes		(4)	*Bryophyta*	NON-VASCULAR PLANTS
17	Microphyllophyta	Club Mosses		(1 000)	*Pteridiophyta*	VASCULAR PLANTS
18	Arthrophyta	Horse-tails		(25)	*Pteridiophyta*	VASCULAR PLANTS
19	Pterophyta	Ferns		(9 500)	*Pteridiophyta*	VASCULAR PLANTS
B	PHANERGAMAE (seed bearing plants)					
20	Cycadophyta	Cycads		(100)	*Gymnosperms*	VASCULAR PLANTS
21	Ginkgophyta	Maidenhair tree		(1)	*Gymnosperms*	VASCULAR PLANTS
22	Coniferophyta	Conifers		(550)	*Gymnosperms*	VASCULAR PLANTS
23	Gnetophyta	No common name		(71)	*Gymnosperms*	VASCULAR PLANTS
24	Anthophyta	Flowering plants		(286 000)	*Angiosperms*	VASCULAR PLANTS

Table 10 Main divisions (1–24) of plant kingdom with approximate number of species and some of major groupings. Vascular plants are those with clearly differentiated and specialised tissues for the transport of water and nutrients from roots to leaves, and from leaves to other organs respectively.

over-all similarity of appearance. Morphological diversity is, in fact, greatest in methods of reproduction and in the form and anatomy of the reproductive organs, be these the spore-containing capsules of mosses or the cones and flowers of the seed-plants. As a result the reproductive organs have long been used as the most important diagnostic characters in the classification of plants and the basic criteria by which species are identified and arranged into a hierarchical succession of genera, families, classes and divisions (see Table 10 and Fig. 17).

Two factors, however, aggravate the problem of species definition. First is the variation of form that occurs even among populations of morphologically similar plants. Second is the fact that a species is not, as was first assumed by early taxonomists, a fixed,

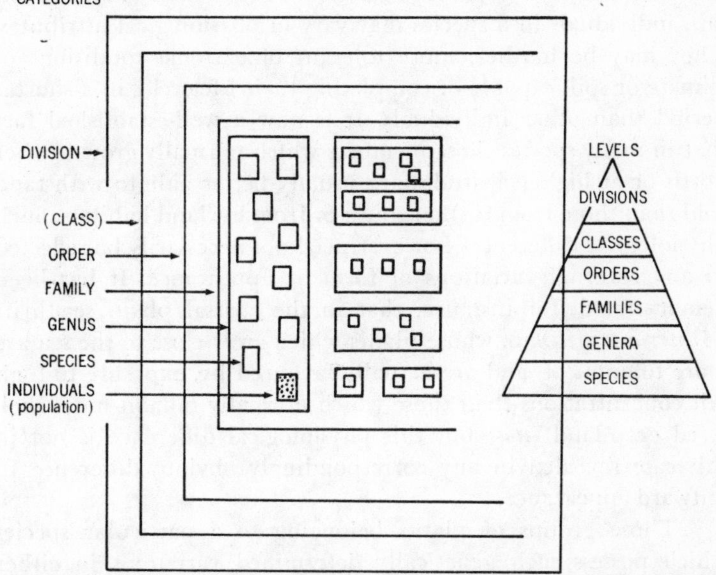

Fig. 17 Schematic representation of plant categories (Taxa) and levels recognised by International Code of Botanical Nomenclature. Species can be composed of a few individuals or small populations (rare) or many individuals or large populations (common). Genera contain one or more species; families one or more genera. (Adapted from Heywood, 1967)

static, permanent unit; in the process of evolution some species have become extinct and new species are continually emerging. Absolutely identical twins are a comparatively rare phenomenon. The individual members of any plant species which, to superficial inspection, may seem to have exactly the same appearance are rarely exactly the same in every respect. Slight variations usually exist between them in characters such as height, number of branches, size and shape of leaves for instance. These variations can and do arise as a result of differences in the environmental conditions under which the plants are growing; those in more favourable habitats will tend to be larger, taller and more vigorous than those of the same species on poorer sites; the stunted low-growing or even prostrate form of trees and shrubs on sea coasts or at high altitudes *may* be a direct result of the inhibiting effect of high wind-force on their rate and direction of growth. On the other hand, variations in form of coastal or mountain representatives of a particular species may be an inherent, genetically determined characteristic which will be retained even if the plants, or seeds from them, are transplanted in different and more favourable habitats. Also, because of variations in genetical make-up, certain individuals in a species may vary in physiological attributes. They may be hardier, more tolerant of extreme conditions of climate or soil, capable of completing their life-cycles in a shorter period than other individuals. It is now a well-established fact that in many species those members which normally grow further north or at higher altitudes are usually better able to withstand cold than those from further south or from lowland habitats. Such physiological differences, however, need not necessarily be reflected in any external variations of form or appearance. It has been demonstrated, for instance, that in the coastal plant, sea-thrift (*Armeria maritima*), whose plants which grow close to the sea are more tolerant of, and are actually favoured by, exposure to high salt concentrations than those which normally inhabit more sheltered or inland sites; but this physiological difference is not in this case revealed in any correspondingly obvious difference in outward appearance.

Those groups of plants belonging to a particular species which possess such genetically determined variations in either form or physiology under varying habitat conditions are known as *ecological races* or *ecotypes,* and may warrant the division of the species into a number of sub-species or varieties. Where there is a marked and discontinuous variation in form, or the populations of plants occur in two contrasting and widely separated habitats (i.e. their distribution is discontinuous) the recognition and classification of ecotypes may be relatively easy. On the other hand, the change in morphology or physiological tolerance of individual plants within one species may be so gradual and continuous that, particularly when associated with a gradual and continuous change in habitat conditions (i.e. an *ecocline*), a precise subdivision may be more difficult to establish. Such is frequently the case with the members of those populations which range from sea-level to high altitudes, or from wetter to drier soil conditions. Many species which occur over a wide range of varying environmental conditions must inevitably be composed of a varying number of ecological races or ecotypes, each adapted to a slightly different habitat. The Scots pine (*Pinus sylvestris*) for example occurs naturally over an area which extends from Southern Italy to beyond the Arctic Circle. The existence of races or ecotypes within this species is not therefore surprising, and many are recognised and distinguished as sub-species or varieties. It is thought that even in Scotland this pine (sometimes referred to as *P. sylvestris var. Scotia*) may be composed of two 'strains' or ecotypes which reflect the climatic variation between the drier east and the much wetter western areas of the country.

Such ecotypes (be they sub-species, varieties, races or strains) probably represent the most recent products of evolution. They will tend to emerge when a species spreads into a slightly different habitat or where for some reason, such as climatic change, conditions may alter in all or part of the existing habitat. The survival of those members of the original species with either morphological or physiological characteristics which make them better fitted or adapted to the new conditions may be favoured more than others. By this process variant plants, which are the normal product of constant gene exchange in an inter-breeding species population, may persist and be perpetuated. This will be more likely to happen when cross-fertilisation between the variants and the other members of the species population is inhibited; such may be the case when they establish themselves in a habitat that is beyond the range over which agents of pollination can operate.

Variants within an existing species population can also arise by mutations; sudden or gradual changes in either the number of the hereditary bodies, the chromosomes, or the composition of their constituent genes. Mutations can be artificially induced by external stimuli and probably also, it is thought, by unusual or sudden natural changes in environmental conditions. The resulting mutant offspring may or may not differ in appearance from its parent. It will not necessarily give rise to a new species unless it is better adapted to the environment than the parent. One particular type of mutation which appears to play an important part in the evolution of new species is that of *polyploidy*. This occurs when offspring are produced which possess a greater number of chromosomes in their cells than in those of the parent plants. It may be the result of a genetic mutation within a particular species or of hybridisation between two different species. Although polyploidy is not necessarily accompanied by any significant change in appearance, it would seem that the polyploid individuals often possess much greater vigour and a much greater adaptability to external conditions than their parents. What is known of their geographical distribution would tend to confirm this; as far as is known, polyploid members of a certain species are particularly prevalent in the cold and arid regions of the world where, presumably, they have also been the source of new species better adapted to more extreme climatic conditions than their parents.

Each species or ecotype varies in its essential requirements for successful reproduction and growth, and in its capacity to survive in face of the competition or depredations of other organisms. The ability of the individuals of a species population to exist and maintain themselves in a particular area will depend on the presence of those conditions necessary for, and the absence of those detrimental to their growth. Plants vary, not only in their essential requirements but in what is called their *tolerance* of environmental conditions. In respect of every habitat factor there are optimum conditions when the reproductive capacity, vigour and competitive ability of a particular species of plant will be greatest. The limits within which a given type of plant can exist are referred to as its *range of tolerance*. Any environmental condition beyond that which the plant can tolerate—either a deficiency or excess of any habitat factor, climatic, edaphic or biotic—will limit its growth and hence the area it is capable of occupying. The wider the range of tolerance the more extensive the area a species can occupy. Tolerance, however, varies not only between species, but within a particular species in respect of different habitat factors (Fig. 18). A species may have a wide range of tolerance to

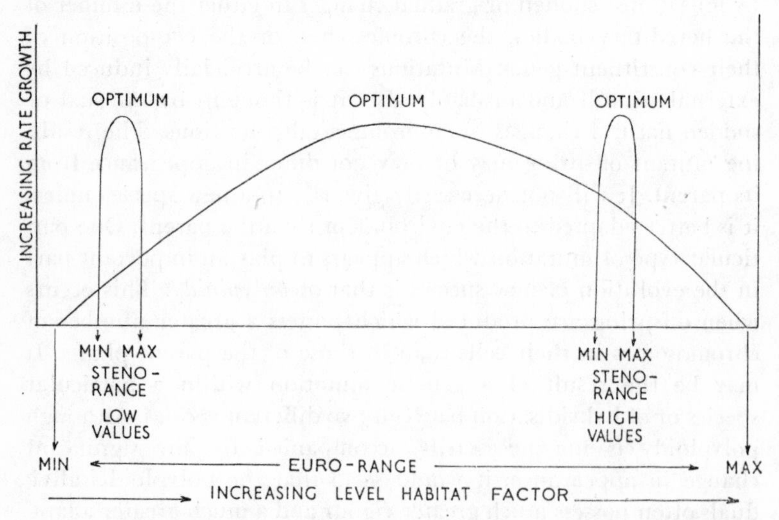

Fig. 18 Diagrammatic comparison of *eury* (= wide) and *steno* (= narrow) ranges of tolerance of organisms to habitat factors. (From: Odum and Odum. 1959)

temperature conditions (i.e. it is *eurythermal*) while its tolerance of soil acidity or alkalinity may be very narrow indeed (i.e. it is *stenoedaphic*). In the latter case the plant's tolerance may be so limited as to provide a sensitive *indicator* of particular and precise environmental conditions. Obviously the decisive factor determining and limiting the distribution of a species will be that habitat factor for which it has the narrowest range of tolerance.

Under natural conditions, however, it is not always easy to isolate the principal *limiting factor* or condition for a particular plant in a given area. It is even more difficult to analyse the exact level or stage at which a particular factor actually inhibits growth. As has already been stressed, habitat factors do not operate independently. They interact with each other in such a way that the condition or operation of one is inevitably affected by that of all the others. The level at which a given amount of rainfall becomes limiting will, for instance, depend not only on the range of tolerance of a particular plant but also on the associated temperature and soil conditions. The latter will determine the amount of rainfall which will become available to the plant. High wind-force can intensify the effects of either low or high temperatures; with increasing altitude it cannot always be ascertained with certainty whether exposure to wind or low temperature is the principal factor limiting tree-growth. The effect of soil, water or temperature conditions below the optimum may reduce a plant's competitive ability and thereby exclude it from a given habitat, without any one of these factors *alone* being below the limit for successful growth or reproduction. Also because of the complex interaction of habitat factors, one may *compensate* for the lack or deficiency of another. Under climatic conditions which might be expected to limit the growth of a particular species a sheltered position, a south-facing exposure or a 'warm' soil may allow its survival. Conversely, in another area, the presence of a sandy soil, or a heavy water-logged clay may reduce water availability and temperature conditions respectively to levels which would become limiting for certain plants that might otherwise grow more successfully under the prevailing climatic regime.

A particular species can only occupy those parts of the earth's surface (those habitats) where environmental conditions are within its range of tolerance. The geographical limits of the area within which a particular species actually occurs is known as its *range*, the maximum possible area it could occupy its *potential* range. All other factors being equal, the limits of a plant's potential range are determined by climatic conditions. Few types of plants are really *cosmopolitan,* that is to say so widely distributed throughout the world that they can exist in every type of climate; temperature conditions alone limit many to arctic, temperate or tropical ranges. A particular species however rarely occupies the whole of its potential climatic range; factors of soil and competition tend to restrict its distribution even further. However, the cultivation of crops, and the introduction and 'protection' by man of exotic species has demonstrated very clearly that, relieved of competition from other plants, many species can grow over a wider range of soil and climatic conditions than they normally do in the wild. The *actual range* of many plants over the earth's surface is, in fact, often less than their *potential range*. The presence of a plant in a particular area will mean that conditions are suitable—it is growing within its range of tolerance. Its absence, however, does not necessarily or invariably mean that the ecological conditions are unfavourable. It may be that, for one reason or another, it has been unable to reach the area. Failure to colonise a suitable habitat may be caused by the presence of some obstacle to migration into the area; by insufficiency of time since the particular species evolved or since the area became ecologically suitable. The present range of any species (plant or animal) can rarely be explained in terms of present conditions alone.

The distribution of plants has resulted from their 'migration' over time into ecologically suitable areas. Many species can spread and increase their area by vegetative propagation. This method of 'movement', though quite effective, is slow and has the additional disadvantage that new shoots from the parent stock do not possess the genetic variability of seeds; hence they lack the potential to adapt themselves to the variations in habitat that may be encountered as they spread. Seed dispersal, which allows greater mobility and a more rapid dispersion over a wider area, is the principal means of plant migration. The effectiveness of seed dispersal, in colonising new areas and in extending the range of a particular plant, is dependent on a number of factors among which some of the most important are: (1) the ability of the seed to germinate and the new plant to establish itself successfully; (2) the number and size of seeds and the frequency of their reproduction; (3) the means by which they are transported. A large number of small, light seeds may facilitate rapid and widespread dispersal, an advantage counteracted by their susceptibility to a high mortality rate. Other plants produce a few large, heavy seeds which, though more difficult to disperse, have a greater chance of survival by reason of their greater food reserves. All other things being equal annual plants which produce, at frequent intervals, large quantities of small, easily transported seeds will have, potentially, a more rapid rate of spread than trees which not only may take many years to reach the seed-producing stage, but also produce fewer, larger seeds at much longer intervals. Speed and distance of dispersal is also affected by the particular means or agency of transport to which the seed is best adapted—be it by wind, water or animals. The relative speed and distance of dispersal by these agents is a matter of some conjecture, and the relative ability of any of them to transport seeds over exceptionally great distances is debatable. In this respect none has been so powerful in increasing the speed and range of seed dispersal in recent times as man. (His repercussions on plant migration will, however, be considered in the next chapter.) Very little, however, is known definitely about the relative effects or significance of the various means of seed dispersal on the rate and distance of plant migration. Many types of plants that would seem to be best-equipped for dispersal are much less widely distributed than those more poorly endowed. In many ways this is not at all surprising. Migration and spread of plants does not depend alone on either high seed output or efficacy of transport. Ultimate success depends on the ability of the particular plant to establish itself in the area into which it is dispersed. The greater its range of tolerance the greater will be its chances of survival and the more widely it is likely to be distributed. On the other hand among seeds with similar ecological requirements efficiency of dispersal will affect their relative speed of migration into a suitable area, and their spread within it.

The migration of a plant from one ecologically suitable habitat into another may be prevented by intervening areas unsuitable for germination and so extensive that they cannot be crossed by any normal means of seed dispersal. In this respect one of the most effective obstacles to plant migration are the oceans. Relatively few land or fresh-water plants have seeds that could survive the long periods of immersion in salt water that would be necessary for their transport over such long distances. Likewise high mountain ranges and extensive deserts present formidable, and for many plants insuperable, climatic barriers to migration. The distribution of land and sea, and of the major relief features of the earth's surface must, then, be included among those factors which have affected plant distribution. Together with the major world climatic zones, they have accentuated the fragmentation and separation of ecologically similar environments. The obstacles they present (or may have presented in the past) to plant migration may help to explain why certain plant species which occur in the equatorial area of South America do not exist to-day in other similar environments; why other species may be confined exclusively to either North America or Europe; why many plants in north-west Europe are not found in equally suitable habitats in the British Isles or why Ireland lacks many species common to Great Britain.

However, there are many features of species distribution that cannot be explained in terms solely of tolerance-range and/or migrational barriers. The latter actually raise more problems than they solve—not least of which is that of the markedly *discontinuous* (or *disjunct*) range of many plants (see Fig. 19). Few wild species are so continuously distributed that they occupy any one piece of ground to the exclusion of all others. Even within a relatively small area variations in local relief, soil and competition will cause discontinuities in the distribution of a particular plant. However the same, or closely related, species frequently occur in two or more areas so widely separated as to preclude the possibility, at present, of inter-migration by normal means of dispersal. This is illustrated in the Arctic-Alpine range of species, like *Salix herbacea* and *Saxifraga oppositifolia,* common to both lowland arctic regions and to the high-altitude habitats on mountain ranges which occur quite far south in Europe; in the markedly discontinuous range of similar or closely-related species which occur in eastern North America and Europe, or in South America and Africa. Some may have even more puzzling distributions—as for instance the tulip tree (*Liriodendron* sp.), and other members of the Magnolia family, found in eastern U.S.A. and Asia but not present in Europe or Africa today; while others occur only in the north temperate and south temperate parts of the earth's surface.

It is most unlikely that long-range dispersal has been so widespread, effective and selective as to account for such marked discontinuities in the range of closely related species. Furthermore, there is little evidence to support the suggestion that similar species have had a completely independent origin in different and widely separated parts of the earth's surface. The most commonly accepted hypothesis is that similar or genetically related types of plants must in the past have evolved from a common ancestral stock and have spread by migration from their orginal centres of origin into the areas they now occupy. The present discontinuities must then be the result of the subsequent creation of the 'barriers' that now separate their present areas of occupation.

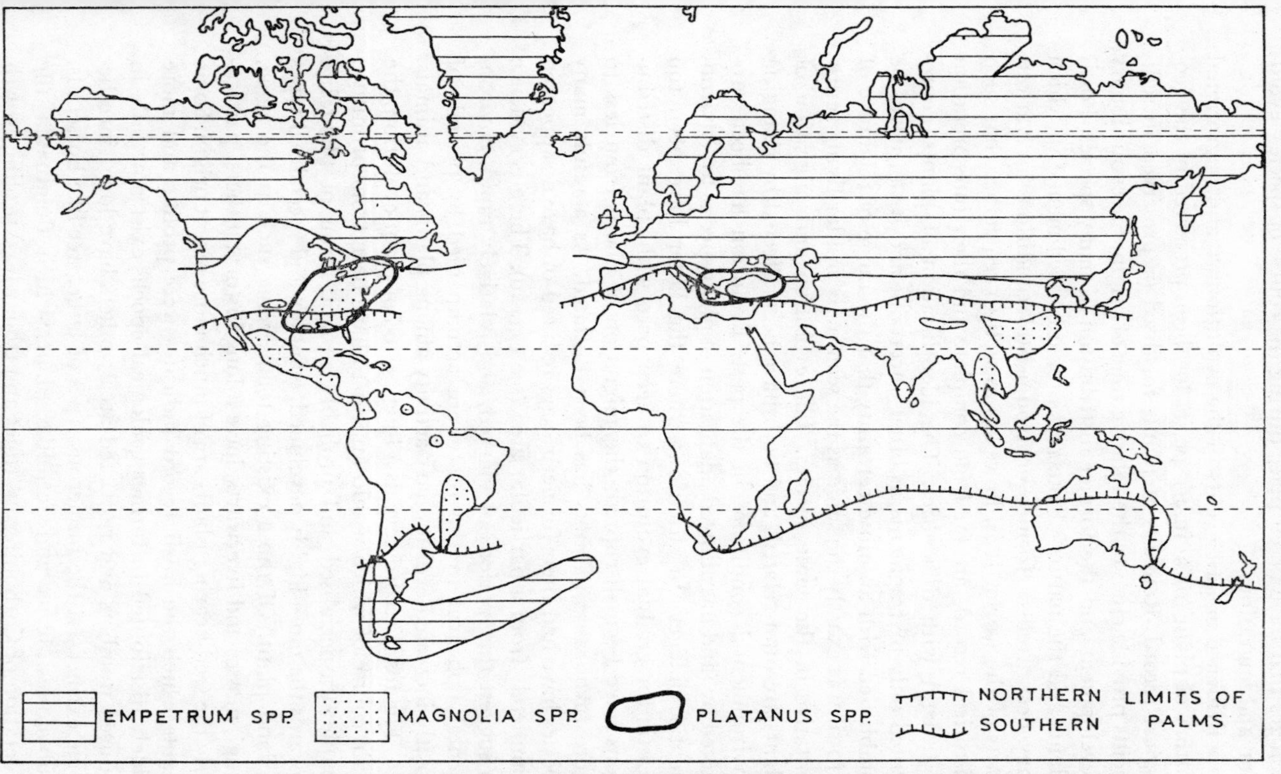

Fig. 19 Examples of *continuous* (e.g. palms) and *discontinuous* or *disjunct* (e.g. *Empetrum* spp., *Magnolia* spp. and *Platanus* spp.) ranges of selected plant genera. (Compiled from Good, R. 1953)

The present distribution is a result presumably of environmental changes that have taken place during the course of species' evolution and migration.

Evidence of past events in the evolutionary and migrational history of plant (or animal) species has been preserved in the geological record. Remains of plants fossilised or preserved in sediments provide clues to the relative time of origin, the evolutionary development and the former distribution of many species. Fossil plants and the nature of the rocks (or other sediments) in which they occur reflect former environmental conditions in various parts of the world. It is, however, as well to remember that such historical records are far from complete and their interpretation is fraught with difficulties. The identification of plants, represented only by fragments, is often uncertain; and it can never be established with absolute certainty that the range of tolerance of a fossil was exactly the same as the seemingly similar living representative of the same species. The evolution and migration of plants are inter-related, and both have been affected by changing environmental conditions in the past. Evolution has been promoted by (and has allowed adaptation of plant species to) environmental changes. In situations where the latter have been too drastic for gradual evolution to keep pace with, plant distributions have been disrupted; the migration of plants from less to more favourable habitats has been initiated. As a result many species have had their former ranges reduced or have disappeared completely from a particular area (see Fig. 20). There is evidence to suggest that periods of marked, and relatively rapid, environmental upheaval in the past were accompanied by periods of greatly increased evolutionary activity among plants and animals.

The fossil record reveals a long period of plant evolution, the main course of which would appear to have been from primitive, simple, unspecialised and probably aquatic forms of plant life towards the more highly developed, complex and specialised types of land plant. Of those existing today, some such as the algae, fungi, mosses and liverworts, have a long history: others, particularly the seed-bearing plants are of much more recent origin. Some plants which the fossil record indicates were prolific and more widely distributed in former geological epochs exist now only in reduced numbers and restricted areas; others have long since become extinct. Of the numerous types of large, tree-like horsetails, club-mosses, ferns and fern-like plants which dominated the vegetation of Carboniferous times and from whose prolific growth the coal seams of that era have been formed, only relatively few,

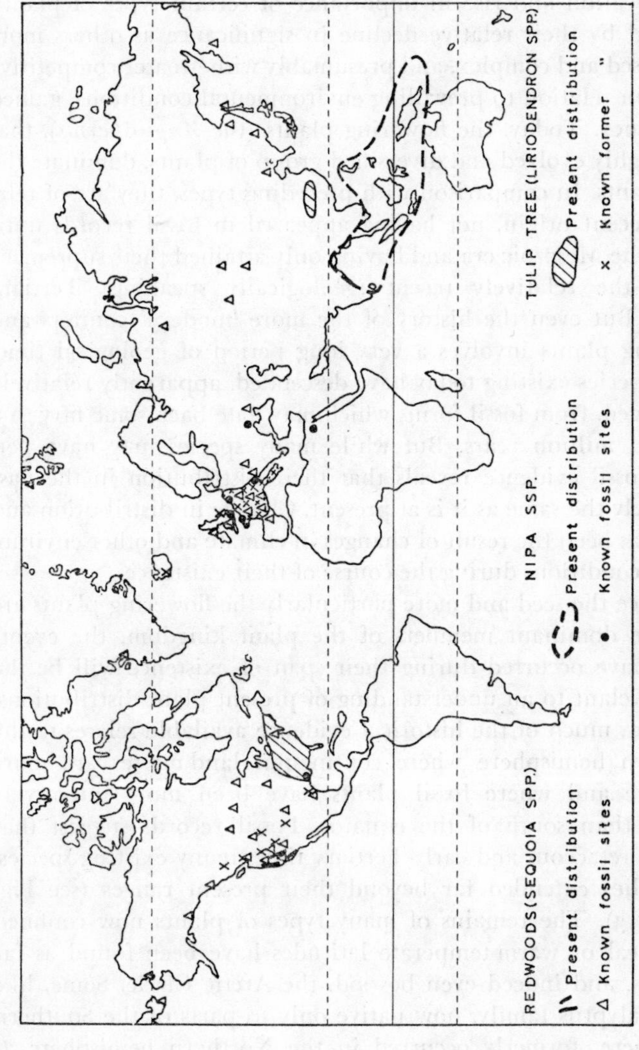

Fig. 20 Present and known fossil distributions of selected plant genera (Compiled from Good, R. 1953 and Cain, S.A. 1944)

insignificant and frequently herbaceous descendants exist today. Of the early seed-bearing plants—the Gymnosperms—which attained such prominence in the succeeding Mesozoic era, only the conifers retain an important place in the present world vegetation cover. Evolution has been accompanied by the development, diversification and rise in importance of certain types of plants, followed by their relative decline in significance as others more specialised and complex, and presumably with greater competitive ability in relation to prevailing environmental conditions, gained ascendancy. Today, the flowering plants (the *Angiosperms*), that most highly evolved and diversified group of plants, dominate the land plants. In comparison with preceding types, they are of relatively recent origin, not having appeared in fossil records until late in the Mesozoic era and having only attained their supremacy within the relatively recent (geologically speaking) Tertiary period. But even the history of the more 'modern' conifers and flowering plants involves a very long period of geological time. Many species existing today have descended, apparently relatively unchanged, from fossil forms which may date back some fifty to a hundred million years. But while many species may have persisted, fossil evidence reveals that their distribution in the past was rarely the same as it is at present. Change in distribution and range has been the result of changes in climate and other environmental conditions during the course of their existence.

Since the seed and more particularly the flowering plants are now the dominant members of the plant kingdom, the events which have occurred during their span of existence will be the most revelant to an understanding of present plant distributions. However, much of the historical evidence available relates to the Northern hemisphere where continental land-masses are more extensive and where fossil plants have been more intensively studied than south of the equator. Fossil records suggest that in late-Cretaceous and early-Tertiary times many existing species, or families extended far beyond their present ranges (see Fig. 20, p. 113). The remains of many types of plants now confined to tropical or warm-temperate latitudes have been found as far north as, and indeed even beyond, the Arctic Circle. Some, like the Eucalyptus family, now native only to parts of the Southern hemisphere, formerly occurred in the Northern hemisphere as well. Cretaceous deposits in Greenland and Spitzbergen contain the remains of plants similar to existing tropical species of banana and bamboos. Fossils resembling such warmth-demanding plants as the tropical palm (*Nipa fruticans*), the Californian redwood

(*Sequoia sempervirens*) and the Maiden Hair tree (*Ginko biloboa*) have been identified from early-Tertiary (Eocene) deposits in Britain; of those associated with the London Clay deposits, for instance, seventy per cent. belong to families which exist today only in south-east Asia. In North America also, tropical and warm-temperate tree species appear to have extended far north of their present range into Alaska and, in addition, to have flourished in areas such as central Oregon where today climate permits but a semi-arid sage-brush scrub. From such evidence it is very tempting to postulate the prevalence of more universally and uniformly warm, humid climatic conditions than at present—lacking the marked temperature gradient from equator to poles, and the uneven rainfall distribution, characteristic of existing world climatic conditions.

Later plant fossils suggest a progressive diminution in the number of tropical and sub-tropical species and a proportional increase in those now characteristic of cool temperate and cold latitudes (see Table 11). By the end of the Tertiary period the plants existing in Southern England were similar to those now present there.

Flora	Number of species	Per cent. whole flora compared	Per cent. exotic and extinct species	Per cent. Chinese-N. American species	Age
Cromerian (E. Anglia)	135	89	5	0·74	Top of Pliocene
Teglian (Holland)	100	75	40	16	Upper ,,
Castle Eden (Durham)	58	55	64	31	Middle ,,
Reuverian (Holland)	133	46	88	54	Lower ,,
Pont-de-Grail (S. France)	17	35	94	64	Base of ,,

Table 11 The relationship of five Pliocene seed-floras. (From: Matthews, J. R. 1955)

The subsequent changes in the composition of fossil floras have been interpreted as indicative of a general deterioration of temperature conditions in temperate and high latitudes. From evidence in North America it would appear that marked variations in rainfall distribution also began to emerge, the initiation of which is thought to be associated with the great orogenic movements of the later Tertiary period and the resultant creation of climate barriers. Such climatic changes must have effected a drastic re-distribution of plant species. Of those less tolerant of cold and drought, some became extinct over wide areas and as a result had their ranges reduced and fragmented, while a general southward migration of the more warmth-demanding species was initiated. The progressive cooling in the northern hemisphere culminated in the Pleistocene period when, for a period of at least a million years, continental ice-sheets advanced and retreated over much of North America and Europe, and glacial conditions alternated with relatively warm, perhaps even near sub-tropical, inter-glacial periods. The spread of these continental ice-sheets was accompanied by a modification of climatic conditions beyond their southern margins. At their maximum extent much of central and even southern Europe must have experienced conditions not dissimilar to those prevailing in the present arctic tundras. Both geomorphological and biological evidence points to the existence of pluvial periods in now completely arid areas in both Africa and North America. The southward migration of species initiated in the late Tertiary period must have been intensified, if not actually accelerated, in the van of advancing ice-fronts.

Over those areas directly subjected to advancing ice-sheets most, if not all, the pre-existing plant life was obliterated. On this point, however, the evidence is conflicting and is greatly complicated by the composite nature of the glacial retreat and advance. The extent to which plants present in pre-glacial times may have survived within or around the margins of formerly glaciated areas has been a subject long-debated, not least in Britain. There are botanists who maintain that the climatic conditions, particularly at that stage of maximum glaciation, when the front of the ice sheet extended from the Severn to the Humber, were so severe that all the pre-glacial plants were obliterated; all existing species had then to re-establish themselves *de novo* by migration from further south following a gradual amelioration of climate. Others believe that many of our hardier, arctic and sub-arctic moorland species, already existing before the onset of glaciation, must have been able to survive in the ice-free tundra zone in the south of England. The existence in south-west Ireland of certain frost-sensitive 'Mediterranean' plants (of which the strawberry tree (*Arbutus unedo*) and certain members of the heath family (*Ericaceae*) are the classic examples), however, raises problems that have never been satisfactorily resolved. Their survival as pre-glacial 'relict' species in this area is hard to substantiate; yet in terms of present geography their now very discontinuous range is just as difficult to explain.

The presence of certain, and often rare, Arctic or Alpine species in coastal areas of north-west Scotland and at high altitudes in England, Wales and Scotland has been attributed to their survival in areas which, for one reason or another, escaped glaciation. The possibility that certain higher mountain areas remained as ice-free 'nunataks' above the general ice-level for some or the whole of the glacial period has long been a question which has intrigued both botanists and geomorphologists alike. But it is a thorny problem which has given rise to views as conflicting as the available evidence. Certainly many of the coastal and high-altitude Arctic and Alpine species may be explained in terms of their habitat requirements; probably more widespread in the severe tundra conditions which prevailed in immediately post-glacial times, their area has been drastically restricted to such bare and open conditions where they have been able to survive the competition of more aggressive and demanding plants which re-instated themselves later as climatic conditions became less severe.

Conflict of opinions about the extent of glacial survival does not, however, invalidate the fact that over the greater part of the formerly glaciated areas both in Europe and North America plant life had to be re-established by migration from beyond those areas actually covered by the ice sheets. For such areas floral re-stocking has taken place only within the relatively short span of the eight to ten thousand years that has elapsed since the final stages of the last glacial retreat. The immigration of plants and their re-establishment was consequent on the progressive amelioration of climate. The composition of the flora that could be re-established was, however, dependent on the availability of a seed-stock and the existence of routes by which migration could take place. It is noteworthy, for instance, that the vegetation of eastern North America and eastern Asia is much richer in numbers of different plant species than that of Europe. In the latter area many Tertiary species did not survive the pre- and full-glacial climatic deterioration. Just as the transverse barriers of the Sahara desert and the Alpine mountain ranges blocked their escape routes southwards, so in the post-glacial period they cut Europe off from the reservoir of plant species that existed further south. In contrast no such transverse barriers existed in North America, so many pre-glacial species were able to find refuge further south, particularly in the south-east. Thus the reservoir of possible colonists was much richer and eventual migration northward in post-glacial times was unimpeded. Such events help to explain the discontinuous range of the Magnolia family (previously noted) which was presumably obliterated in the southward plant migrations in Europe, but survived in south-east U.S.A. and Asia. They perhaps also help to explain why tropical Africa is much less rich in species than similar areas in South America or south-east Asia. The effect of post-glacial migration routes on the composition of the existing flora is also strikingly demonstrated in British Columbia. In terms of present climatic conditions the dominance of coniferous trees and the relative paucity of broad-leaved deciduous and evergreen species is somewhat anomalous. It has been suggested that the longitudinal and very effective barrier of the Western Cordillera cut the area off from the rich plant reservoir of the south-eastern U.S.A. Restocking of the north-west then came from the mountain ranges further south where conifers in particular had been able to survive.

Although of relatively short duration, the climatic and geological events of the post-glacial period have also had a considerable influence on the migration of plants into areas formerly glaciated and on the present composition of the flora and vegetation. A record of these events is provided by plant remains, particularly pollen grains, preserved in peat which has accumulated in ill-drained habitats left in the wake of the receding ice-sheets. From these an increasingly detailed picture of the post-glacial vegetational history in both Europe and North America is beginning to emerge. Further, dating of plant remains found at different peat levels has been aided by correlation with associated geomorphological (such as *varves*—layers of clay and silt in post-glacial lake deposits) and archaeological features, and has in recent years been made even more precise by techniques of radio-carbon dating.

The relative abundance of remains, particularly of the more easily identifiable tree-pollen trapped and preserved at various levels in peat deposits, has revealed the course of post-glacial plant immigration and vegetation development in lowland England. The progression from open tundra and birch-scrub heath to forest cover, in which the earlier hardy colonisers, birch and pine, give way to the eventual establishment of broad-leaved deciduous forest, reflects the gradual establishment of the existing climatic and natural vegetation patterns. The relative proportions of tree pollen from various peat-levels suggest, however, a period of climatic fluctuation rather than one of continuous amelioration; drier continental (Boreal) alternated with more humid oceanic (Atlantic) phases. The early Boreal period, somewhat warmer and drier than at present, is marked by a predominance in England of pine and hazel. This period, sometimes referred to as the 'xerothermic period', was paralleled in Europe by the northward migration of many characteristically Mediterranean plants beyond their present range; and, in North America, by a movement eastwards of prairie grassland species. Post-glacial climatic conditions appear to have attained their 'optimum' in the succeeding Atlantic period when temperatures remained higher than at present but were accompanied by increased humidity. Deciduous forest reached its maximum extension and greatest development in north-western Europe; in Britain elm became more prevalent than at any other stage and lime attained its maximum northward range; while in France the beech markedly increased its area. In many places, higher humidity favoured the increased development and extension of peat bogs, particularly in northwest Britain. There is evidence that these became drier and were recolonised by birch and pine in the ensuing somewhat drier sub-Boreal phase—another more 'continental' climatic period when pine and birch ascended to their highest altitudes in the Scottish Highlands. The latest Atlantic phase, though subjected to minor climatic fluctuations, has been one of cooler, wetter, oceanic conditions such as prevail today. During its course, beech and hornbeam became established in the south of England, lime and elm decreased in amount, birch became more widespread in the north-west and the accumulation of peat mosses was again accelerated (see Table 12).

The migration of plants into Britain in the early post-glacial period was undoubtedly facilitated by the existence of land connections with the continent. Remains of forests now submerged around the coasts and *moorlog* dredged from the bed of the North Sea testify to sea-levels lower than at present. The exact date at which the presumed link with Europe was severed is debatable; some would put it in Atlantic times, others later. While hardly a formidable barrier, the eventual establishment of the English Channel may well have retarded plant migration and hence have contributed to the relative paucity of plant species in Britain in comparison with France, which has three times as many. Only four species of conifers (the Scots pine, juniper (2) and yew) are indigenous to Britain. The Norway spruce (*Picea abies*) present in pre-glacial times failed to re-instate itself, while many species now

	Approx. Date	Blytt & Sernander's Periods	Pollen Zone	Vegetation	Climate
	B.C.	Sub-Atlantic	VIII	Decline of mixed forest. Beech, Hornbeam	Cold and wet, more oceanic
	500	Sub-Boreal		Mixed forest	Warm and dry, continental
			VII		
	3 000			maximum	climatic optimum
Post-glacial		Atlantic		Alder, Oak, Elm, Lime	Warm and wet, oceanic
	5 500				
		Boreal	VI	Oak, Elm, Pine, Hazel	oceanic
	7 500		V	Pine, Birch	Warm and dry
	8 000	Pre-Boreal	IV	(Hazel), Birch Pine	Cold : sub-arctic
	8 300	Younger Dryas	III	Birch scrub tundra	Sub-arctic
Late-glacial		Alleröd	II	Birch tundra heath	Milder
	13 000	Older Dryas	I	Open tundra	Sub-arctic

Table 12 Sequence of late-glacial and post-glacial vegetation and climatic changes in Britain. (Adapted from: Goodwin, H. 1956)

common (such as the horse-chestnut and sycamore) owe their presence to introduction by man. Ireland with only sixty-seven per cent. of the number of plant species present in Britain, reflects an even greater isolation.

Fluctuations in the relative levels of land and sea have been a universal occurrence throughout geological time. In those areas of more extensive, shallow, epi-continental seas overlying the wider parts of the continental shelf, a lowering of present sea-levels by only thirty to sixty metres would connect Britain again with the continent, the East Indies with Asia and Australia, and would create an almost continuous land connection, via the Bering Straits, between Alaska and Siberia. The formation and disruption of such land connections, perhaps several times in the past, must have influenced plant migration and, as in Britain, resulted in similarities and differences between adjacent land areas. However, neither geomorphological changes of this magnitude nor past climatic fluctuations can satisfactorily explain the often striking similarities between the plants in areas now widely separated by oceans. The earlier theory that inter-connecting land masses—the 'land-bridges'—once existed and have, like the 'lost continent of Atlantis', foundered beneath the sea has little evidence to support it. On the other hand there is much to support the later theory of *continental drift*. For example, the similarities between the plants in trans-oceanic areas, the occurrence of comparable geological strata and structures, together with other physiographic features on now widely separated continental masses, and finally the suggestive outline of such opposing coastlines as those of South America and Africa. Basic to this theory is the assumption that existing continents are fragments of a formerly continuous, or almost continuous, ancient land-mass which have gradually drifted apart. Modern geophysical and palaeomagnetic work is producing further evidence which is tending to reinforce and substantiate this theory. As critical, however, as the fact of continental drift is the period when it was initiated. Biogeographical distributions suggest that this must have been sometime during the Cretaceous period.

Whatever the causes, changing climatic and geological conditions over a long period of time would appear to have effected a continual sorting and redistribution of plant species throughout the world. Many, formerly in closer communication, have become isolated; others formerly more widely, if not universally, distributed have been confined to one particular area which often represents only part of their potential world-range. The fragmentation of land masses and the consequent isolation has tended to increase variation of and differentiation between the types of plants now existing in widely separated though ecologically similar habitats. The evolution of new or slightly different species from the same ancestral stock has been facilitated by isolation. This has been even more striking in the case of animals than of plants. A great many different types of plants and animals had already evolved before the environmental disruption of the Cretaceous and Tertiary periods commenced. Most of our modern flowering-plants were already in existence, as were many present-day types of insects. The mammals, birds, and fresh-water fishes were, however, as Charles Elton points out, 'poised on the edge of a tremendous bout of evolution' at this time. The divergent evolution of different types particularly of mammals *since* the period of continental fragmentation and drift, has resulted in even more marked differences in animal than in plant distributions.

Physical and ecological isolation has given rise to the phenomenon of *endemism*. Endemic plants are those (whether members of a family, genera or species) peculiar or exclusive to a particular area. They are plants which have either evolved within a particular area; or plants which, because barriers have been created subsequent to their migration, have become confined within a particular area. The degree of endemism, i.e. the number of endemic species (expressed as a percentage of the total of all species in an area), depends largely on how effectively and for how long an area has been isolated. Areas surrounded by oceans, extreme climatic conditions or extensive mountain ranges—limited by well-marked obstacles to migration—have been most favourable to endemism. Off-shore 'continental' islands, within easy reach of the mainland, and often (as in Britain) only recently separated from it, normally have few endemic species. Areas of characteristically high endemism are long-established and isolated oceanic islands (see Table 13) and mountain areas. The latter can, in fact, be considered as isolated 'climatic islands' which, in

Canary Islands	45
Corsica	58
Madagascar	66
New Zealand	72
Hawaii	82
St. Helena	85

Table 13 Percentage endemic species in selected island floras. (From: Wolff, E. V. 1950)

addition, because of marked variations in local relief and climate often contain a great number of small endemic areas within them.

The particular kinds of plants, irrespective of their abundance or relative importance, which occur in any area are known collectively as its *flora*. The composition and diversity of any flora is an expression partly of the present and partly of the past environmental events which have influenced plant evolution and migration. Within any area the species which comprise its flora can be grouped into components or elements characteristic of their mode or time of entry, their distribution within the area, or the source from which they migrated. The distinction made between native or indigenous species, established by normal, 'natural' means, and *aliens*, deliberately or accidentally introduced from other areas by man, is not always easy to establish for all species. Many 'weeds', for instance, long-considered aliens in Britain, have since been identified from late and post-glacial deposits. Nor is it possible to determine with *certainty* the time of establishment of every species, particularly in those areas where the extent of glacial survival has not been completely resolved.

The grouping of species according to their present ranges or distribution within and beyond a particular area is more feasible. The former will reflect the influence of existing environmental conditions, the second the migrational history of the existing flora. In Britain the distribution or range of a species was originally established on the basis of its recorded occurrence within vice-countries, areas of roughly comparable size corresponding to existing small counties or sub-divisions of large counties. A more recent inventory of the British flora has used kilometre-squares (within the existing National Grid) as the recording unit to plot the distribution of species and produce the *Atlas of the British Flora,* edited by E. H. Perring and S. M. Walters. Apart from those species which have a widespread distribution throughout Britain and those of extremely localised occurrence, the general coincidence in the range of certain species has permitted their classification into broad 'range or distribution types', designated according to their geographical distribution within the area. The varying distribution of such groups of species reflects physical, and particularly climatic, variations within the country. In Britain, for instance, we can recognise those species with a more northerly (Scottish) range, those of a southerly (English) type, or those with a characteristically south-easterly continental (Germanic) distribution (see Fig. 21).

Finally, the species which comprise the flora of any area can be grouped according to their range not just within but beyond it. Some may be endemic (and hence their range will be restricted to the area in question) others, however, may extend, and the greater part of their range lie well beyond its limits. The latter are often referred to as extraneous, or 'foreign', floral elements; they represent species or groups of species that have migrated from beyond the area rather than having originated within it. In Britain, where the number of endemics is extremely small, the present flora is essentially an extension of that in Europe. A classification by J. R. Matthews of those species in Britain which share similar ranges beyond the country reveals the source of the principal 'geographical elements' that have contributed to the British flora. Those species which have a very wide range throughout and even beyond

Europe account for more than half of the total flora. The remainder can be grouped into a dozen geographical elements. Half comprise species whose main centre of distribution is towards the south and south-west Mediterranean and central Europe (Fig. 22),

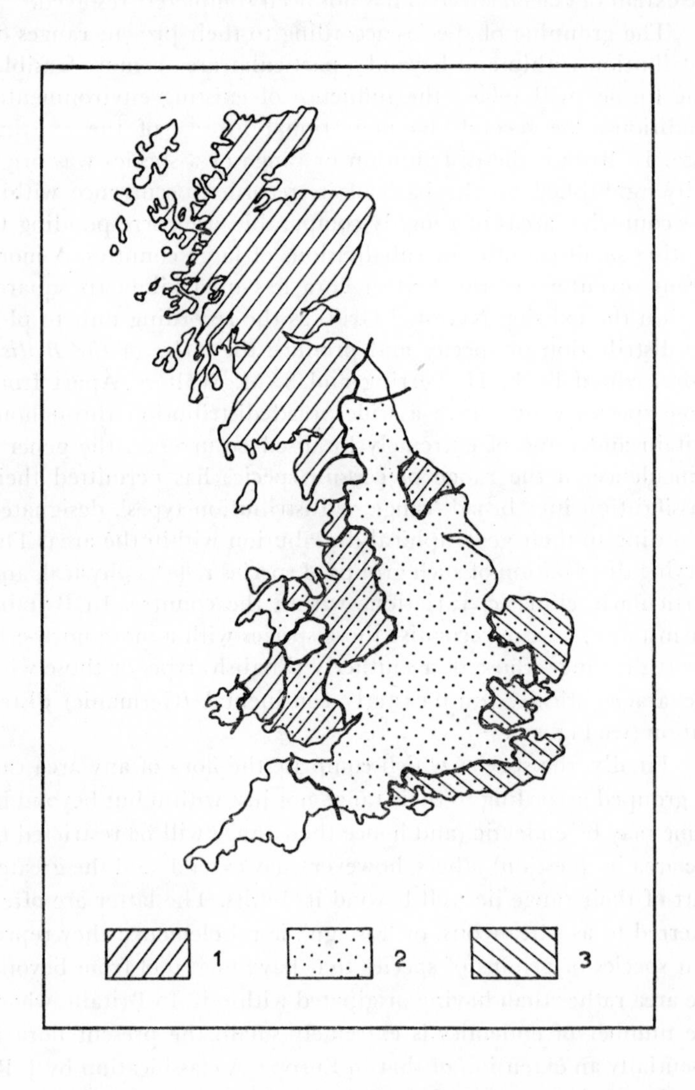

Fig. 21 Types of species-range or distribution in Britain: 1. *Trollius europeaus* (globe flower) = Scottish type: 2. *Bryonica dioica* (white bryony) = English type; 3. *Frankenia laevis* (sea-heath) = Germanic subtype. (Compiled from Turrill, William B. 1958)

the other half is either northern, north-western European, including Arctic areas, or the Alpine areas of the mountains in west, central and south-east Europe (see Fig. 23). J. R. Matthews points out that the northern elements include a high proportion of aquatic, marsh, moorland and mountain species, many of which have been identified from glacial and late-glacial deposits. These he suggests represent the oldest members of our flora, a number of which very probably survived the last glaciation. The species of

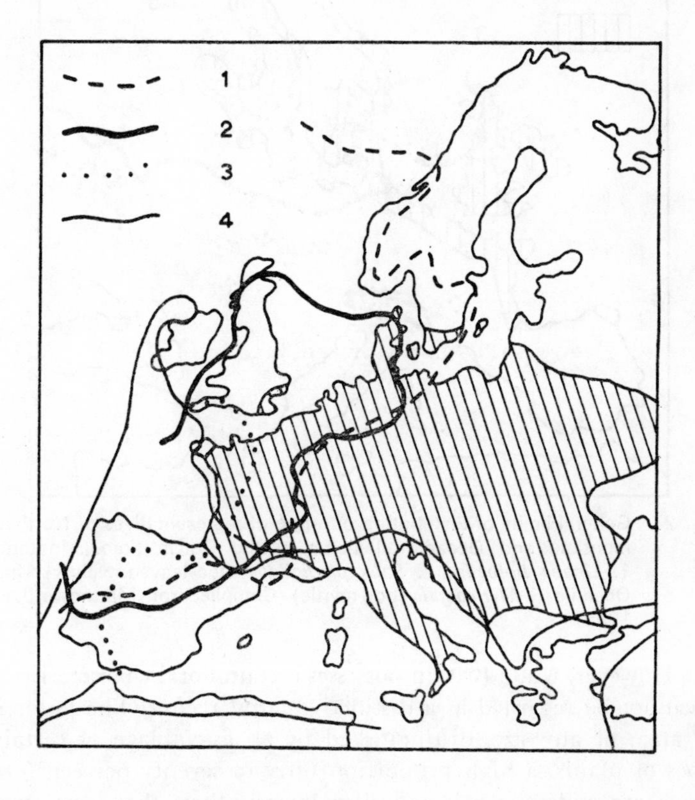

Fig. 22 General limits of distribution or range of representatives of (A) Oceanic West European and (B) Continental elements in the British flora: (A) 1. *Erica tetralix* (cross-leaved heath); 2. *Genista anglia* (needle furze); 3. *Erica ciliaris* (ciliated heath); 4. *Arbutus unedo* (strawberry tree); (B) shaded areas = *Carpinus betula* (hornbeam), example of a 'continental' element. (From: Matthews, J. R. 1955)

the southern elements would be those which did not enter the country until warmer post-glacial times. In the words of Matthews 'although the British area is small and its flora limited . . . it epitomises the whole story of northern temperate regions and their plant life since the Pleistocene'.

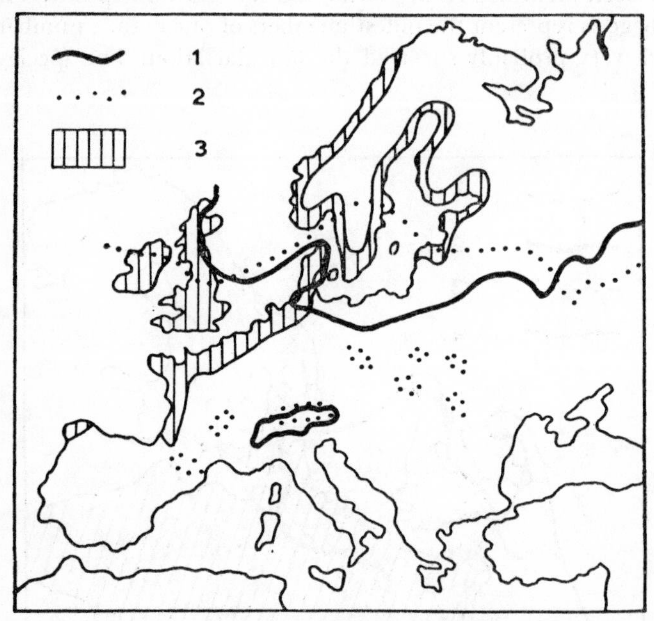

Fig. 23 General limits of distribution or range of representatives of Northern montane and Oceanic elements in the British flora. Montane: 1. *Linaea borealis*; 2. *Salix phylicifolia* (tea-leaved willow); and Oceanic: 3. *Myrica gale* (bog myrtle) (Compiled from Matthews, J. R., 1955)

However, while Britain possesses certain floral characteristics, it cannot be regarded as a distinctive *floral region*. The latter is an area, of any size, distinguished by an assemblage of certain types of plants, a high proportion (fifty to seventy per cent.) of which are endemic to it as well as many others that occur predominantly within its boundaries. Several attempts have been made by botanists to delimit floral regions of varying rank on the basis of groups of associated taxa, a large number of whose ranges coincide. But neither the classification nor delimitation of such regions is easy; the problems are similar to those attendant upon the regionalisation of any phenomena. The characteristics and size of floral regions depend very much on the criteria used to identify and delimit them. For instance, French plant geographers

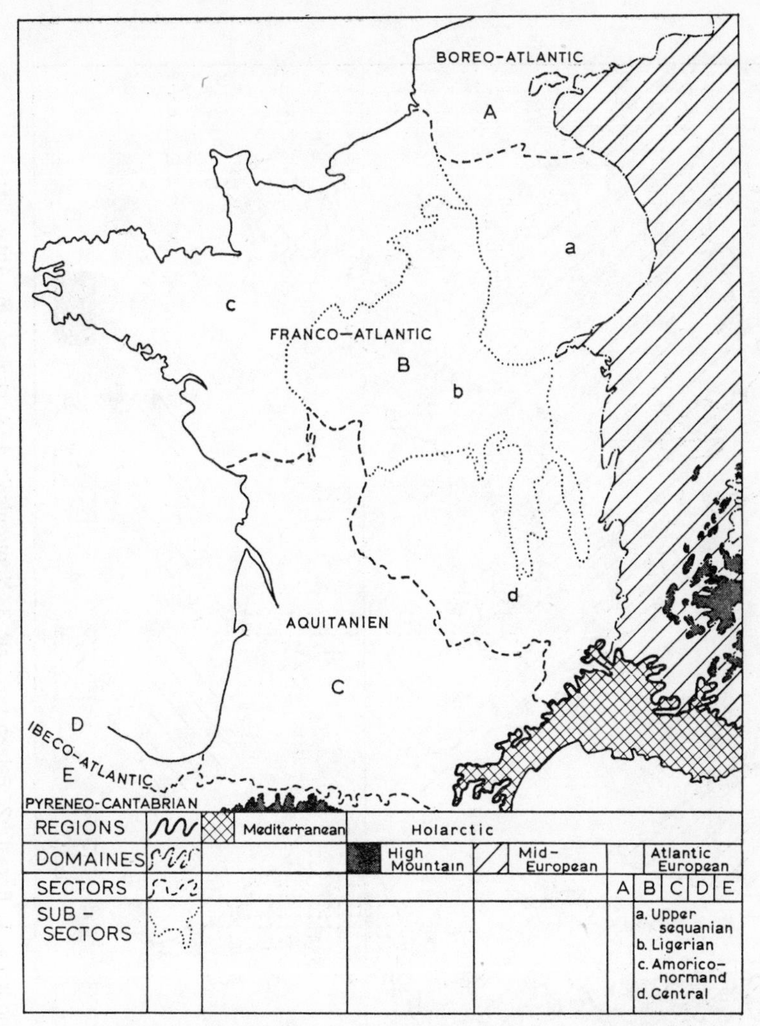

Fig. 24 Floral regions of France (From: *Atlas de France*)

have distinguished a hierarchy of floral regions (see Fig. 24). They are, (1) *regions*: with a high proportion of endemic genera subdivided into, (2) *domaines*: with a high proportion of endemic species, and within these (3) *sectors* and (4) *districts* on the basis of the presence of sub-species or ecotypes. The English plant geographer R. Good has attempted a division of the land areas of the world into thirty-seven provinces whose boundaries are determined by a marked coincidence of plant ranges (see Fig. 25). However the location of regional boundaries is difficult. The

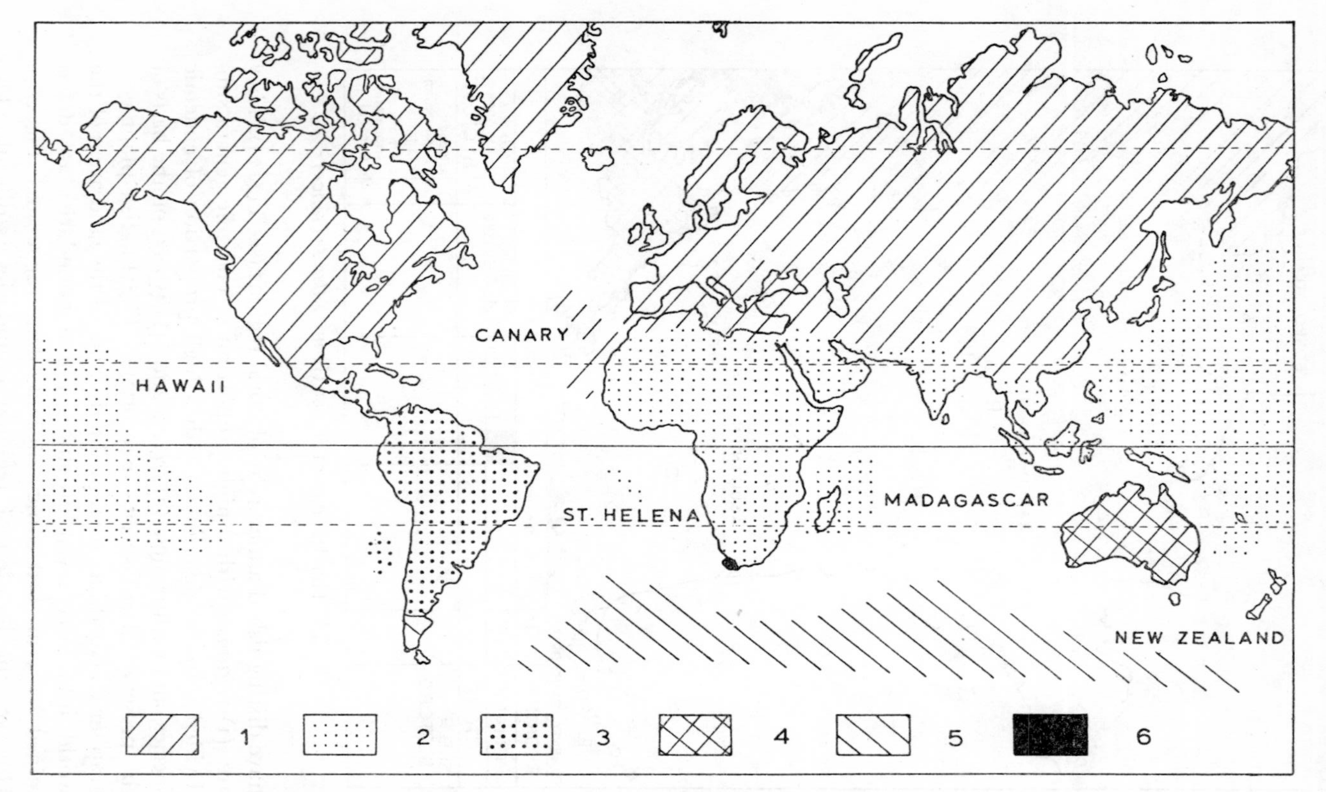

Fig. 25 Floral kingdoms. 1. Boreal; 2. Palaeo-tropical; 3. Neotropical; 4. Australian; 5. Antarctic; 6. South African. (After Good, R. 1953)

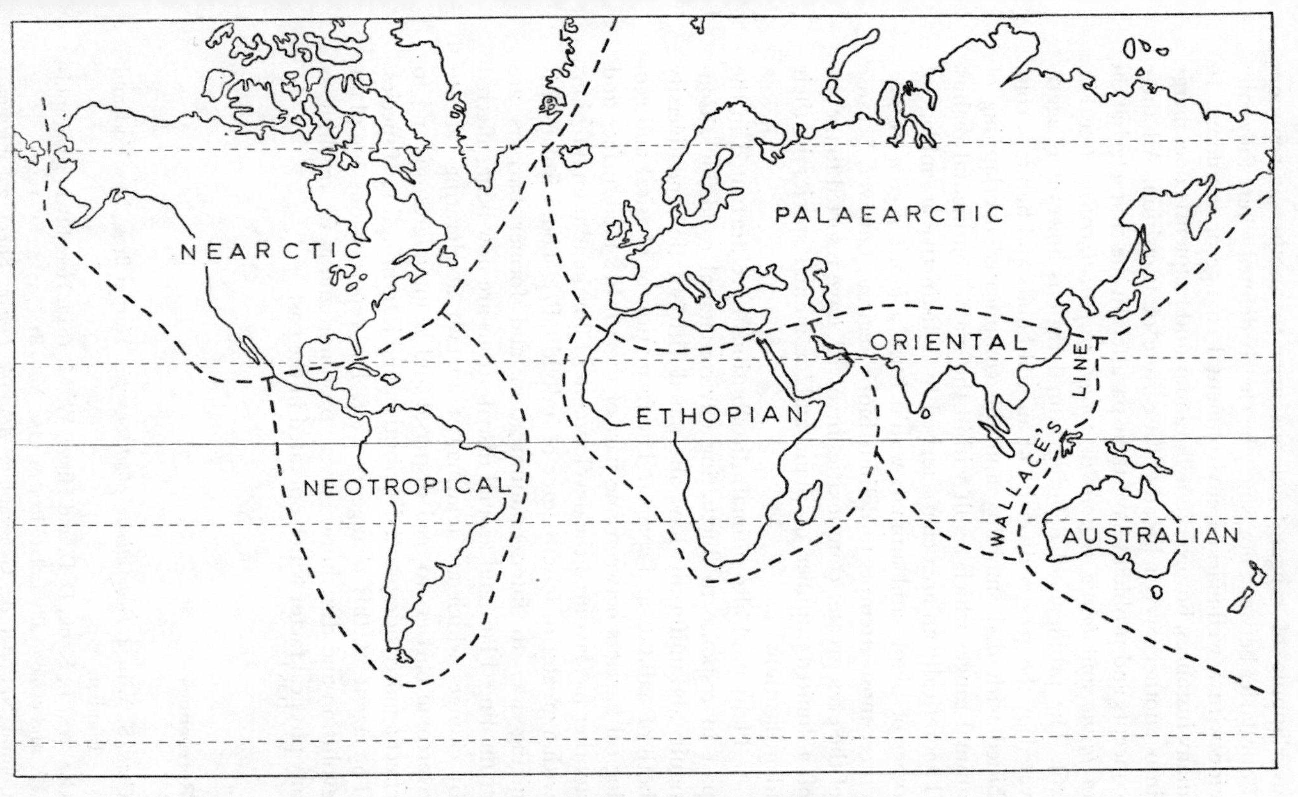

Fig. 26 Wallace's zoo-geographical regions.

actual ranges, even of endemics, rarely coincide precisely and overlapping of ranges from one area to another is common. Boundaries between floras may be clearly defined where they coincide with a well-marked environmental change or boundary. In many instances, however, one flora may and frequently does merge into another over a broad diffuse zone of transition. All these concepts and problems have their parallels in the characterisation of *fauna* and *fauna regions* or *zoogeographical realms* (see Fig. 26). The problems of defining faunal regions however is aggravated by the greater diversity of animal habits and habitats, combined with their mobility and the complicated overlapping of animal ranges which result within the major continental realms. Theoretically an area characterised by a distinctive assemblage of types of plants and animals would form a *biotic region* or *province*; most attempts to delimit biotic regions, however, because of the even greater complexities involved, have resorted to the use of a limited number of plant and/or animal species in their identification.

Plant (and also animal) distributions have fluctuated in the past in response to fluctuating environmental conditions. Contiguity and uniformity have alternated with isolation and diversity both of habitats and species. The formation of physical and ecological barriers between one part of the world's land-surface and another has favoured the development of floral and faunal regions with a greater or lesser degree of individuality. Today isolation is giving way to inter-communication and former barriers are crumbling. Plant and animal distributions are now being affected by changes which are tending, at an ever accelerating rate, to overcome barriers to seed dispersal and plant migration, and to effect a more widespread distribution and intermingling of species. The agent of this change is *man*. His repercussions on plant evolution and distribution are as profound and as far reaching as any physical factor which operated in the past.

References

Cain, S. A. 1944. *Found ationsof plant geography*. Harper Bros., New York and London.

Clausen, J., Keck, D. D. and Hiesy, W. M. 1947. Heredity in geographi- and ecologically isolated races. *Am. Nat.*, **81** (797): 114–133.

Comité National de Géographie. *Atlas de France*, Sheet 27. Eléments floristiques et limites d'espèces végétales.

DEEVY, E. S. 1949. Living records of the ice age. *Scient. Am.*, May, **180** (5): 48–50.

DEEVY, E. S. 1952. Radio-carbon dating. *Scient. Am.*, February, **186** (2): 24–42.

ELTON, C. S. 1958. *The ecology of invasions by plants and animals.* Methuen, London.

FURON, R. 1958. *Causes de la répartition des êtres vivants: paléogéographie et biogéographie dynamique.* Masson et Cie, Paris.

GAUSSEN, H. 1954. *Géographie des plantes* (2nd ed.). Armand Colin, Paris.

GLEASON, H. A. and CRONQUIST, A. 1964. *The natural geography of plants.* Columbia Univ. Press, New York and London.

GOOD, R. 1953. *The geography of flowering plants* (2nd ed.). Longmans Green, London.

GODWIN, H. 1956. *The history of the British flora.* Cambridge Univ. Press.

HESLOP-HARRISON, J. 1953. *New concepts in flowering plant taxonomy.* Heinemann, London.

HEYWOOD, V. H. 1967. *Plant taxonomy.* Studies in Biology No. 5, Institute of Biology. Edward Arnold, London.

JONES, M. 1968. Biological square-bashing. *New Scient.* 25th July, **31**: 193–196.

LOUSLEY, J. E. 1953. *The changing flora of Britain.* Botanical Society of the British Isles, Arbroath.

MARTIN, P. S. 1967. Pleistocene overkill. *Nat. Hist. N.Y.*, **76** (10): 32–38.

MATTHEWS, J. R. 1955. *The origin and distribution of the British flora.* Hutchinson's University Library, London.

ODUM, E. P. and ODUM, H. T. 1959. *Fundamentals of ecology* (2nd ed.). W. Saunders & Co., London.

OLDFIELD, F. 1967. Some notes on the implications of pollen-analytical evidence. In *Liverpool essays in geography*, Ed. by R. W. Steel and R. Lawton. Longmans, London.

PENNINGTON, W. 1969. *The history of British vegetation.* English Univ. Press, London.

PERRING, E. H. and WALTERS, S. M. (Eds.) 1962. *Atlas of the British flora.* Botanical Society of the British Isles, and Nelson, London.

POLUNIN, N. 1960. *Introduction to plant geography and some related sciences.* Longmans Green, London.

RIDLEY, H. N. 1930. *The dispersal of plants throughout the world.* Reeve & Co., Ashford, Kent.

SEWARD, A. C. 1931. *Plant life through the ages: a geological and botanical retrospect.* Cambridge University Press.

STEENIS, C. G. G. VAN. 1964–5. On the origin of island floras. *Advmt. Sci. Lond.*, **21** (89): 79–92.

TURRILL, W. B. 1946. The ecotype concept. *New Phytol.*, **45** (1): 34–43.

TURRILL, W. B. 1958. *British plant life* (2nd. ed.). Collins, London.

WILLIS, J. C. 1922. *Area and age: a study in geographical distribution and origin of species.* Cambridge Univ. Press.

Wulff, E. V. 1950 *Introduction to historical plant geography*. Chronica Botanica, Waltham, Mass., U.S.A.

Zeuner, F. E. 1959. *The pleistocene period: its climate, chronology and faunal succession*. Hutchinson, London.

7
Effect of Man on Plant Evolution and Distribution

The recent course of plant evolution and distribution has been no less eventful than that of the remote geological past. Rather, it has been more dramatic and spectacular. Within the brief time-span since the climatic fluctuations of the Pleistocene period man has evolved. With increasing numbers and developing techniques he has emerged as the major, most powerful and universal instrument of environmental change in the biosphere today. He has triggered off a new phase of organic upheaval which is continuing with increasing intensity. During this phase the *rate* of plant evolution and migration has far outstripped that in any comparable period of time in the past. As the dominant animal he has affected, either directly or indirectly, accidentally or deliberately, every other type of organism. He has altered the balance of plant and animal species. He has increased the numbers and range of some at the expense of others which have, as a result, been partially or wholly destroyed. He has become a major agent in the dispersal and redistribution of species throughout the world. He has intentionally or unintentionally modified the direction and accelerated the rate of plant and animal evolution.

Man's influence on plant evolution has been twofold. He has deliberately produced new species of plants by the selection and propagation of variants and by the cross-fertilisation of different types of plants. In addition he has created environmental conditions that have tended to facilitate the evolution of new species. In relatively stable habitats evolution is inevitably a slow process. The variant or mutant individuals which develop within any species will only give rise to a new 'race' or 'strain' under conditions which favour their survival and allow them to maintain their identity free of continual cross-fertilisation with other members of the same species. The production of new species by hybridisation is also relatively rare under these conditions. The reasons for this vary; some species cannot interbreed, others may but, as is frequently the case, the hybrid progeny is sterile and cannot propagate itself. On the other hand two species that can be crossed under cultivation do not necessarily produce hybrids when they occur in close proximity in the wild, or if they do the hybrids fail to perpetuate themselves. The failure of hybrids to maintain their identity is, as in the case of variants or mutants, due partly to continual back-crossing or *introgressive hybridisation* with the parent plants, which tends to cancel out marked differences between species and their hybrid off-spring. Also, the optimum requirements and tolerances of the hybrids and variants will vary from those of the parent stock. In the absence of a suitable habitat in which they would have a competitive advantage it is difficult, if not impossible, for them to become established *and* maintain their identity. New or changed conditions to which they are better adapted than other plants will tend to favour their survival and allow the emergence of a new species-population. Variants and hybrids require, therefore, different or what has been referred to as 'hybrid' habitats if they are to maintain their identity successfully.

It is thought that in the geological past, periods of more drastic and extreme environmental upheaval—mountain building, volcanism, the advance and retreat of epi-continental seas, and climatic changes were accompanied by bursts of evolutionary activity as a result of the greatly increased opportunities for selection and, particularly, hybridisation. The migration of species initiated by such changes brought about a greater intermingling of diverse types of plants; while the new or disturbed habitats that were produced provided conditions for the survival and establishment of new species or races resulting from inter-contact. Such must have been the case in the northern hemisphere, for instance, during the marked climatic, soil and vegetation 'disturbance' of the Pleistocene period.

Fresh habitats with open and unstable soil conditions are a natural and continual product of those areas subjected to marked erosion or deposition at present. They are characteristic of areas directly affected by river action, by coastal erosion or deposition, or by slumping and sliding on steep mountain slopes. But the extent and diversity of such naturally disturbed habitats is infinitesimal in comparision to those which have been, and continue to be produced by man. From the time man began to make his first impression on the vegetation and soil cover he has created an ever-increasing variety of new habitats. Some are associated with his place of habitation—from prehistoric camp-sites and kitchen middens to modern gardens and back-yards—others with his place of work, whether farm, rubbish-dump, mine or quarry. Yet again his movement by primitive trail or modern road or rail lines and widespread clearing of the former vegetation cover by burning, deforestation or cultivation have helped to create new habitats. These are the 'hybrid' habitats born of the interaction of man and his physical environment. They are characteristically open, exposed to full sunlight and not infrequently enriched in organic and mineral matter as a result of man's activities.

Such open, disturbed habitats, have provided a foothold for those species of plants tolerant of such conditions. Their vigorous growth and survival is favoured by the absence of competition. Of those which occupy these sites, some were originally native only to naturally open and exposed habitats; others were formerly inconspicuous and often minor species repressed by the competition of the other plants with which they originally grew. A great many more have arisen from variants or hybrids to which these disturbed habitats gave an opportunity for establishment. Numerous species have become associated with man and through him achieved an increase in abundance or even evolution, and are dependent on him for their continued existence. These are the 'camp followers', the types of plant often designated as *weeds*. Sometimes defined as plants growing 'out of place' or 'where they are not wanted', they can perhaps be better described as those which man unintentionally (and often to his detriment) permits to grow with greater abundance and vigour than they would otherwise do. Many are herbaceous plants, frequently though not invariably annuals, and most, in the absence of competition, are vigorous and aggressive. Prolific seed production combined with a wide range of tolerance of varying environmental conditions has aided and accelerated their widespread dispersal and distribution by man himself. Among the few species of plant which have attained an almost world-wide range—in both tropical and temperate latitudes—are some of the notorious weeds of cultivation, such as the dandelion (*Taraxacum officinale*), shepherd's purse (*Capsella bursa-pastoris*), plantain (*Plantago spp*) and chickweed (*Stellaria spp*). The very early association of weeds with man is testified by the appearance of pollen derived from such species as the rib-wort plantain (*Plantago lanceolata*) and fat-hen (*Chenopodium spp*) in peat deposits of Neolithic age in Northern Europe. In many areas weeds persist as evidence of man's former presence in an area. The American Indian called the plantain 'the White Man's footprint'. Dense stands of nettles (*Urtica dioica*) frequently remain on or near sites of former habitation where the soil is particularly rich in phosphorus. The bracken fern (*Pteridium aquilinum*) will grow with continued vigour for many years on land once cultivated or heavily manured by animal droppings. But deprived of the open, disturbed conditions which favour their rampant growth weeds can rarely maintain their dominance indefinitely in face of competition from more demanding species.

The simultaneous development occurred not only of species which were adapted to and could tolerate the disturbed habitats man created but also of those plants which he selected for his own particular uses. By the conscious or unconscious selection of certain species (and moreover of particular variations and strains within them) best suited to his needs, he has created by the process of domestication, forms of plants (and animals) quite distinct from those that exist in the wild. These are the cultivated crops, the so-called *cultigens* or *cultivars*—the products of man's direct influence on plant evolution. The uses for which man has selected certain types of plants are as varied as the types of cultigens he has produced. As C. D. Darlington remarks, 'Within cultivated plants there is an endless variety of types according to their antiquity and mode of origin and improvement . . . At one extreme there are the original grain crops few in number and of a remote and undocumented origin. These are the plants to whose cultivation and improvement man himself owes the origin of settled society. At the other extreme are the ornamental plants, the frills and superfluities of civilisation, plants existing in an enormous number of species'. However, while the diversity of cultivated crops is great, they have in fact been developed from a relatively small number of existing families. This is particularly striking in relation to the food crops. By far the largest amount of man's carbohydrate requirements is supplied by a few select members of one large and very varied, cosmopolitan family—that of the grasses (*Gramineae*); while the legume (*Leguminosae*) family provides the bulk of his plant-protein supply. As the evolution of cultivated plants has been dependent on man, so too is their continued existence. Some have no known or recognisable wild ancestors. Few could exist independently in the wild. The process of domestication has involved selection, planting and propagation by man. His selection has involved not only certain species, but particular variants within or hybrids between them. He selected for protection and propagation those individual plants which were best-suited to his needs but which frequently were those least-fitted for their survival in face of competition under natural conditions. The majority of cultivated plants lack the ability to spread or mantain themselves independently of man. Most (with the exception of such tree-crops as coffee, tea, cacao) are heliophytic, intolerant of shade. Apart from those which are propagated vegetatively, the majority are annuals; some were originally annuals, while others have developed this habit under cultivation. A classic example in this respect is cotton, originally a perennial shrub confined to frost-free tropical areas. The selection of forms that fruited early permitted cultivation in areas where summers were warm and long enough for fruiting but where winters were cold and liable to frost, and this was accompanied by a change from a perennial to an annual habitat in the plants.

Many of our food crops in particular are dependent on man for their propagation. Some have been propagated vegetatively for so long that their seed-producing capacity has been drastically reduced. The extreme example is the edible banana whose continued propagation is entirely dependent on man. The origin of this highly domesticated cultigen was the selection of a sterile hybrid able to produce the fruit so attractive to man but unable to develop seeds necessary for its perpetuation. Characteristic of the cereal crops is the lack of effective or efficient means of seed dispersal—an advantage for harvesting but a decided disadvantage for survival and spread. Many of the wild grasses and primitive species from which the modern grain-crops evolved have heads whose central ear-bearing stem is brittle and fragile. When ripe this breaks up (or 'shatters') and allows the seeds to be easily dispersed. In the more evolved cultivated forms, selection has favoured those plants whose flower stems were tough. Their heads would remain intact when mature and could hence be easily gathered. Indian corn, or maize, has become so highly domesticated that it is incapable of reproducing itself without the aid of man. Unlike the other grains whose kernels, or ears, are separate, those of maize occur as a densely packed cluster of seeds enclosed in a tightly wrapped sheath or husk. When this drops to the ground there is no means whereby the seeds can be released and few survive the excessive competition in the cob or the depredations of animals.

In addition, cultivated plants have in the course of domestication tended to 'lose' those characteristics which might make them unattractive to man, for example, hairiness, thorns, toughness or an unpleasant taste, but which in the wild afford a measure of protection from grazing or browsing. Indeed the increased size and palatability of herbaceous crops make them more liable to attack. Furthermore the concentration of plants of one kind under conditions favourable for their growth creates optimum conditions for the multiplication and evolution of animal pests and disease organisms. The more highly evolved and domesticated, the greater is their dependence on man's protection. It is somewhat ironic that in his cultivated fields man creates disturbed habitats in which his weeds, with their greater aggressiveness and competitive capacity, are better fitted to survive than the cultigens he has selected to grow there!

All the major crops and most of the minor ones are ancient in the sense that they have been domesticated since pre-historic times. Forms of cultivated cereals almost identical to those grown today are known to have been in existence at least seven thousand years ago. Recent archaeological evidence has revealed the existence of agricultural communities in Turkey nearly nine thousand years ago. As Carl Sauer notes, 'historic man has added no plant or animal of major importance to the domesticated forms on which he depends, and none are important food plants.' Over very considerable areas of the world man's cultigens and his weeds are now the dominant type of plants; it has been noted that the number of wheat plants alone may outnumber those of any other wild or cultivated seed plant! In view of the extremely brief period of man's existence, this represents a rate of evolution probably greater than in any other type of organism at any other time in the history of the biosphere. The place of origin, the wild or primitive ancestors, and the evolutionary development of cultivated plants are however difficult to trace. They involve a field of investigation in which the biologist and the archaeologist meet and look to each other for mutual assistance. For the biologist the study of the origins of cultivated plants is fundamental to an understanding of their present characteristics and to that of evolutionary processes, as well as to the discovery of varieties and strains which might provide possibilities for future development and use. The imprints and carbonised remains of seeds preserved in pre-historic sites (even in the faeces and stomachs of pre-historic man) form part of the evidence as to the types of plants grown in a particular place at a particular time. The evolution of human societies and cultures and their dissemination is inextricably linked with that of plant and animal domestication. The pre-historian, the archaeologist, the anthropologist, and the human geographer are deeply involved in these problems; they, in turn, must look to the biologist for an understanding of how the 'organic artefacts' of man evolved, where they originated and the directions in which they spread.

In the period between the two world wars Russian plant geographers under the guidance of I. N. Vavilov (one of the outstanding biologists of his day) conducted an extensive survey of the main crop plants of the world. Their primary object was to study the distribution of existing species or varieties of cultivated plants and their closely related wild relatives. Vavilov discovered that for a large number of crop plants (and their associated weeds) the greatest number of species and varieties within them, or in closely related wild species, tended to be concentrated in certain areas. These areas or, as they become known, 'centres of diversity', which he identified occur mainly in the Old World, (see Fig. 27). But here, and in the New World, they are characteristically upland or mountainous regions—areas of marked physical diversity in tropical or sub-tropical latitudes. On the assumption that a particular type of plant—a particular genus—would reveal the greater diversity of species and varieties (as well as the presence of wild ancestors) near its centre of origin, Vavilov argued that the centres of diversity which he had identified were probably the 'centres of origin' of the particular crop plants. This being so, and since many of his areas were centres of old civilisations, he thought they may well have been the foci where domestication and agriculture first developed. He later modified his ideas to allow for the fact that while some of his areas could be regarded as the *primary* centres of crop-plant evolution, others were probably secondary. The latter would have developed later when already domesticated plants migrated or were carried by man into other areas. In contact with different plants and under different physical conditions, the original cultigens would then give rise to a new set of hybrids and variant strains, and a secondary centre of diversity would thus be created. In conjunction with this idea he distinguished between *primary* and *secondary* herbaceous crops (see Table 14).

The primary crops were those which man originally selected and many of which, as C. D. Darlington suggests, may have first come to his notice as 'habitation weeds'. These included wheat, barley, flax, soy and maize. It is probable that many of the secondary crops, such as rye, oats, mustard, rape and other cruciferous plants, would have made their first appearance as 'weeds of cultivation'. Under primitive methods of agriculture and rudimentary cultivation, a great variety of other plants must have accompanied

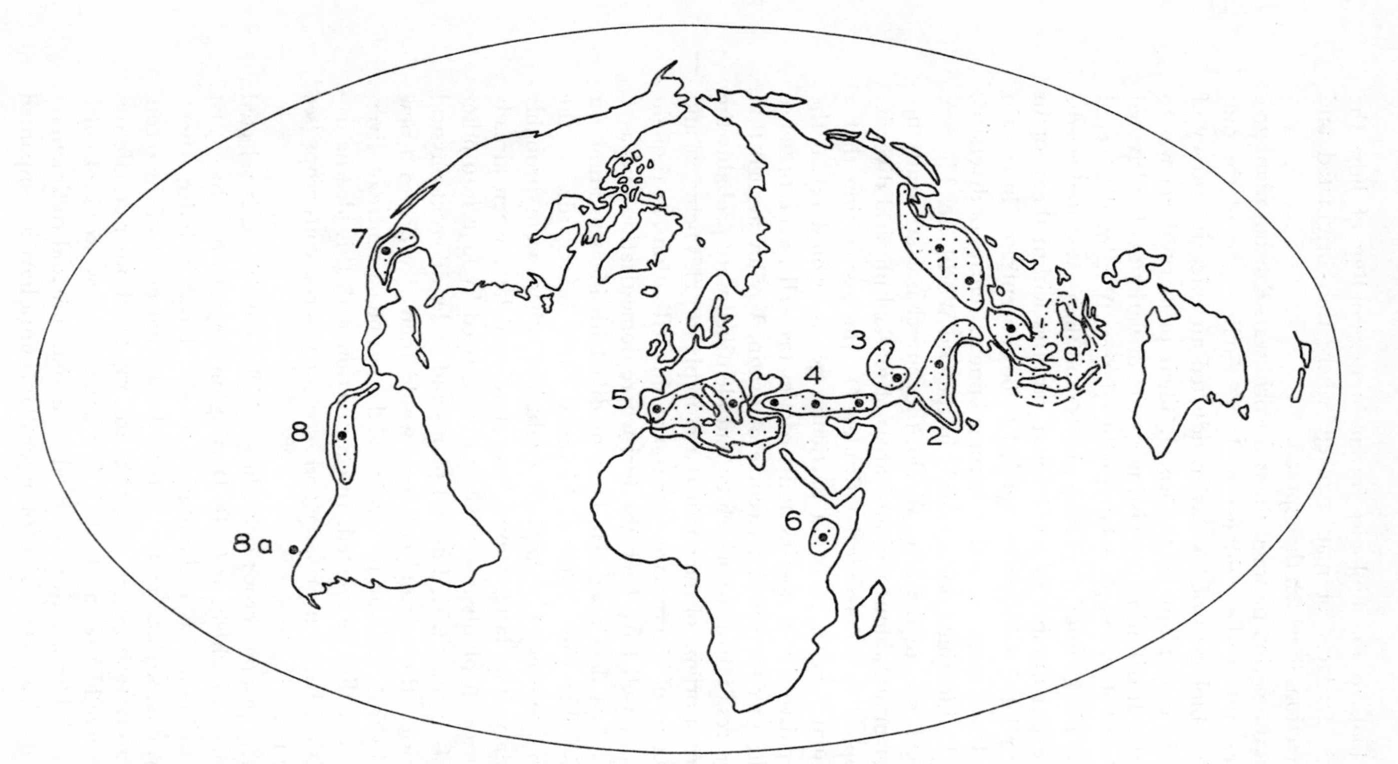

Fig. 27 World centres of origin of cultivated plants according to Vavilov: 1. Chinese; 2. Indian; 2a. Indo-Malayan; 3. Central Asian; 4. Near East; 5. Mediterranean; 6. Abyssinian; 7. Southern Mexican and Central American; 8 and 8a. South American; Black dots indicate centres of origination of form of principal cultivated plants. (From: Vavilov, I. N. 1949–50)

Primary Crops	Secondary Crops
Common wheat (*Triticum vulgare*)	Rye (*Secale cereale*)
Common barley (*Hordeum vulgare*)	Oats (*Avena sativa*)
Emmer wheat (*T. diococcum*)	
Flax (*Linum usitatissum*)	Spurry (*Spergula linicola*) Mustard (*Brassica campestri*) (*Eruca sativa*) (*Camelina sativa*)
Buckwheat (*Fagopyrum esculentum*)	F. tatarium
Cereals	Common vetch (*Vicia sativa*) Field pea (*Pisum arvense*) (*Coriandrum sativa*) (*Cephalavia syriaca*)

Table 14 Origin of secondary crops as weeds of cultivated primary crops. (After: Darlington, C.D. 1963)

man's crops. His fields and plots (as in tropical 'garden' cultivation today where a great variety of different plants are grown together) must have been 'hot-beds' of hybridisation and introgression. The weeds of cultivation—many of them hybrids—would have accompanied man as he moved from one area to another. As in the case of rye and oats, they eventually became the dominant crop under climatic conditions to which they proved better adapted than the primary crop. That the areas which Vavilov identified are centres of marked species-diversity for many crops is an established fact. The development of such diversity must undoubtedly be closely associated with the tremendous variety of habitats characteristic of these mountainous areas; it may also, as has been suggested, have been accentuated by the mingling and intermixing of peoples from different areas. The interpretation of these centres of diversity has, however, been disputed as over-simplified. Edgar Anderson, for instance, would maintain that the existence of a great diversity of often primitive varieties of particular crops in isolated mountain areas may be due to the persistence of ancient and conservative methods of agriculture—they are regions of survival rather than of origin. Vavilov's survey (on a scale that has never been repeated) was of its nature a pioneer study—an overture rather than a finale. It is not, therefore, surprising that many of his preliminary ideas, especially about the centres of origin of particular crops, have been and will no doubt continue to be revised in the light of new archaeological and botanical evidence.

The origins and development of the two major cereal crops, wheat and maize, have been subjected to the most intensive of genetic and archaeological study. The earliest archaeological evidence of agriculture comes from sites in Asia Minor—in the belt of mountains which form the north-western flank of the 'Fertile Crescent' in Iraq and Syria. Here in the ruins of pre-historic settlements of Jarmo (Iraq) and Tepe Saras (Iran)—dated about 6 750 to 6 500 B.C. respectively—remains of primitive barley and of two forms of domesticated wheat have been identified. Of the latter one is almost identical with an existing wild wheat (wild emmer) the other identical with a still cultivated but very primitive variety (emmer). These are the earliest known ancestors of our modern wheats, wild forms of which do not exist. Modern work in plant genetics has, in this case, confirmed the archaeological evidence that the original centre of wheat domestication was located in Asia Minor. Further it has thrown much light on how the different wheat species must have evolved. Existing wheats can be classified into three groups distinguished by their chromosome numbers (Table 15). The einkhorns, with the lowest chromosome numbers are considered to be the most primitive. The second group were produced by the hybridisation between einkhorn and a closely related wild grass. The third and most recently evolved, which contain the modern bread wheats (common, shot, club), represent a still later stage of hybridisation. This was not deliberately controlled hybridisation (which had to await the scientific knowledge of the twentieth century) but a result of contacts between crops and weeds, and the survival of the resulting hybrids that man unconsciously facilitated at a very early stage.

Maize, however, does not exist in a wild form today. The problems of its ancestry and place of origin have been more difficult to solve. Pollen—identified as that of maize—has recently been found in deposits thought to be some eight thousand years old in Mexico. Corn cobs dated at 5 000 B.C., which may be those

Latin Name	Common Name	Chromosome number	Distribution	Earliest Evidence
T. aegilopoides	Wild Einkhorn	7	Asia Minor, Greece, S. Yugoslavia	Pre-agricultural
T. monococcum	Einkhorn	7	Asia Minor, Greece, C. Europe	4750 B.C.
T. dioccoides	Wild Emmer	14	Near East	Pre-agricultural
T. diococcum	Emmer	14	India, C. Asia, Europe, Abyssinia	4000 B.C.
T. durum	Macaroni	14	C. Asia, Asia Minor, Abyssinia, U.S.	100 B.C.
T. persicum	Persian	14	Georgia, Armenia, Turkey	None
T. turgidum	Rivet	14	Abyssinia, S. Europe	None
T. polonicum	Polish	14	Abyssinia, Mediterranean	17th century
T. timopheevi	—	14	W. Georgia	20th century
T. aestivum	Common	21	World wide	Neolithic period
T. sphaerococcum	Shot	21	C. & N.W. India	2500 B.C.
T. compactum	Club	21	S.W. Asia, S.E. Europe, U.S.	Neolithic period
T. spelta	Spelt	21	C. Europe	Bronze age
T. macha	Macha	21	W. Georgia	20th century

Table 15 Existing species of wheat (*Triticum*): all but wild Einkhorn and Emmer are still cultivated. Only the bread wheats (common, shot and club) and the macaroni wheat are of commercial importance today. (From: Mangelsdorf. 1953).

of wild or a primitive cultivated variety of maize, have been discovered in a prehistoric site, also in Mexico. Such findings make it clear that this cultigen originated in the New World and evolved from a wild maize which has long since disappeared. Although today the greatest diversity of types of corn is found in the Andean region of north-western South America, the centre of origin of domesticated maize may, in fact, have lain much further north in Central America.

The earliest known archaeological evidence of agriculture in both the New and Old World dates from about 7 000 B.C. However, as Carl Sauer, in his stimulating work, *Agricultural Origins and Dispersals,* points out, this is a relatively late date in the history of the human race. Even the earliest remains of agricultural communities in the Near East must represent a fairly advanced stage of plant domestication and of human culture. Domestication was a long, intricate process which must originally have involved more primitive methods. It started much earlier and probably originated in areas far removed from the known and 'archaeologically blessed' cradles of civilisation. Sauer suggests that the early ancestors of the Neolithic farmers were sedentary peoples whose main source of food would have been fish and other animals of fresh-water lakes and rivers. The first plants to be domesticated may well have been those used for their fibres and fish-poisons. Be that as it may, it would seem reasonable to expect, as Sauer does, that man learnt to plant the crops he first selected for food and other purposes and to propagate them vegetatively by cuttings and shoots, *before* he began to collect and sow seeds. He was a vegetative planter before becoming a seed planter.

Furthermore, it is unlikely that man's first attempts at agriculture would have been made in areas where environmental conditions were such that highly developed skills and techniques were needed in order to live successfully. Settled agriculture in areas with a long dry or cold season, in unstable flood plains or in grasslands where the tough, matted sod is hard to penetrate, demands a knowledge of fairly advanced farming techniques adapted to these particular conditions. Sauer maintains that the selection and domestication of plants and the development of agriculture would have been more likely to originate in areas where it was easy rather than difficult for primitive man to live. He envisages this early home as characterised by a warm mild climate with sufficient rain to ensure a reasonably long growing season; by easily cleared and dug soils—light forest and shrub, rather than dense tropical forest or grassland and by a diversity of plants with which man could 'experiment'. For these reasons Sauer has proposed South-East Asia and north-western South America as the areas—'the hearths'—of original domestication of plants and animals (see Fig. 28). Climatically favourable, these are both highly dissected mountain areas of marked diversity of physical habitat and, as Vavilov and other botanists have confirmed, areas of exceptional plant and animal variety. Both are areas where opportunities exist for settled fishing communities, and both are strategically placed

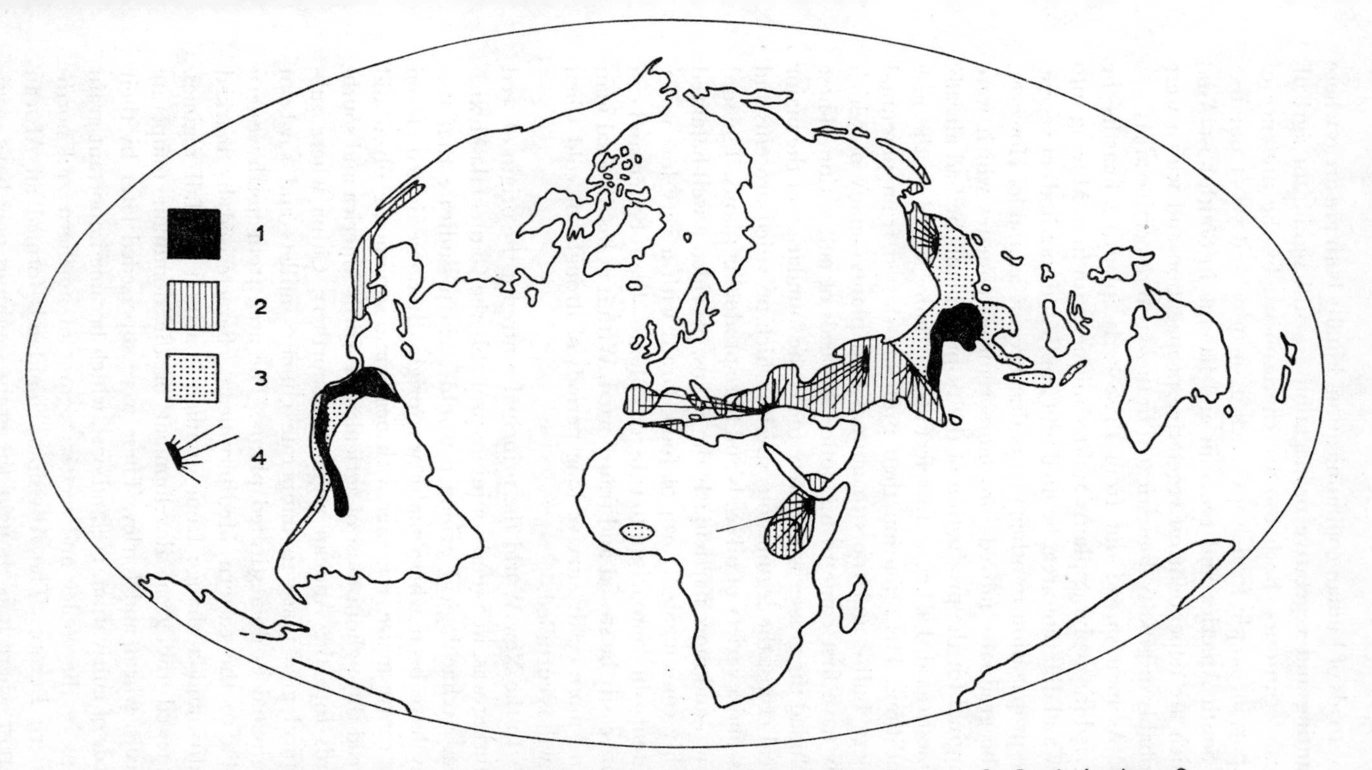

Fig. 28 Sauer's 'hearths of domestication': 1. Vegetative-planting hearths; 2. Seed-planting; 3. Colonial extensions from 1 and 2; 4. Centres of dispersal of seed-planting. (Adapted from: Sauer, C. O. 1952)

crossroads of human communication. Finally both are areas where planting and vegetative propagation are still significant agricultural techniques. Both too are characterised by the presence of important staple food-crops (such as manioc and sweet potatoes in South America and the yam and banana in South-East Asia) which are dependent on vegetative propagation and which were probably originally domesticated in these two respective areas.

As man spread out from these early 'planting hearths' he carried his evolving plants and techniques with him. Moving into new and different areas he no doubt applied his methods of vegetative propagation to other plants that became available. However, as he gradually moved into more rigorous climates which were marginal for the production of the tropical plants he had already domesticated, he began to select species more suited to the new conditions. His attention then probably shifted from perennial tubers, bulbs, and roots to annual seed plants—many of which may have first come to his notice as weeds of cultivation. These included the grasses which were to lay the foundation of the major cereal crops; the leguminous species which provided proteins and fats, and a variety of oil-seeds and fibre-producing plants. The seed plants, and more particularly the grains, provided a well-balanced highly concentrated form of food. Grain with a very low water content—in comparison to tuberous plants—keeps better, and can more easily be stored and transported. With it at his disposal man could more easily survive long periods of drought or cold when growth is curtailed.

In the New World the principal centre for the origin of seed planting was, in Sauer's opinion, probably the Gautemala-Mexico area. Archaeological evidence would seem to indicate that this may have been where maize was originally domesticated. From this centre maize, together with beans and squashes, evolved and spread throughout most of agricultural North America and southwards into the Andean region. In northern China where subtropical give place to temperate climates millets and soy-bean originated as domesticated plants. The area from north-western India to the eastern Mediterranean, characterised by marked aridity and a change from summer to winter rainfall regimes, witnessed the great development of the dominant temperate cereals, wheat and barley. These were superseded later by their weeds of cultivation, rye and oats, which became important grain crops in the cooler and wetter regions of northern and north-western Europe. The Abyssinian highlands formed an African outpost where it is thought the grain sorghums may have originated and in which Vavilov found a great diversity of primitive forms of wheats, and, more particularly, barley.

The process of domestication of cultivated crops was one in which man's conscious selection of the plants best suited to his needs in different areas was combined with his unconscious, unintentional stimulation of their evolution by his methods of cultivation and his migrations from one region to another. He accidentally intermixed plants that otherwise might not have come into contact. He inadvertently produced new species and varieties, and either deliberately or unintentionally favoured their survival. So effective and rapid was this process that the staple food-crops of today had already been determined at least 7 000–8 000 years ago, if not earlier. The number of additional types of plants that have been brought into the domestic fold within historic times is relatively few and, as has already been stressed before, includes no important food crops. The more recently evolved cultigens are ornamental plants, animal fodder crops; in particular the forage grasses and legumes which at present are in a very active and relatively early stage of domestication, and plants such as the South American rubber-producing tree (*Hevea*) and *Chinchona* (quinine) whose products, though formerly known, have assumed a commercial significance only since the latter part of the nineteenth century.

The evolution of cultivated plants has, however, continued with increasing rapidity within the early-established groups of domesticated plants. Its direction, within historic time, has been greatly influenced by an increasing human population, by the development of more specialised farming techniques and by the replacement, particularly in temperate latitudes and modern industrialised societies, of subsistence by commercial agriculture. From the main types of domesticated plants modern man has, in fact, selected only relatively few of the available species and varieties for commercial development. Of the fourteen known species of wheat in existence the three species of bread wheat (common, shot and club) account for about ninety per cent. of the wheat crop in the world today. Man has so increased the range and distribution of the favoured commercial species that their main centres of production now lie well beyond their original areas of origin. In addition modern farming methods and commercial agriculture have stimulated, and in their turn have been stimulated by, the revolutionary advances that have occurred in crop breeding within relatively recent years.

The new phase in the evolution of cultivated plants began

when man started deliberately to create, to breed, plants for particular purposes. The pioneers in this field were the Dutch bulb breeders who as early as the end of the seventeenth and beginning of the eighteenth centuries were already producing standardised seeds for food crops. The contributions of Charles Darwin and Alfred Wallace to an understanding of the principles of evolution and of Mendel to the laws of inheritance, laid the foundations for the development of new scientifically based techniques of breeding and propagation. By rigidly controlled selection, 'pure-line' varieties of crop plants could be isolated and maintained. By deliberate hybridisation new varieties with particular characteristics could be produced. The development of these new techniques has been mainly within the last sixty years and their application has been more particularly concerned with the major commercial food crops and ornamental plants of temperate regions. Modern advances in plant breeding have been most spectacular in countries or regions of recent agricultural colonisation, e.g., the prairie regions of North America and Russia, and in New Zealand and Australia, where crop plants have been introduced from elsewhere. They have been particularly striking, too, in areas of highly developed commercial agriculture where new varieties of cultivated plants have either allowed the application of new methods of farming, or were necessitated by them.

The main aim of modern plant breeding is to produce varieties of crops that will give, most economically, most efficiently, the maximum production in terms either of quantity or quality, under the conditions in which they are grown. Man is more able than his ancestors not only to select the plant best suited to his needs, but also that plant best adapted to produce them under given environmental and farming conditions. He endeavours to produce ecotypes; agricultural races of cultivated crops, these have been termed 'agro-types' as they are adapted not just to natural environmental conditions but to soils modified by drainage or fertilisation and to mechanical methods of cultivation. They are ecological races adapted to particular agricultural habitats.

Early-maturing varieties of grain are bred that can be grown in regions where either drought or frost may curtail the growing season; they are bred for 'hardiness' or drought resistance. The over-riding objective in plant breeding, however, is to increase *yield*—to produce varieties of crops that will, if well cultivated and heavily fertilised, build up plant tissues most efficiently. Such highly-bred, highly-specialised plants of a uniformly standardised genetic composition have, however, little resistance to the conditions to which they are not adapted. A severe drought or prolonged frost may destroy a whole crop, since it will not contain some individuals with a slightly greater tolerance to these conditions, as in the case of more variable natural or less highly domesticated species. For the same reasons they are particularly susceptible to disease. A result of man's creation of specialised inbred crops is the necessity to breed disease- or pest-resistant strains, and to continue doing so as ever more virulent pathogens evolve. In breeding for high yield, man produces plants which grow so vigorously that they can hardly stand upright and are hence susceptible to 'lodging' under the onslaught of high wind or heavy rain. Efforts must therefore be made to produce strains that have a good 'standing ability' as well as high yield. Since efficiency of farming is becoming more and more dependent on a high degree of mechanisation, man must either design machines suited to handle a particular crop or breed crops suitable for mechanisation. The successful mechanisation of cotton harvesting in the U.S.A. was dependent on the breeding of varieties that would ripen evenly and were of a particular uniform height. The introduction of drought-resistant grain sorghums into Texas and Oklahoma from Africa necessitated the production of short forms; the African varieties, some two and a half to four metres in height were too tall for machine harvesting. The production of such 'machine crops' is assuming an ever-increasing importance in plant breeding today.

The evolution of varieties of domesticated crops under modern methods of controlled, directed plant-breeding has been particularly rapid in the major cereal crops, and above all in the wheats. It has been noted, for instance, that in the great wheat-growing regions of Canada and the U.S.A. there is hardly a region where the principal varieties of wheat grown today are the same as those used fifty years ago. This has been consequent upon the rapid development of the so-called hybrid wheats, which in their turn contributed greatly to the economic development of the prairie regions of Canada and the spring wheat belt of the U.S.A. The first of these, the Marquis wheat (a hybrid produced by crossing the Hard Red Calcutta (India) and Red Fife (Poland) varieties in the first decade of this century) matured earlier than spring wheats which were previously planted. They combined high yield with grain that gave flour of a superior baking quality. Marquis wheat has since been used to produce a wide range of hybrids adapted to different regions and for a variety of uses. A comparable, though even more dramatic, revolution has occurred within the last thirty years with the production of hybrid corn (maize). This came into commercial production in 1933; by 1946 over three-quarters of all the corn grown in the U.S.A. was of hybrid origin. Within the same period average yields of corn increased by fifty per cent. In the richest parts of the great corn-growing areas of the Middle Western States yields have doubled, or even trebled. As Mangelsdorf remarked in 1951, 'hybrid corn may prove to be the most far-reaching development in applied biology in the last quarter-century.' Since 1950 new breeding techniques have been developed which permit the mass production of hybrid corn seed without the necessity of detasselling by hand to prevent self-pollination.

It is not beyond the bounds of possibility that in the future man may produce not only new varieties of cultivated cereals but a new cereal cultigen by hybridisation between existing types of grains or between these and wild grasses. So far such attempts as have been made have yielded nothing superior to existing grains. The most important development to date in this respect is probably the recent production, in Canada, of the 'new' grain species, *Tricale*. A cross between durum wheat and rye, the initial problems of hybrid sterility have been solved by drug-induced chromosome changes; the result is a viable plant capable of reproducing itself true to type. *Tricale* has a higher protein content but lower yields than existing strains of bread wheat, which it resembles. It is, however, expected to provide an economic source of animal fodder and alcohol. The continued breeding of existing crops, however, has an important, indeed vital, part to play in increasing food production, not least in the under-developed countries of the world (many of them areas where the major food cultigens originated!) and in meeting the ever-growing food demands of the whole world. One of the most spectacular recent biological 'break-throughs' has been the mass production of true self-producing hybrid wheats of increased vigour and potentially higher yield than ever before. Breeding techniques have overcome the major barrier—that of self-fertilisation—to easy hybridisation. Wheat is to-day on the verge of an 'explosion' comparable to that experienced by corn thirty years ago. The new Mexican dwarf-wheat, developed within the last fifteen years, is a self-pollinating hybrid which combines a potential for high yields with exceptional disease resistance. A short stalk which diminishes the risk of lodging make it particularly suitable for use in irrigated areas. It has already revolutionised wheat production in Mexico;

and within the last seven years it has been introduced into India, Pakistan and Turkey where it is beginning to make a significant contribution to increased food production in these areas. As important as increasing wheat yields has been the production of protein-rich crops. Rice, the staple died of a large percentage of Asia's population, has an average protein content (five to seven per cent.) less than half that of wheat (ten to fourteen per cent.). The recent discovery of high-protein varieties has stimulated breeding programmes designed to increase protein content. Already some third-generation plants had more than ten per cent. The attainment of these levels, however, required heavy nitrogen fertilisation and was accompanied by increasing susceptibility to disease.

It is, however, little use producing 'new' and 'better' crop varieties unless soil moisture and nutrient conditions are suitable for their cultivation. A particular variety of crop may have a potential to give a high yield but this will not be realised unless its water and nutrient supply is correspondingly high. As Sir Joseph Hutchinson points out, '. . . it is no accident—and not merely a misguided pre-occupation with cash crops—that plant breeders have achieved little in the improvement of African food crops in the past half century. The husbandry of these crops is virtually unchanged and the existing races are thoroughly well adapted to it'. New types of plant breeds need new 'environmental' conditions, new methods of cultivation; alone they cannot increase yield.

As well as being an evolutionary catalyst man has also become an ever-more powerful agent of plant and animal dispersal and distribution. The intermingling of plants as a result of man's migrations has, with the consequent development of ecological races, allowed species to become more widely distributed. Man's effectiveness as an agent of dispersal has been gathering momentum particularly during the last four to five hundred years with exploration, the growth of trade and the development of more and faster means of inter-continental communication. The latter have bridged the barriers that formerly restricted the range of many species. One of the outstanding consequences of this has been the increased range of cultivated plants formerly confined to one continent or hemisphere. The Irish potato, maize and tobacco, for example, have been introduced into Europe, rubber and chinchona into South-East Asia from the American continents. Wheat, oats, rye, barley, flax, sorghum, rice, sugar and bananas are among the important crops which man has carried

F

to the New World from the Old. This spread of crop plants has been paralleled by that of domestic animals. Before discovery and European colonisation few existed in either North or South America, while Australia and New Zealand contained no indigenous mammals other than bats. Man has extended the range and number of some plants and animals but has drastically reduced many others in the process.

The domesticated plants and animals, however, that man has introduced from one region to another represent but a minute fraction of the total number of plants and animals that he has carried and continues to carry around the world. As his means and speed of transport have increased so too have the number of plant and animal species accidentally distributed by him. They travel as 'stowaways' in grain and feedstuffs, in plant and animal fibres, in timber, ballast or packing materials; many more are carried attached to man, to living animals or the vehicles in which they travel. Only a relatively small proportion of those species accidentally transported and dispersed actually become established in a country or region foreign to them. Some obviously will not survive the journey; others will be dispersed into areas climatically or otherwise unsuitable. Under wild conditions it is difficult for an alien plant-intruder to find a foothold and survive competition in an already well-established and well-adapted vegetation cover. That some can, often with disastrous results, is largely a result of man himself. The invasion and establishment of alien (or *adventive*) species is and has been most successful in man-modified habitats. Nowhere has it been more dramatic than in those formerly isolated areas such as New Zealand or remote oceanic islands such as Hawaii, where the existing number of plant and animal species before the arrival of man was comparatively small. Most aliens whose introduction is accidental or which may occasionally 'escape' from cultivation join the company of weeds; some may in time become naturalised members of the existing vegetation in undisturbed habitats.

Many alien plants, which in their native homes are only minor or inconspicuous 'citizens', often become much more vigorous, aggressive and abundant in a foreign environment. They may even spread so rapidly and effectively as to compete, not only with native weeds but with natural vegetaion. The classic, oft-quoted example is that of the prickly pear cactus which, deliberately introduced from South America into Australia as a possible fodder plant, overran and seriously reduced the quality of natural pasture land. Somewhat comparable in the U.S.A. is kudzu (*Pueraria thumberglana*), a vine introduced from Japan and used to control gulley-erosion in the south-eastern States, which has become a rampant and often smothering weed in woodlands. Nearer home the common purple-flowered Rhododendron (*R. ponticum*), introduced from the Mediterranean region in the mid-eighteenth century, has successfully escaped from gardens and invaded sandy and acid soils. Casting a heavy shade and contributing very poor humus, it has become a very troublesome weed in woodlands and newly established Forestry Commission plantations, where it checks tree-seedling growth and is difficult and expensive to eradicate. The aggressiveness and rapid proliferation of alien weeds which do become established, may be due partly to lack of competition, especially in open, man-modified habitats and partly to the absence of predators or pests to which they are normally subjected in their native habitats. Other aliens may cross-fertilise closely related native species and produce hybrids of exceptional vigour. This has been the case with the hybrid rice or cord grass (*Spartinia townsendii*), a fertile, polyploid hybrid resulting from the cross between the original British and an accidentally introduced American species of this grass. Within the last half-century the hybrid has not only suppressed its parents but has colonised with astonishing rapidity considerable areas of tidal mudflats around the coasts of southern England and northern France.

The changing distribution and balance of plants affects, and is in turn affected by man's dispersal of animals and micro-organisms. The introduction of larger animals from one region to another has mainly been deliberate; but in many instances it has been accompanied by a tremendous increase in the number of the introduced species to the detriment of cultivated and wild plants alike. The rapid proliferation of the rabbit and the red deer introduced into New Zealand from Europe has contributed to the destruction of natural vegetation and accelerated soil erosion. The grey squirrel and the muskrat are but two of the animals introduced from North America that have multiplied alarmingly in Europe. Also very considerable repercussions have followed the transport and introduction from one region to another of insects, other invertebrate animals and micro-organisms —the pests and diseases of plants, animals and man. Pests have been called the 'animal counterparts of weeds'. These (and the disease organisms) are like weeds in being organisms which have so increased in numbers as to become harmful pests as a result of favourable conditions created for them by man. Insect pests and fungus diseases of plants introduced into areas where the existing wild or cultivated plants do not possess a natural resistance to attack can multiply and spread rapidly, helped by an abundant source of food and the lack of their natural predators. The fluted scale-insect introduced to California from Australia in the late-nineteenth century became a serious threat to citrus orchards, until its natural enemy was also brought into the infested areas. Both the dreaded Colorado beetle and the fungus which causes potato blight occur naturally among wild potato species in North America without being particularly harmful. They assumed the proportions of a devastating pest and disastrous disease respectively when introduced to Europe where the cultivated potato crop provided the conditions for their rapid increase and spread. Similarly the aphid, Phylloxera, which decimated French vineyards at the end of the nineteenth century was a natural but inoffensive companion of wild vine species in America which proved fatal when brought into contact with the unresistant European cultivated varieties. Many of the worst insect pests or plant diseases, in fact, are those which have been transported accidentally from one region to another, or native species which have become pests on unresistant 'alien' crops.

The cumulative effect of man (which continues with increasing intensity) has been to favour, either deliberately or accidentally, the evolution and distribution of some types of plants and animals at the expense of others. In Tertiary times, one region or habitat was isolated from another by physical barriers. These barriers allowed an independent evolution of plants and animals within different parts of the world. Now these barriers have been broken down by man. Over great areas of land which were once the habitat of a rich variety of different plants and animals have been grown large numbers of a few specialised species which man deliberately protects or which can tolerate the modified habitat conditions he has created.

References

ANDERSON, E. 1948. Hybridisation of the habitat. *Evolution, Lancaster, Pa.*, **2** (1): 1–9.

ANDERSON, E. 1967. *Plants, man and life.* University of California Press, Berkeley and Los Angeles.

ANDERSON, E. and STEBBINS, G. L. 1954. Hybridisation as an evolutionary stimulus. *Evolution, Lancaster, Pa.*, **8** (4): 378–381.

BENNETT, E. 1965. Plant introduction and Genetic conservation: Geneological aspects of an urgent world problem. *Scottish Pl. Breed. St. Rec:* 27–113.

Braidwood, R. J. 1960. The agricultural revolution, *Scient. Am.*, September **203** (3): 131–168.

Baker, H. G. 1964. *Plants and civilisation.* Macmillan, London.

Burkhill, I. H. 1951–2, Habits of man and the origins of the cultivated plants of the Old World. *Proc. Linn. Soc. Lond.*, **164**: 12–42.

Byrd, C. C. and Johnston D. R. Hybrid wheat. *Scient. Am.*, May, **220** (5): 21–29.

Crane, M. B. 1940. The origin and behaviour of cultivated plants. In *The New Systematics*, Ed. Julian Huxley. Clarendon Press, Oxford.

Darlington, C. D. 1963. *Chromosome botany and the origins of cultivated plants*, 2nd ed. Allen and Unwin, London.

Edlin, H. L. 1967. *Man and plants.* Eldus Books, London.

Elton, C. S. 1958. *The ecology of invasions by plants and animals.* Methuen, London.

Frankel, Sir O. 1967. Guarding the plant-breeder's treasury. *New Scient.*, 14 September, **30**: 538–539.

Guyot, A. C. 1942. *Origine des plantes cultivées.* Coll. 'Que sais-je', Paris.

Harlan, J. R. and Zohary, D. 1966. Distribution of wild wheats and barley. *Science, N.Y.*, **153** (3740): 1074–77.

Harris, D. R. 1966. Recent plant invasions in the arid and semi-arid Southwest of the United States. *Ann. Ass. Am. Geogr.*, **56** (3): 408–22.

Harris, D. R. 1967. New light on plant domestication and the origins of agriculture: a review. *Geogr. Rev.*, **57**, (1): 90–707.

Helbaeck, H. 1959. Domestication of food plants in the Old World. *Science N.Y.*, **130** (3372): 365–371.

Holdgate, M. W. and Wace. N. M. 1960–61. The influence of man on the floras and faunas of southern islands. *Polar Rec.*, **68** (10): 475–93.

Hutchinson, Sir J. 1965. *Essays on crop plant evolution.* Cambridge University Press.

Lousley, J. W. (Ed.) 1953. *The changing flora of Britain.* Botanical Society of the British Isles. Arbroath.

McNeish, S. R. 1964. The origins of New World civilisation. *Scient. Am.*, November, **211** (5): 29–37.

Mangelsdorf, P. C. 1950. The mystery of corn. *Scient. Am.*, July, **183** (1): 20–29.

Mangelsdorf, P. C. 1951. Hybrid corn: its genetic basis and its significance in human affairs. In *Genetics in the 20th Century*, Ed. L. C. Dunn. Genetics Society of America, New York.

Mangelsdorf, P. C. 1951. Hybrid corn. *Scient. Am.*, Aug, **185** (2): 39–47.

Mangelsdorf, P. C. 1953. Wheat. *Scient. Am.*, July, **189** (1): 50–59.

Polunin, N. 1960 *Introduction to plant geography*, Longmans Green, London.

Salisbury, Sir E. J. 1961. *Weeds and aliens.* Collins, London.

Sauer, C. O. 1947. Early relations of man to plants. *Geogr. Rev.* **37** (1): 1–25.

Sauer, C. O. 1952. *Agricultural origins and dispersals.* American Geographical Society, New York.

SAUER, C. O. 1963. The theme of plant and animal destruction in economic history. In *Life and man*, Ed. John Leighly. University Press, California.

SCHWANITZ, F. 1966. *The origin of cultivated plants*. Havard University Press, Cambridge, Mass.

THOMAS, W. L. (Ed.). 1956. *Man's role in changing the face of the earth.* Chicago Univ. Press, Chicago.

VAVILOV, I. N. 1949–50. The origin, variation, immunity and breeding of cultivated plants. *Chronica Botanica*, **13** (1–6).

WILSIE, C. P. 1962. *Crop adaptation and distribution.* Freeman & Co., London.

WRIGHT, S. W. 1946. *Plant invaders.* Year book of the Royal Society of Edinburgh.

8 Vegetation

The species of plants that occur in any area constitute its flora. But whether the term flora is restricted to wild plants (as is most usual) or includes man's cultigens, it refers to the types of plants present irrespective of their prevalence or relative abundance. A plant of a particular species, however, rarely grows in complete isolation; it grows in company with others of the same or different species. Each species is composed of a varying number of individual plants which, in any one place, constitute a *population*. Such a population may be continuously or discontinuously distributed and in any one habitat may be sparse or abundant. Take a very simple example. The common daisy is, locally, discontinuously distributed, occurring in some fields and gardens but not in others. In two adjacent lawns, the daisy population may be dense in one and very sparse in the other. Areas, even of limited extent, occupied by members of one species to the exclusion of all others are exceptional in nature. Those which approach most closely to this condition are gardens and fields devoted to a cultivated crop, and from which other plants are excluded by assiduous weeding.

All the plants which grow together in any area form its *vegetation,* the character of which depends not just on the different species present but on the relative proportions in which their members are represented. It is quite possible, and indeed not unusual, for two habitats to have similar floras but different sorts of vegetation or indeed similar types of vegetation but quite different floras. For instance, the species heather (*Calluna vulgaris*), moor mat grass (*Nardus stricta*) and bracken (*Pteridium aquilinum*) may all be present in three separate habitats but in such varying proportions as to form three quite different types of vegetation. While a distinction is frequently made between natural and semi-natural vegetation, and between these and cultivated crops, this is in many ways an artificial convention. Cultivated crops and forest plantations form the principal components of the vegetation cover over a significant percentage of the earth's surface. Areas where vegetation can be considered 'natural', in the usually accepted sense of having been completely unaffected by the direct or indirect activities of man, represent a relatively small and continually dwindling proportion of the total cover. Indeed one might go so far as to question whether any such vegetation still exists.

The vegetation of any part of the earth's surface is composed of a collection—an assemblage—of a variable number of plants belonging to a few or many different species. They are not necessarily the only plants that could, given the opportunity, grow there; nor is their association together merely a fortuitous coincidence. They represent those species that have been able firstly, to reach, and secondly, to establish and maintain their populations in the area. The various species that grow together, however, do so not only because the physical habitat satisfies at least their minimum requirements for growth and reproduction, but also because they are able to live together. As that eminent plant sociologist, J. Braun-Blanquet, has expressed it, they are compatible 'table companions'—capable of feeding from the same table! The requirements and/or life-cycles of the various species are sufficiently different so that each can make slightly different uses of the soil and atmosphere, and sometimes of one another, on the same site. They are not in direct competition since each occupies a slightly different 'niche' or micro-habitat. Where two species have very similar requirements and are in direct competition for light, water and nutrients, the more demanding, aggressive and larger can exclude the other, although the physical conditions might otherwise be quite suitable for its existence. Hence competition 'selects' from the available plants those which can co-exist in a given habitat.

In any particular habitat, then, the vegetation is composed of a group of plants which, though differing in species and in form, are *ecologically related*—in the sense that they are capable of occupying the same 'home'. Some, because of their greater competitive ability and/or range of tolerance, may be able to occupy a wide diversity of habitats, while others will have a more restricted distribution. However their successful co-existence in one particular place stems from their ability to cope with the same physical environment and to tolerate each other. Those plants which grow together in a particular habitat are referred to as a *plant community*, by which something more than a mere collection or assemblage is implied. It suggests some degree of organisation and integration of the component members. Among ecologists, and more particularly plant ecologists, the 'community' has been as contentious a concept as has that of the 'region' among geographers. On the one hand, some have considered the organisation and integration of groups of ecologically related plants to be analogous to one living organism or to a human community; while on the other hand, doubts have been expressed as to whether the concept has any valid basis in reality. The definitions of a plant (or for that matter any other organic) community have been expressed with varying degrees of ambiguity and in almost as many ways as there are ecologists who have committed themselves to paper. For some a plant community is any collection of plants occurring together which possesses a certain degree of unity or individuality; for others it is an assemblage of plants which grow together in a particular physical habitat or site. Yet others define plant communities as groups of species distinguished by a characteristic composition and structure in association with a distinctive physical habitat. The term can also be used in a *concrete* sense to mean one particular community or in the *abstract* sense of implying a *type of community*. However, implicit to a greater or lesser degree in most definitions of a plant community are three characteristics: (1) it is composed of two or more different species of plants—as distinct from a *society* which is a collection of contiguous plants all belonging to the same species; (2) it is composed of species capable of growing together in a particular habitat, i.e. its members are ecologically related; (3) it is organised in the sense of having a composition and structure which result from the interaction between the component plants and between them and their environment through time.

Competition is considered by many ecologists to be a major factor in the 'integration' of plant communities, in the determination of their composition and structure. The former depends on the number of species and the relative proportions of their populations. The presence, and the sparsity or abundance of any one species is largely determined by how successfully its members can compete with those of others in a given site. Competition influences not only the relative importance of a particular type of plant but also the place or *niche* it will occupy in the community. One of the most important results of competition is reflected in the vertical layering or *stratification* of the component species. It is the most obvious structural characteristic of plant communities. It reflects the ability of plants of varying potential height-growth to live in the shade of taller ones and to tolerate the particular conditions of light, temperature and humidity so created. Each stratum will be composed of plants which can either make use of different parts of the same habitat or can make use of it at a different time. Smaller shade-tolerant species can exist beneath taller more light-demanding ones. In deciduous woodlands many herbaceous plants can complete their growth-cycle rapidly in early spring or in autumn when light conditions in the community are more favourable than when trees and shrubs are in full leaf. Stratification above ground is also paralleled below the surface where the root systems of plants of varying size avoid excessive competition by drawing on different soil levels (see Fig. 29).

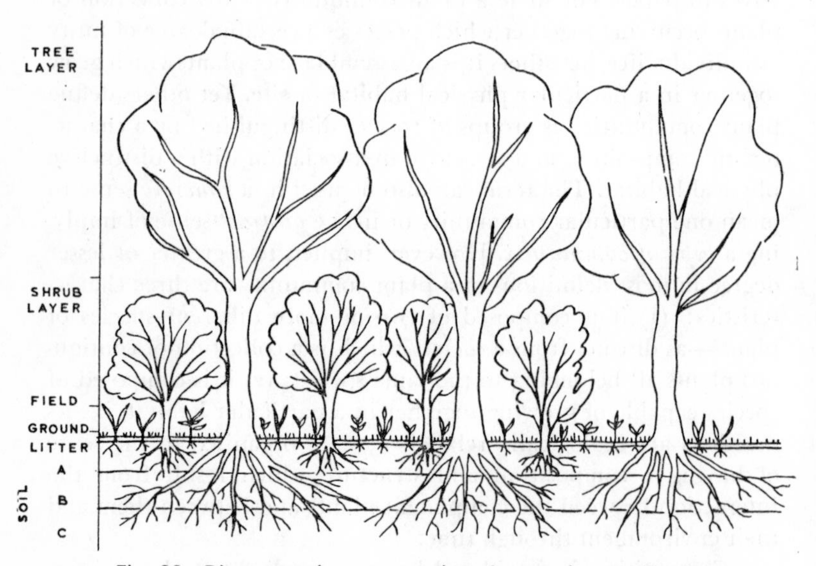

Fig. 29 Diagrammatic representation of vegetation stratification.

To a very great extent, both the composition and structure of plant communities are controlled by those plants which can compete most successfully for the resources of a particular habitat. These are the *dominant* plants, which, because of their greater numbers and/or size, comprise the largest percentage of the total bulk or volume of the living plant matter (the so-called plant biomass) in the community. They need not be the most numerous individuals; usually they are the largest and tallest. In an oakwood the number of individual bluebell plants will often far exceed the number of oak trees: the latter, however, are not only taller and larger but their total bulk and volume will far surpass that of the bluebell population. Because of their proportionately greater biomass the dominants exert the greatest influence on both the physical habitat and the other plants that can grow with them. They contribute most to and take most from the soil; by virtue of their size, bulk and the shade they cast, they modify light, temperature and humidity conditions and thereby create their own particular micro-climate. In so doing they play a major rôle in determining those species that will be able to accompany them (those that will be 'socially acceptable') and the place they will occupy in the community. In fact, a hierarchy of dominance exists within a plant community, as each stratum will be dominated by those plants which can compete most successfully under the shade of those above.

A. G. Tansley has commented that 'fundamentally the vegetation of the world is made up of a mosaic of plant communities whose distribution is determined by a corresponding mosaic of habitats'. Variations in habitat from one part of the earth's surface to another are reflected in varying combinations of 'ecologically related' plant species. The resultant plant communities can theoretically be of any size: ranging from those associated with a restricted micro-habitat, such as the group plants that grow on a narrow rock-ledge, localised sand-dunes or peat bogs, to those capable of existing together under a particular climatic regime, and of considerable extent as in the tropical rain forest. Where abrupt variations of habitat occur such as are associated with a sharp break of slope or a marked difference in rock- or soil-type, the change from one plant community to another may be correspondingly abrupt and the boundary between them clearly defined and easily identifiable. Nowhere is this more marked, for instance, than in man-created or man-modified landscapes. Among the most sharply defined vegetation boundaries are those between cultivated and uncultivated land, between burned and unburned vegetation. Changes in physical conditions, whether of slope, soil or climate are, more often than not, gradual and continuous. Under these circumstances one assemblage of plants will grade gradually into another through a zone of transition (an *ecotone*) whose width will depend on the steepness of the environmental gradient involved. In ecotones there is a greater number of species than in the adjacent communities. They are composed of an intermixture of species which include, on the one hand, the most tolerant members of the neighbouring communities and, on the other, *edge* species peculiar to the ecotone itself. However, abrupt changes in vegetation composition are not infrequently associated with rapid but nevertheless continuous environmental gradients. This is revealed in the well-marked zonation of communities in mountain, coastal or lacustrine habitats. The reasons for this apparently anomalous situation are not fully understood and may well vary from one area to another. In some cases, the rapid replacement of one group of plants by another is the result of the narrow range of tolerance of the component species to such environmental factors as temperature, salinity, degree of inundation. In other instances it has been demonstrated that certain species live together in stable combinations only in definite proportions. Some botanists maintain that competition naturally gives rise to communities more sharply defined than might be expected given the environmental gradient.

Plant communities vary in diversity of their composition and the complexity of their structure. The simplest are those composed of a few species of similar requirements and form, as represented by moss or lichen 'mats' on rock surfaces or by a regularly burned heather-dominated moorland. The most complex are those composed of a great number of species of different forms and of varying requirements, which have a highly developed vertical stratification, as in many forest communities. Variations in the complexity of plant communities can be broadly related to three groups of factors: (1) the nature of the physical environment; (2) the environmental history; (3) man. In (1) where *the environment* is limited by the climate and condition of the soil fewer species can exist and hence the resulting communities tend to be simpler in composition and structure. An example of this is in arid and very cold habitats where a one-layered, often discontinuous or 'open' vegetation cover may be dominated by a few highly specialised types of plants adapted to these extreme conditions. It is also reflected in the difference between the Boreal forest of high latitudes and the tropical rain forest of equatorial regions. The former is dominated, over extensive areas, by a relatively small number of hardy coniferous species, with correspondingly simple sub-strata. The latter is characterised by a tremendous diversity of trees—as many as fifty to a hundred different species per hectare have been noted—and a complex stratification of these into as many as three tiers according to varying light tolerances. Generally speaking the more favourable physical conditions are for growth the greater the diversity of composition and complexity of structure that will develop. (2) *The environmental history*: within comparable physical habitats the existing nature of the plant communities may vary because of past events. Plant communities of northern Europe, and particularly Britain, are less rich in species-composition and often less complex in structure than those in ecologically similar areas in America or Asia. In the former, as has already been noted, the effect of the Pleistocene glaciation and barriers to subsequent plant migrations have left a legacy of relative floristic poverty. Also the time that has elapsed since a particular habitat has become available for plant colonisation (as on newly created mud-flats or volcanic islands) may not have been sufficient for the most diverse and complex plant communities possible to become established. (3) *Man* has been and continues to be one of the most powerful factors in the simplification of plant communities. As a result of his intensification of burning and grazing, he has in many areas drastically reduced the number of species capable of maintaining themselves and has disrupted a formerly more complex structure by removing and inhibiting tree and shrub growth. Among the simplest types of plant communities, in fact, are those composed of a particular type of crop or planted tree.

Plant Formations

Two of the most important features by which plant communities are commonly characterised, and which result in observable variations in the vegetation cover are *physiognomy* or *form* and *floristic* or *species composition*. The latter is dependent on the proportion of various species, the former on the proportion of plants of a particular growth-form which make up the vegetation cover. Growth-form (for which life- or vegetative-form are commonly used synonyms) means simply the whole form or structure of a plant body in terms of the size, shape and structure of its various organs, its general habit of growth and the changes of form that take place during its life- or growth-cycle. The individual plants belonging to a particular species have a generally similar form which is genetically determined. Differences in appearance between closely-related species may be, in many cases, so slight as to be difficult for the untrained eye to detect easily. Some families of plants, containing several genera and many species may be characterised by some particularly obvious common morphological feature—as in the blade-like leaf of the grass family or the form of the trunk and leaves of the palm family. Among the more primitive, less highly evolved groups of plants, such as the algae, mosses, lichens and liverworts (and to a lesser extent the ferns) whose tissues and organs are relatively simple and undifferentiated, many different kinds of plants may share a general overall similarity of form. Complexity and variation of form is, however, greatest in the more highly evolved Gymnosperms and Angiosperms. Particularly in the latter, closely related species belonging to the same family may have a quite different appearance, while a generally similar growth-form may be common to a number of genetically unrelated types of plants. Oaks for instance are all woody, perennial plants. However some species of the genus (*Quercus*) may be broad-leaved and deciduous, while others have small, tough, evergreen leaves, some are trees of considerable height and others, such as dwarf or scrub oaks, are no more than three to four feet in height. Members of the legume family contain some species—clovers and vetches—which are small creeping herbs and some which, like laburnum and acacias attain the proportions of trees. Even one particular genus may contain species of highly contrasted form. That of *Euphorbia,* which in Britain is represented by the inconspicuous spurges, contains species such as the large, succulent tree-like Euphorbias of East Africa. Conversely many genetically unrelated species may possess in common such features as fleshy, succulent leaves or stems, spines and thorns instead of leaves, or a climbing habit which, as with vines and lianas, makes them dependent on other plants for support.

Growth-form or physiognomy has been one of the aspects of plants and vegetation that has long attracted the attention of geographers. It has been, and indeed still is, the basis of their description of the world's vegetation cover. The significance of growth-form for the geographer has been linked to the possible relationships which may exist between the form of plants and the particular environmental conditions with which they are associated. The ability of plants to modify the physical habitat in which they grow is partly dependent on their size, structure and the effect of these features on such conditions as water and light. The competitive ability of plants is, as has previously been stressed, frequently a function of their physical form, particularly their size. The structure or morphology of the plant communities that comprise the vegetation cover is dependent on the predominance of certain growth-forms (such as trees for instance) and the arrangement of similar growth-forms (trees, shrubs, herbs, mosses, etc.), into strata (*synusiae*). The possible significance of growth-form as a direct expression of the adaptation of plants to their physical environment has long engaged the geographer's attention and has influenced his approach to the study of vegetation even more profoundly.

To what extent is the growth-form of plants, and the resulting physiognomy of the vegetation cover an index of environmental conditions? We need not become too involved here with the problem of how plants become adapted to survive in different habitats. Suffice it to note that the majority of biologists would still support the view of adaptation as the result of natural selection—the survival and perpetuation of those genetic variations (which arise among and between species) best fitted to cope with a given environment. There was, however, an early tendency to assume that plants survived in an area primarily because their particular form endowed them with the necessary means to do so; that plant-forms evolved as a direct result of the selective effect of environmental conditions. However while adaptation is the result of evolution, evolution does not necessarily ensure adaptation or survival. It has in the past led as often to extinction. The interpretation of the significance of plant-forms as adaptations to environment has been beset with teleological explanations. The former tend to suggest, often without adequate proof, that all or certain morphological features have a positive adaptational survival value. The latter implies a direct, even conscious 'human' reaction of the plant to the physical environment in the use of such terms as 'drought evaders' or 'shade tolerance', which now are accepted descriptive expressions. Attempts to explain the relationship between plant-forms and the physical environment in which they occur have tended to imply, wittingly or unwittingly, a direct causal relationship between the two and more particularly between plant forms and climate. Over-generalisation and simplification have often unconsciously reinforced the idea that, for instance, trees are deciduous because of particular climatic conditions.

That there exists a general degree of correlation between the occurrence of plants of similar growth-forms and similar types of environment cannot be denied. The prevalence of such forms as needle-leaved conifers, winter- or summer-deciduous trees, succulent cacti forms or broad-leaved evergreen trees in particular climatic regions is almost a geographic axiom. However, neither the nature of the correlations, nor the reasons for them, are simple or even fully understood. The attributes of plant form are many and not all necessarily have a special adaptational value under existing conditions. It has, for instance, been suggested that some species which possess such features as succulent tissues or small sclerophyllous leaves (which *may* have evolved as a result of adaptation to formerly existing drought conditions) now exist in areas where such apparently 'xeromorphic' characteristics have little obvious value. Investigations have also revealed that the form of a plant is not always a reliable indication of the way in which it functions. Further, some doubt has been cast on the significance of form as an indication of adaptation to environment. In some cases a morphological characteristic (such as deciduousness in trees) has an obvious adaptational advantage in climates with a marked cold or dry season. In other instances a species' ability to tolerate low temperatures, prolonged drought, short day-length, acid or saline soils need not be a result of any particular morphological features. It has already been noted in another context that all plants adapted to dry habitats do not necessarily possess obvious xeromorphic features. Indeed such plants demonstrate very forcibly that there is no one morphological feature common to all xerophytes; rather there is a variety of forms equally well fitted to survival in arid or semi-arid regions. It should also be remembered that plants of a similar form do in fact exist in a wide variety of environmental conditions; the needle-leaved pines range from arctic to tropical, and from humid to arid climates. Furthermore, there are few habitats occupied by plants of a particular growth-form to the exclusion of all others. Plant communities are, in fact, composed not only of different species but also of those of different forms which can exist together in a given situation.

The description and classification of plants on the basis of growth-form, although useful, is particularly difficult. The attributes of form are so varied as to allow an almost unlimited number of possible categories. Many methods have been proposed. Some have been based on the most obvious features, others on those which are assumed to be of the greatest significance for the adaptation of plants to their environmnt. Of the latter, probably the most frequently employed by botanists is Christian Raunkiaer's classification of *life-forms.* He grouped plants according to the nature, and position relative to the ground-surface, of those organs (particularly the resting buds) from which growth is renewed after dormancy. For him this basic criterion indicated the way in which a plant survived these periods unfavourable to growth, and was therefore the most significant outward and visible expression of its adaptation to climatic conditions. His original classification of land plants included five primary life-form classes: (1) (Th) *Therophytes;* annual plants which, having completed their life-cycles, survive periods of cold or drought as seeds or spores; (2) (Cr) *Cryptophytes*: including all those perennial plants whose resting buds are below soil or water surfaces (these can in turn be sub-divided into (G) *Geophytes*: perennial, herbaceous plants with underground 'food-storage' organs such as bulbs, tubers, rhizomes, etc., and (HH) *Hydrophytes*: marsh and water plants); (3) (H) *Hemicryptophytes*: mainly herbaceous plants with resting buds located on the soil surface (these include many mosses and lichens as well as plants with a tussock or rosette arrangement of their leaves); (4) (Ch) *Chamaephytes*: herbaceous or small, low-growing woody plants with resting buds carried on stems up to, but not exceeding, twenty-five cm above the ground surface; (5) (Ph) *Phanaerophytes*: woody perennials—trees and shrubs with resting buds on upright perennial stems twenty-five cm or more above the ground surface. Raunkiaer further sub-divided this final class on the basis of height, duration of leaves and bud-protection; later he included classes comprising such special forms as succulents and epiphytes. There have been many modifications of this system which have involved the sub-division, in greater or lesser detail, of Raunkiaer's primary classes, as well as the inclusion of the micro-organisms of the sea and soil.

The main purpose of Raunkiaer's scheme was to analyse the relationship between plant-form and climate. The percentage of species in the *flora* of a particular region belonging to each of his life-form classes formed what he termed the *biological spectrum*.

Locality and phyto-climate	No. of Species	Percentage Life-form classes Ph.	Ch.	H.	Cr.	Th.
Phanaerophytic climate: (warm, humid, tropics)						
St Thomas and St Jan	904	*61*	12	9	4	14
(Seychelles)	258	*61*	6	12	5	16
Therophytic climate: (tropical and sub-tropical arid and semi-arid)						
Death Valley (California)	294	26	7	18	7	42
Argentario (Italy)	866	12	6	29	11	42
Hemicryptophytic climate: (mid-latitude cool and warm regions)						
Altamaha (Georgia)	717	23	4	*55*	10	8
Denmark	1084	7	3	*50*	22	18
Chamaephytic climate: (high latitude (tundra) and high altitude)						
Spitzbergen	110	1	*22*	*60*	15	2
Alaska	126	0	*23*	*61*	15	1
Normal spectrum	1000	46	9	26	6	13

Table 16 Life-form spectra of four major phyto-climates, (According to Raunkiaer. 1934.)

A comparison of biological spectra from different parts of the world with the *normal spectrum* (an approximation of the percentage life-form composition of the world's flora) revealed that significant variations could be correlated with climatic variations. On this basis, Raunkiaer distinguished four main type of phytoclimate (Table 16) characterised by the most significant types of life-form associated with them and their degree of deviation from the normal spectrum. The biological spectrum has also been used in the analysis of plant communities as illustrated in Table 17.

Locality in Scotland	Community type	Proportion of species in each Raunkiaer's Life-Form groups as % total number of Phanaerophyte species			
		N	Ch	H	G
Orkney	Calluna-Erica cinerea	14·3	28·6	57·1	—
Inverness-shire	Calluna-Vaccinium	11·1	22·2	55·6	11·1
Kincardine-shire	Calluna-Vaccinium	11·5	15·4	65·4	7·7
Aberdeen-shire	Calluna-Arctostaphylos	9·1	22·7	68·2	—

Table 17 Representation of life-forms in selected examples of heath communities in Scotland. N = nanophanaerophytes, dwarf shrubs 25 cm–2 m tall. (From: Burnett, J. H. 1964)

The main limitation of the spectrum analysis is that it is a measure of the flora of a region or community and not of the composition of the vegetation. It gives no indication of the relative abundance of plants of a particular life-form. In fact, the most prevalent life-form class may not, as indicated in Tables 16 and 17, necessarily be the *dominant* forms. This is obvious from the spectra for Denmark and Georgia where—although there is a greater number of hemicryptophytic species—trees (phanaerophytes) are the normally dominant plants in the 'natural' vegetation cover. Also in the examples of Scottish heath vegetation although there are a greater number of hemicryptophytic species (particularly mosses and lichens) the dwarf shrub and chamaephytic heath plants named comprise the highest proportion of the plant biomass. However while Raunkiaer's classification is still employed in the analysis of biological spectra, the significance of these as expressions of adaptation to climate alone has been subjected to much criticism.

Raunkiaer's classification, however, has the merits of clarity, simplicity and conciseness by reason of the consistent use of one easily definable primary criterion. Of the other features that have been employed in the classification and description of growth-forms, the most important are: (1) the general overall form or 'architecture' of the plant in terms of height, structure (composition of tissues) and duration (by which a distinction is made between annual, biennal and perennial; woody and herbaceous; trees, shrubs, herbs and moss forms, etc.); (2) the shape, structure, size and duration of its green parts (leaves and some stems). Together these categories include those features which produce the most obvious physiognomic contrasts and which, in addition, determine the plant's competitive ability under varying external conditions. The particular feature or combination of features chosen varies according to the purpose or scale of description and classification. For instance, two recently proposed classifications of growth-forms for the purpose of description are those of Dansereau and Küchler. The former suggests a combination of five criteria: (1) height or size; (2) life-form; (3) function or habit; (4) leaf-size and shape; (5) leaf texture. The latter makes a primary distinction between woody and herbaceous plants and sub-divides these into groups according to height, size and duration of leaves, and other special features.

As has already been pointed out, the physiognomy and structure of plant communities are determined by the growth-form of the plants of which they are composed. A community distinguished by the dominance of a particular growth-form (or a combination of growth-forms) is commonly referred to as a *plant formation,* what in more general terms would be called a 'type of vegetation'. While formations can be readily observed in the changes in the general appearance (and colour) of the vegetation cover of an area, their description and classification suffers to an even greater extent from the same problem as those posed by the form of the individual plant. Formations may be described in lesser or greater detail by one or more attributes of the dominant growth-forms, as for instance forest, broad-leaved forest, or broad-leaved evergreen or deciduous forest. Alternatively, they may be characterised by the predominant growth-form in association with a distinctive feature of the physical habitat in which they occur, as in acid grassland, desert scrub, monsoon forest. There is also a variety of colloquial, descriptive, and often 'regional' terms which are indicative of a particular type of vegetation and associated environmental conditions which include such categories as heath, moorland, prairie, steppe, garrigue, llanos and savanna.

A. G. Tansley has defined a plant formation as 'a unit of vegetation formed by habitat and expressed by distinctive life-forms'. Comparable formations associated with similar, though widely separated habitats make up a *formation-type* (i.e. group or class). On the basis of the most obvious contrasts of appearance, the world's vegetation cover can be sub-divided into a few major formation-types of which the most commonly recognised are forest, grassland, wooded-grassland (savanna), scrub and desert, each of which can be broken down into smaller units according to difference in form of the dominants. However the term formation has been reserved by some authors for those major units of vegetation, such as the deciduous forests of north-west Europe, the broad-leaved, evergreen forests of equatorial Africa or the grasslands of central North America which comprise the characteristic and most prevalent type of vegetation over extensive areas. The formation type is then synonymous with what we usually think of as world-types of vegetation. Tansley, however, distinguishes between climatic-formations associated with a major climatic region (and hence the largest type of plant community possible) and local formations which are related to soil conditions and/or the activities of man. These major formations together with the associated animals are designated *biomes*.

A general coincidence between the distribution of major formation-types and climatic regions does exist. The predominance of comparable types of vegetation in similar, though widely separated, climates is a distinctive feature of world vegetation distribution that can hardly be ignored. The dominance of a particular growth-form often common to a great number of different species in the major plant formations of the world is usually attributed to *convergent evolution* under similar climatic conditions. That is to say, during the course of evolution the species which have become dominant are those with a form that has best fitted them to compete successfully under the prevailing climate. In other words, the environment has favoured the survival of plants of a similar form. The degree of convergence of growth-forms in any region varies and is by no means complete. It is a feature of the tropical rain forest formations of South America, central Africa and south-east Asia which is stressed by many students of world vegetation. Each is composed of a tremendous number of different species but at the same time is characterised by a marked predominance of evergreen broad-leaved trees often of remarkably similar appearance. It has been suggested by plant geographers that this high degree of convergence is the result of a long uninterrupted period of unchanged climatic conditions. Under conditions of temperature and humidity which are optimum for growth, the struggle for existence is so severe and the competitive advantage of the tropical broad-leaved evergreen tree-form such, that the evolution and survival of deviations from it have been inhibited. Convergence of growth-form is illustrated both in the predominance of needle-leaved conifers in the Boreal forest (*taïga*) formations of America and Eurasia, and in the broad-leaved winter- or summer-deciduous forest in temperate and tropical climates which have a marked seasonal regime of temperature or rainfall respectively. Convergence is probably *least* developed in arid and semi-arid scrub formations. Although here drought-resistance is a characteristic of the dominant plants, this may be expressed in a great variety of different growth-forms such as deciduous or evergreen sclerophyllous shrubs and many varied forms of succulent plants occurring under similar climatic conditions. Other 'climatic' regions are characterised by a mixture of contrasted forms, as in the mixed broad-leaved deciduous and coniferous forests of north-east America or by a combination of trees and grasses in tropical savanna. Nor would it appear that the dominance of one particular growth-form is restricted exclusively to one type of climatic region. As S. R. Eyre points out, needle-leaved coniferous formations exist under a wide variety of climatic conditions ranging from cold to warm, from humid to semi-arid. Similar climatic regions are not invariably distinguished by comparable formations. N. C. Beadle has noted that in similar climatic conditions in central California and south-east Australia grassland is the natural formation in the former, eucalyptus woodland in the latter. He suggests further that the former has presumably remained grassland because of the absence of a suitably adapted woody species. The currently very successful spread and regeneration of eucalyptus, introduced as shade trees into California, gives point to this observation. In the broadly comparable climatic regions of north-west Europe, British Columbia, and southern Chile the major vegetation formations are broad-leaved coniferous, mixed evergreen, broad-leaved and coniferous forests respectively.

One is, in fact, tempted to speculate on the value of attempting to characterise world vegetation in terms of its dominant growth-forms. Just how significant is the general uniformity of appearance of vegetation over extensive areas anyway? Is it perhaps not due as much to the basic structure of plants as to convergent evolution in comparable environments? The anatomy and morphology of plants, their nutritional requirements and physiology is much simpler than that of animals; they exhibit much less variation of form (there are, at least, four times as many species of animals as there are plants). The *gross,* structural organisation of a great many different species of plants is very similar and adaptation to a wide range of environmental conditions can be effected without marked variations in say the basic tree-form.

Nevertheless the broad and general correlation which, in spite of certain anomalies, exists between the distribution of formation-types and climatic regions has profoundly influenced the development of plant-geography, ecology, biogeography and climatology. It has been the basis of the concept that climate is the dominant, if not controlling, factor determining the major features of the world's vegetation cover. As S. R. Eyre says 'it seems that each formation-type occupies a region which possesses certain climatic characteristics to which a particular life-form is most suitably adapted'. The conviction that vegetation of a particular form is the outward and visible 'expression' of climatic conditions is maintained by the Americal ecologist, H. J. Oosting in the following statement, 'the climate of a region controls the kinds of plants that may grow there. The general vegetation-type or growth-form, such as grassland, desert or forest is a product of the complex climatic factors effective in a region and can be used as a generalised basis for evaluating the climate. For example, knowing something of the growth-forms able to survive under the extreme conditions of moisture and temperature associated with a desert one may automatically accept a repetition of these growth-forms anywhere else in the world as indicative of desert conditions. The scrubby broad-leaved evergreens (chaparral) that cover much of southern California are a product of the climatic conditions peculiar to the area. The same growth-form is repeated in a few widely separated regions of the world where, although made up of quite different species, it exists in a similar complex of climatic conditions. In the same way, the vast expanse of deciduous or coniferous forest in the temperate regions of the world each occur where climatic characteristics fall within definite limits, similar throughout'. That the 'ultimate' type of natural vegetation which can exist, under stable conditions and undisturbed by man, will be determined by climate is basic to the originally propounded concept of climax vegetation.

Students of world vegetation, both botanists and geographers, have long been preoccupied with its relationship to climate. Many attempts have been made to express in quantitative terms the characteristics of desert, grassland and forest-climates and the climatic limits of these major-formation types. Such efforts have been based almost exclusively on temperature and rainfall (or evaporation expressed as a function of temperature) as being the main climatic factors limiting plant growth. The climatic definition of vegetation boundaries have, in the main, been concerned with those between the major formations of tundra, forest, grassland and desert. By inspection of climatic data available, attempts have been made to assess what quantity (or duration) of either temperature or rainfall coincided most closely with these particular vegetation boundaries. For instance the general but by no means exact coincidence of the 10°C isotherm for the warmest month and the boundary between forest and tundra in the northern hemisphere was early noted. It has been suggested however, that a closer 'fit' is obtained if a limit of three months with a mean temperature of 6°C (then generally accepted threshold for plant growth in temperate regions) is taken. Similarly, by the inspection of temperature and rainfall values along the margins between forest and grassland and between grassland and steppe, many efforts have been made to calculate the amount of 'effective' rainfall (expressed in terms of the relationship between rainfall and temperature or rainfall and evaporation) which coincided with these boundaries. The assumption that vegetation provided the 'faithful reflection' of climate has also influenced many systems of climatic classification. It was considered by biologists and geographers alike that, in the words of A. A. Miller, 'a satisfactory classification of climate must reflect the climatic control of vegetation: that climatic provinces must coincide as closely as possible with the major vegetational regions of the globe'. Efforts were made to define the boundaries of climatic types in terms of those rainfall and/or temperature parameters which revealed the highest degree of correlation with the limits of major vegetation types. Types of climate were designated by vegetational characteristics, as for instance tropical rain forest, tropical savanna, steppe, tundra climates, etc. Conversely, the classification of the major types of world vegetation was constructed on the basis of climatic characteristics!

One of the most commonly used systems of climatic classification—that of the German biologist, W. Köppen—was developed from an intensive study of the climatic limits of the major types of world vegetation. On the basis of mean annual and monthly rainfall and temperature data Köppen delimited five major climatic regions which were intended to correspond to major vegetation regions, as summarised on page 174.

CLIMATE	VEGETATION
A Tropical: no cool season; temperature of coldest month $>18°C$.	Megathermal plants; Tropical rain forest and savanna
B Dry climates; excess of evaporation over rainfall, semi-arid and arid.	Xerophyllous plants; Steppe and desert
C Warm temperate rainy climates with mild winters; average temperature warmest month $<18°C$: average temperature coldest month $>10°C$.	Mesothermal plants Temperate forests
D Cold climates; snowy; severe winter; average temperature warmest month $>10°C$; coldest month $<-3°C$.	Microthermal plants Boreal forest
E Polar climates; no warm season; average temperature warmest month $<10°C$.	Hekistothermal plants Tundra vegetation Permanent ice caps

While the main humid forest climates were delimited on the basis of temperature, the boundary between these and dry climates was defined by 'effective rainfall'. From a study of rainfall and temperature data available along the forest-steppe boundary, W. Köppen constructed formulae which expressed the effective rainfall in terms of mean annual temperature, mean annual rainfall and its seasonal distribution. This gave a value which showed a consistent fit with the vegetation boundary, e.g. mean annual rainfall = 0·44 mean annual temperature − K (3 when rainfall maximum in summer; 14 when in winter). It is worth noting that other climatologists and biologists have proposed alternative R : T ratios which they purport give a more accurate measure of the same boundary.

A later classification by the American climatologist, C. Warren Thornthwaite, also attempted to define climatic boundaries in terms of their coincidence with the limits of major vegetation types. His system differs from that of Köppen in being based on the combination of three criteria which aimed at providing a more realistic measure of climatic efficiency for plant growth. The criteria used were precipitation-efficiency (expressed as monthly rainfall divided by monthly evaporation), temperature-efficiency, and seasonal incidence of rainfall. From these Thornthwaite defined humidity and temperature provinces; the limits of the former being those P/E values calculated for the areas which lay along the boundaries of the major types of vegetation. In cold

Humidity Province	Vegetation	P/E Index T/E	Temperature Province
A Wet	Rainforest	> 128	A′ Tropical
B Humid	Forest	64–127	B′ Mesothermal
C Sub-humid	Grassland	32–63	C′ Microthermal
D Semi-arid	Steppe	16–31	D′ Taiga
E Arid	Desert	<16	E′ Tundra
			O′ Frost

Table 18 Thornthwaite's first classification of humidity and temperature provinces: P/E and T/E indices = sum of twelve monthly P/E and T/E ratios respectively.

regions where temperature is the more significant factor limiting plant growth a primary division is made on the basis of temperature provinces. In those with a temperature efficiency index of over thirty-two the basic division is by humidity provinces (see Fig. 30).

These are but two of the most familiar and frequently used of the many systems of climatic classification which, as G. T. Trewartha remarks, are 'really vegetation regions climatically defined'. Their shortcomings and the criticisms to which they have been subjected stem basically from the problems that must inevitably be inherent in attempts to correlate types of climate and vegetation and to define one by the characteristics of the other. Even if formation types are assumed to be expressions primarily of climate, the latter is a complex of variable and interrelated factors among which temperature and rainfall are only two of the many which influence plant growth. In many parts of the world reliable information about even these two elements is still relatively scanty, while the characteristics, distribution and exact limits of the major vegetation types throughout the world have not yet been fully and accurately established. Further attempts to define vegetation or climatic boundaries quantitatively must necessarily tend to be unrealistic; both are zones of transition, of varying width, which may, as in the forest-tundra ecotone be as much as 160 kilometres wide. In addition, Miller has pointed out that

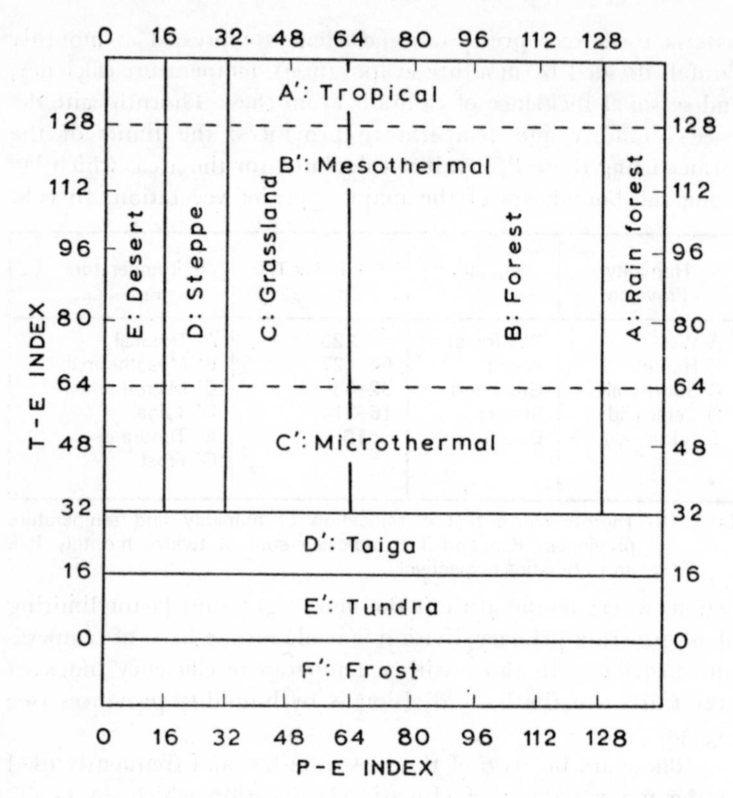

Fig. 30 Temperature and humidity provinces according to Thornthwaite's system of climatic classification (From: Trewartha, Glen T. 1954)

vegetation, particularly in high latitudes in the northern hemisphere, may still be in process of re-adjustment to post-glacial climatic fluctuations and that the arctic timber-line may not be stable. There is evidence in Alaska that the colonisation northwards of trees is still in progress (see Chapter 9).

Finally much doubt has been expressed as to the possibility, indeed validity, of defining the limits of the major vegetation regions by climatic data alone. The vegetation of any area is the expression of the sum total of all environmental factors operating through time. While climate is undoubtedly one of the major factor-complexes, in many parts of the world historical and biotic factors have been equally if not more important in the determination of the presently existing type of vegetation. The influence of past climatic change and of geographical barriers to plant migration have, as has already been noted, resulted in different formations in similar climatic regions. It has become increasingly obvious that man has so modified the natural vegetation cover that in many parts of the world its form cannot be explained in terms of climate alone. In the first place he has removed or modified so much of the natural cover that the distribution of the pre-existing 'natural' vegetation cannot be reconstructed with certainty. In many areas long-continued grazing and burning has been instrumental in favouring some types of vegetation at the expense of others. There is now strong evidence to support the theory that much of the tropical savanna grasslands, as well as the more humid temperate prairie and steppe, owe their existence to burning which, rather than climate, prevents tree growth. As a result the existing boundaries between forest and grassland, grassland and desert are man-made rather than climatically, or otherwise 'naturally' determined. As A. A. Miller pertinently remarks, 'the correlation (between climate and vegetation) remains a will o' the wisp whose presence we may sense but whose outlines remain blurred and indefinite'. The realisation of the problems and pitfalls of assuming a direct relationship between climate and plant formations has motivated more recent recommendations, and attempts, to classify world types of vegetation and climate on the basis of their own characteristics rather than those of each other.

Associations

Plant communities, however, can be characterised not only by their form but by their floristic composition. While discrete units or types of vegetation can be distinguished on the basis of the dominant growth-forms of which they are composed, they are not necessarily—indeed are rarely—composed of the same species throughout. A particular formation or type of vegetation is made up of a variety of communities which may have in common certain characteristics of form but which are distinguished by different assemblages of species reflecting varying habitat conditions. In British and American ecological literature, a fundamental distinction is usually made between plant formations—characterised by the dominance of plants of a particular growth-form, and *associations*—characterised by the dominance of one (consociation) or several (association) species. Hence the broad-leaved winter-deciduous forest formation is composed of a number of associations in which one or more of such trees as oak, ash, beech and elm, may be dominant according to the particular conditions of soil and climate under which it can most successfully compete with other tree species. In the remnants of this formation in Britain, for instance, the common or pedunculate oak (*Quercus robur*) is thought to have attained its greatest development on the heavier damper soils of south-eastern England while the durmast or sessile oak (*Q. petraea*) became dominant on the poorer siliceous soils of the north and west of Britain. This ecological and geographical distinction, however, has been blurred by the ease of hybridisation between these two species and deliberate planting, particularly of *Q. robur*, on Scottish estates in the eighteenth and nineteenth centuries. Beech and ash are common, though not exclusively confined to chalk and limestone habitats respectively. Similarly the dominant species of the semi-natural grasslands of the downs and moorlands vary with differences in rock type, rainfall, soil drainage and base status and with varying burning, and grazing regimes.

But even within plant communities or associations so defined there may be considerable variation according to the types of subordinate species which accompany the dominants. Oakwood communities vary from those with well-developed, species-rich shrub and field layers to those in which the shrub layer may be absent and the ground layer composed of a few species of heath plants and/or mosses. Such subordinate species tend to reflect minor variations in habitat more sensitively than do the more tolerant and competitive dominants (see Table 19). Among the commonest and most widespread types of moorland vegetation in Britain are the *Agrostis-Fescue* dominated grasslands, and the *Calluna*-dominated heaths. In the former, there are communities characterised by the presence of such species as moor mat grass (*Nardus stricta*), heath-bedstraw (*Galium saxatile*), tormentil (*Potentilla erecta*) on poor acid or heavily grazed habitats, and by the appearance of plants such as thyme (*Thymus vulgaris*), clover (*Trifolium repens*) on base-rich soils, and the meadow grass (*Poa pratensis*) on sites heavily manured by grazing animals. Similarly the ubiquitous heather-dominated communities range from those on drier sites, accompanied by varying proportions of bell heather (*Erica cinerea*) crowberry (*Empetrum nigrum*) and blaeberry (*Vaccinium myrtillus*), to those on wet peaty areas where the more common associates will be the cross-leaved heath (*Erica tetralix*), bog-cotton (*Eriophorum spp*) and deer-sedge (*Scirpus/ Trichophorum caespitosus*).

Obviously the type of plant community distinguished will depend on the number of attributes by which it is characterised.

	1	2	3	4
Dominant tree species	oak/birch	oak	oak	oak/birch
average tree height (m.)	13	12–13	12–13	13–17
canopy depth (m.)	5	5	7	7
canopy cover (percentage)	55–60	50	50–75	50–75
Shrub layer	none	none	scattered individuals oak, ash, hazel.	scattered individuals rowan.
Bracken/Bramble (percentage cover)	20	65	30	45
Field layer (number of herbaceous species)	8	3	15	11

Table 19 Variation in even-aged oak (*Q. robur*) wood on Arrachymore Point, Loch Lomond. Sites 1 and 2 on soils derived from acid quartzose schists; 3 and 4 from base-rich serpentine. Differences in number of herbaceous species related to (i) density of bracken/bramble cover and (ii) soil base status. As well as supporting a richer ground flora sites 3 and 4 were characterised by the abundance of species, e.g. *Brachypodium sylvaticum* (False wood brome grass) *Mercuralis perennis* (dog's ermcury) and *Rubus spp* (bramble) not found on sites 1 and 2.

Those distinguished on the basis of dominant growth-form alone will tend to be larger units than those defined by dominant species or a combination of dominant and associated species. The term 'association' has unfortunately been, and still is, used in a number of different senses. On the one hand it may be used to denote an actual assemblage of species, an individual plant-community, existing in a particular place or habitat. In this sense it means an actual 'concrete' association or more correctly a *stand*—a particular bit of vegetation that can be identified from the surrounding vegetation cover by reason of its *relative* homogeneity of physiognomy and species composition. On the other hand, association, in the strict ecological sense means an *association-type*, e.g. similar assemblages of species or stands which may recur in similar though widely separated habitats. The concept of what constitutes an association-type—the criteria used to classify bits of vegetation or stands—has varied according to different schools of ecological thought.

American, British and Scandinavian plant ecologists have tended to distinguish associations (community-types) on the basis primarily of dominant species. Dominance has been expressed in terms of either relative abundance or cover-values; the latter is expressed as the percentage of a unit area covered by the constituent species when viewed from above (see Table 20).

	I	II
10	—	cover 9/10ths to complete (91–100%) total area
9	—	cover 3/4–9/10ths (76–90%) total area
8	—	cover 1/2–3/4 (51–75%) total area
7	—	cover 1/3–1/2 (34–50%) total area
6	—	cover 1/4–1/3 (26–33%) total area
5	cover over 75%	cover about 1/5 (11–25%) total area
4	any number individuals covering 50–75%	cover up to 1/10 (4–10%) total area
3	any number individuals covering 25–50%	occurring as numerous individuals but with cover less than 4% total area
2	very numerous: cover value at least 5%	occurring as several individuals: no measurable cover
1	plentiful but small cover-value	occurring as one or two individuals with normal vigour; no measurable cover
+	sparse or very sparsely present, very small or negligible cover-value	occurring as a single individual with reduced vigour; no measurable cover

Table 20 Comparison of the Braun-Blanquet (I) *cover-abundance scale* and the Domin (II) *abundance-vigour-cover scale*, as modified by Dahl (From: Braun-Blanquet, J. 1932, and Evans, F. C. and Dahl, E. 1955)

While the English-speaking ecologists have classified stands on the basis of primary dominants, the Scandinavians have placed more emphasis on community structure and strata dominance. They have classified associations (or, as they are termed, sociations) on the basis of characteristic combinations of strata-dominants.

Species-dominance is a more characteristic feature of plant communities in northern latitudes, where the flora is much less rich, than in other more climatically favourable regions. In northern Europe, particularly Britain, this relative poverty has been reinforced by the effects of the Pleistocene glaciation and man. Not only are there fewer species adapted to the more severe conditions but competition is less intense, and those species with a wide range of tolerance can maintain their dominance over an extensive range of soil and climatic conditions. This is illustrated in the sub-arctic Boreal forest, for instance, and in the range and extent of heather-dominated communities in Britain. Under these circumstances the same plant may be dominant in a variety of different habitats but be accompanied by varying assemblages of subordinate species in each. The dominant plant is usually less sensitive to environmental changes than the subordinate associated species whose range of tolerance is much less and which, as a result, occur in more 'specialised' conditions. For this reason the classification of communities on the basis of dominance alone is considered by some workers to be too superficial and lacking in ecological significance. On the other hand, the conspicuousness of the dominant species is the main criterion whereby major contrasts in the vegetation cover can most easily be observed. Also the dominant species exert the greatest influence on the habitat and on those species that can exist with it. For instance, under certain circumstances and particularly when subjected to well-controlled burning, heather may be so dense and vigorous as to preclude the existence of strata other than a ground carpet of mosses and lichens. In cool humid climatic conditions its litter provides a raw, acid humus low in nitrogen which only a relatively small number of species can tolerate. The conspicuousness of dominant plants must inevitably be an important factor in the initial visual identification of distinctive stands of vegetation.

In other parts of the world, however, where environmental and particularly climatic conditions are more favourable, the diversity of species is greater and competition more severe, dominance of extensive areas by one or a few species is much less common. This is true for instance in the floristically richer deciduous forests of eastern North America, in the mixed broad-leaved evergreen and deciduous forests, and it is nowhere more strikingly demonstrated than in the tropical rain forest. It has been suggested that because in many parts of the latter the number of different tree species even in one area is so great, and because one combination of species is so rarely repeated elsewhere that the association concept, however defined, is not applicable to this type of vegetation. Recent studies have, however, revealed that in certain areas the dominance of one or a few trees of the same species (or family) and characteristic assemblages, do recur to an

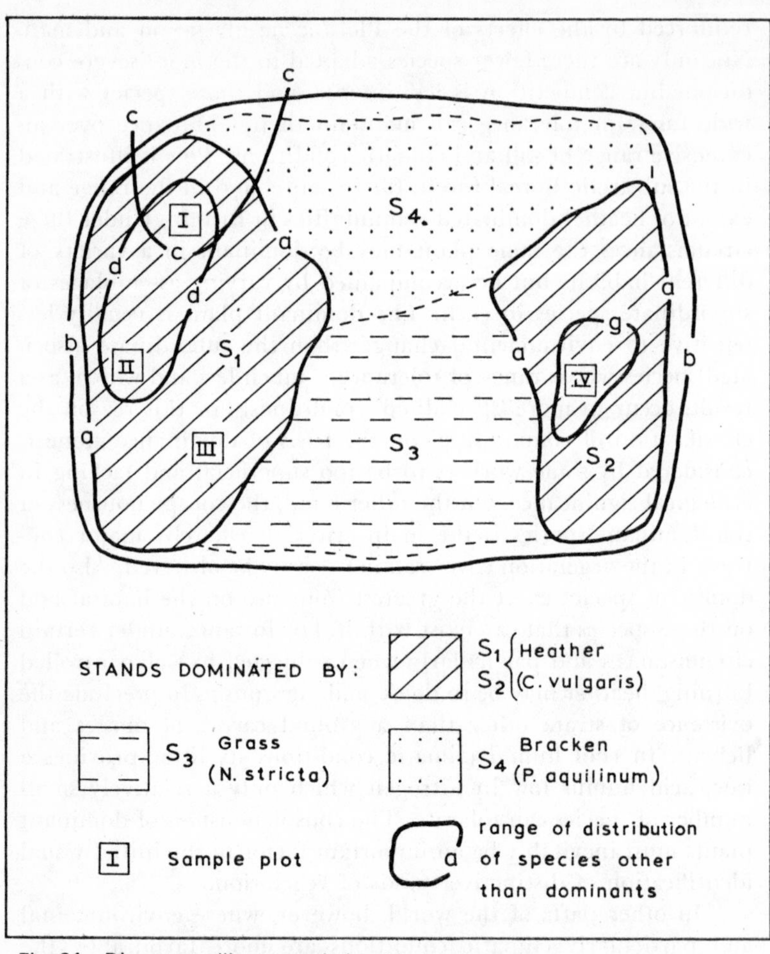

Fig. 31 Diagram to illustrate relationship between stands (concrete associations) and the phyto-sociological characteristics of community types. Certain species are found only in stand 1 or 2 but not in 3 and 4.

Species list from sample plots

STAND 1			STAND 2
(I)	(II)	(III)	(IV)
a	a	a	a
b	b	b	b
c	—	—	—
d	d	—	—
e	e	e	e
—	—	—	g
—	—	—	m

Not all species are shown on diagram. In the heather-dominated community – type a, b, and c are constants; d, g and m are characteristic of stands 1 and 2 respectively (Adapted and simplified from Gaussen, H. 1954) extent that permits the identification of characteristic associations often related to particular soil conditions.

In the floristically richer areas of southern Europe advocates of what have become known as the Zurich–Montpellier school of plant sociology have placed less emphasis on dominance and more on total floristic composition in their attempts to classify vegetation. They have taken into consideration other criteria such as the *constancy* (presence) and *fidelity* (exclusiveness) of species in recognition of related stands (see Fig 31). The former is a measure of the frequency with which a particular species occurs in a given number of stands of vegetation. And on the basis of 'constancy classes' (see Table 21) species can be described as rare or accidental, accessory or constant.

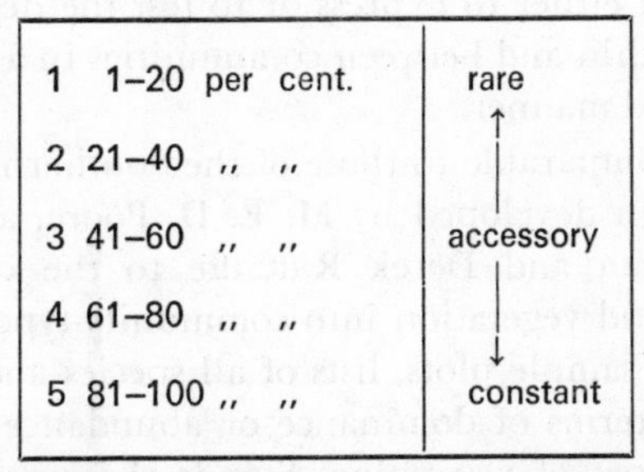

1	1–20 per cent.	rare
2	21–40 ,, ,,	↑
3	41–60 ,, ,,	accessory
4	61–80 ,, ,,	↓
5	81–100 ,, ,,	constant

Table 21 Braun-Blanquet's five-point rating of constancy classes, e.g., percentage of total stands examined in which a given species occurs irrespective of its abundance or relative importance. The number of classes and their limits vary with different workers.

Community-types (or associations) are hence distinguished on the basis of distinctive assemblages of species. The French school of phyto-sociologists, in particular, have attempted to identify associations by characteristic or faithful species. These are most abundant or attain their greatest vigour only in particular stands; they are exclusive to certain communities. To this end Braun-Blanquet classifies species into five fidelity classes:

1 *Strangers,* appearing accidentally
2 *Companions,* indifferent species without a pronounced affinity for any community
3 *Preferents,* present in several community-types, but predominantly in one
4 *Selectives,* present particularly in one community-type but occasionally found in others
5 *Exclusives,* found almost exclusively in one community-type.

Classes 3, 4 and 5 are the so-called characteristic or 'faithful' species. The recognition of community-types on this basis has been subjected to considerable criticism because theoretically it is only possible after an intensive investigation of a particular type of vegetation has been completed.

In addition the description and classification of community-types in Europe has been based, until recently, on more standardised and analytical techniques than in either America or Britain. These involve the analysis of floristic composition of numerous stands of vegetation on the basis of sample areas and the measurement or estimation of such attributes as dominance, abundance or density of the component species with reference to standardised scales such as have been indicated above. In recent years, there has been an increasing application of statistical techniques designed either to express or to test the degree of floristic relationship within and between communities in a more objective and standardised manner.

Methods comparable to those of the Southern European ecologists have been developed by M. E. D. Poore, and adapted by Donald MacVean and Derek Ratcliffe to the classification of Scottish moorland vegetation into community-types. On the basis of standardised sample plots, lists of all species and their relative significance in terms of dominance or abundance were recorded for different types of vegetation. Stands characterised by assemblages of species occurring in much the same proportion (e.g. cover-abundance plus constancy) are classed in the same community type or *nodum*. The latter is an 'abstract' vegetation unit (or taxon) of any category; it summarises the characteristic features common to a group of similar, though not necessarily absolutely identical, stands of vegetation. Hence the nodum concept is to the concrete plant community or stand what the species concept is to the individual plant.

Continuum Concept of Vegetation

During the last two decades the validity and value of classifying vegetation on the basis of community-types have been the subject of increasing debate among plant ecologists and phyto-sociologists. Fundamental to vegetation classification is the occurrence of discrete stands with common characteristics of composition and/or structure distinctive enough to allow them to be grouped into a community-type and to be clearly differentiated from other types. Methods of classification have undoubtedly been influenced by the practical requirements of cartographic representation and the firmly entrenched hierarchical approach to taxonomy. Although classification is essential to a fuller understanding of the nature of vegetation it has, as in so many other disciplines, often tended to become an end in itself directed towards the study of the taxonomy of vegetation *per se*.

Types of vegetation (e.g. formation-types) distinguished by common physiognomic characters such as forest, savanna, scrub or grassland can, particularly on a large scale, be fairly readily identified. In contrast, however, the range of variation in species composition and structure within a given type of vegetation is often such that vegetation change is continuous rather than discontinuous. The recognition of any clearly delimited community-types is therefore difficult, if not virtually impossible. Variation, however, is not entirely random; it is more often than not a function of a complex of interacting environmental variables. Spatial discontinuities do occur in such conditions as rock type, slope and soil drainage between contiguous habitats, while variation in atmosphere conditions is continuous. It can be argued that the number of possible combinations of environmental variables is such that, particularly when anthropogenic activities are involved, no two habitats will be exactly similar in every respect. Hence while generally similar, though not necessarily identical, stands of vegetation may occur within a range of comparable habitats, variation from one to the other will be so gradual (i.e., continuous) that a particular type of vegetation will be composed of a continuous series or sequences of community-types, one overlapping with and merging into the other through gradual changes in species composition. The whole then forms what has been termed a *vegetation continuum* rather than a mosaic of clearly defined community-types; classification is therefore only possible, if at all, on the basis of arbitrarily determined limits.

A vegetation continuum, in the words of R. P. McIntosh, 'Although not formally defined, . . . is plainly described as a gradient of communities in which species are distributed in a continuously shifting series of combinations and proportions in a definite sequence pattern'. The continuum concept is the product of the more rigorous analytical and quantitative studies of vegetation which have developed since the nineteen-fifties. These studies have highlighted the difficulties *and* disadvantages of the hierarchical system of classification as a method of vegetation study. For such reasons it has been suggested that the description and analysis of vegetation in terms of *continua,* i.e. of changes along continuous environmental gradients is more realistic and revealing. This in turn has stimulated the development and application of what are known as *ordination* techniques to vegetation study. These involve 'the arrangement of communities, species or environments in sequences' along one or more axes (co-ordinates). The object may be to examine the degree of association or correlation between species, between communities or between these and one or more environmental factors. Ordination methods directed towards the study of relationships between stands and/or community-types and environmental gradients have, in some instances, been termed *direct gradient analyses*. The familiar transect is a well-known and acceptable method of spatial or geographical ordination (see Fig. 32); it is still a standard method for analysing variations in vegetation across a continuous area of land or water. Non-contiguous stands or communities can be arranged in relation to selected environmental gradients such as altitude, slope, aspect, soil moisture or nutrient status, height above water-table or amount and duration of flood-water or snow-cover.

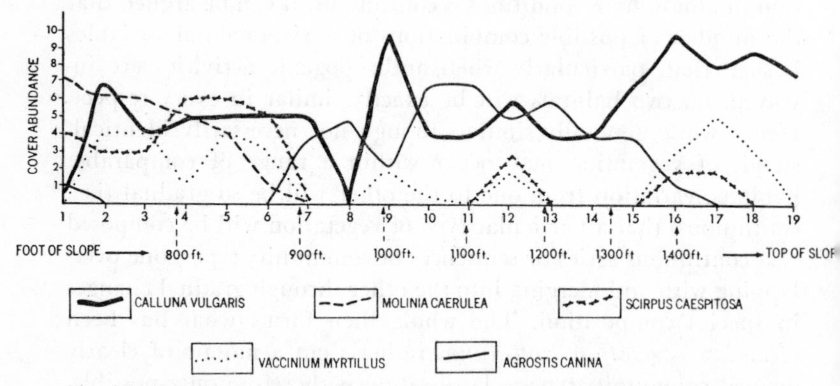

Fig. 32 Spatial or geographical ordination of one-metre quadrats along a continuous transect of a hill-slope near Creetown, Kirkcudbrightshire; showing variation in cover-abundance of selected species. (Reproduced from unpublished work by Mrs Joan Mitchell, Department of Geography, University of Glasgow)

The development of the continuum concept of vegetation and the application of ordination methods owes much to the 'Wisconsin school' of ecologists, among whom J. T. Curtis, R. J. Bray and R. P. McIntosh have been the most outstanding protagonists. The last ten years have seen a rapid development of ordination techniques in the studies of British moorland vegetation. For instance, in their summary of the grass moorlands of the forest and sub-alpine zone in *The Vegetation of Scotland*, edited by J. D.

Burnett, J. King and I. A. Nicholson gave an example of the range and nature of variation in the widespread *Agrostis/Fescue* dominated type of grassland in the Pentland Hills, south of Edinburgh. When groups of sites are arranged in an 'environmental sequence' from the most leached (low pH) to the most flushed (high pH)

Site Group	1	2	3	4	5	6	7
Log. Soil water %	1·35	1·43	1·67	1·61	1·53	1·77	1·69
pH surface soil	5·04	5·50	6·03	6·05	5·73	6·10	6·23
Mean no. species/site	9·3	13·0	18·2	16·7	15·5	14·6	12·7
Agrostis canina ssp. canina	—	3·3	—	—	2·0	—	—
A. canina ssp. montana	9·0	6·7	—	—	—	—	—
A. stolonifera	—	—	—	—	—	5·5	37·1
A. tenuis	91·0	93·3	84·3	98·3	96·0	92·7	65·7
Anthoxanthum odoratum	65·0	86·7	57·1	48·3	70·0	34·5	8·6
Briza media	—	—	42·9	6·7	4·0	1·8	—
Cynosurus cristatus	—	3·3	18·6	35·0	44·0	29·1	28·6
Deschampsia caespitosa	4·0	—	10·0	8·3	16·0	10·9	8·6
D. flexuosa	5·0	—	—	—	—	—	—
Festuca ovina	95·0	86·7	87·1	56·7	44·0	5·5	—
F. rubra	54·0	76·7	94·3	100	100	100	100
Holcus lanatus	—	—	2·8	8·3	6·0	—	—
H. mollis (2n = 21)	—	—	—	3·3	—	—	31·4
H. mollis (2n = 35)	23·0	—	—	—	20·0	14·5	11·4
Poa annua	—	—	—	—	—	3·6	34·3
P. pratensis	22·0	6·7	14·3	23·3	48·0	29·1	25·7
P. trivialis	2·0	—	5·7	15·0	22·0	49·1	77·1

Table 22 Grass species composition in terms of specific frequency of a series of seven groups of sites arranged in order of the soil sequence from the most leached (Group 1) to the most flushed (Group 7) (From: Burnett, J. H. 1964)

(see Table 22) the vegetation associated with these shows the changes in the composition of grass species to be continuous. No one group is clearly differentiated from the next in the sequence. No species is exclusive to any one group. The dominant genera *Agrostis* and *Festuca* are characterised by a high specific frequency in all. The frequency of several grasses tends to increase or decline from one end of the series or continuum to the other: *F. rubra* (red fescue), *Cynosurus cristatus* (crested dog's tail) and *Poa* spp. (meadow grasses) and *Holcus* spp. are more abundant on the wetter and more basic soils; *F. ovina* (sheep's fescue), *Agrostis tenuis* (bent) and *Anthoxanthum odoratum* (sweet vernal grass) on drier and more acid sites. Other species are most abundant in the intermediate habitats. The gradation in species composition is even more strikingly illustrated in an *Agrostis/Fescue* continuum in the

Cheviot Hills. Figure 33 illustrates the gradual changes in cover-abundance of a number of species, common to this type of grassland, in relation to a sequence of sites with soil profiles, showing a continuous variation from the more acid peat-podsols and brown-forest soils to the less acid brown-forest soils and gleys.

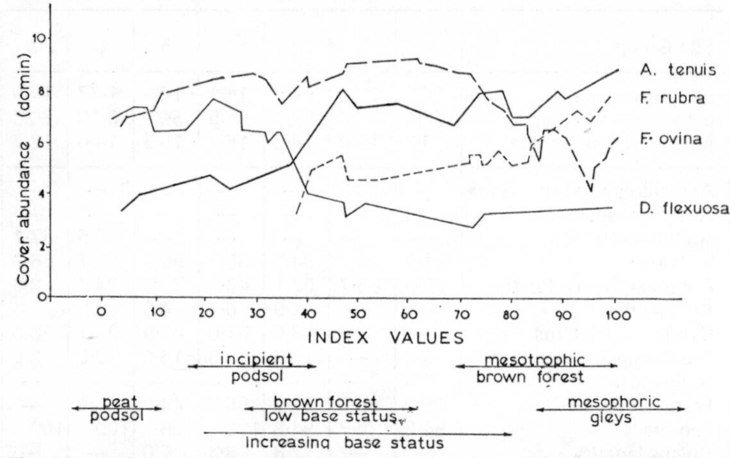

Fig. 33 Variation in cover-abundance of a number of grass species *Agrostis tenuis, Festuca ovina and rubra* and *Deschampsia flexuosa* in relation to a sequence of soil profile types (From: Burnett, J. H. 1964)

Variation in vegetation however is the product not just of one but of a complex of a great number of interacting environmental variables. Simple, uni-dimensional ordinations, while nonetheless valuable, are obviously inadequate to cope with the analysis of multiple correlations. The combination of two or more environmental gradients in a multi-dimensional ordination as in Figure 34 provides a partial solution to this problem. This summarises the interrelationships of the grassland community-types and their relationship to two edaphic gradients. However as J. King and I. A. Nicholson point out, 'the ordination of these types in relation to two environmental axes cannot account for the total floristic variation. In consequence it should not be assumed that adjacent or overlapping types are necessarily as closely related as their proximity in one plane might suggest. For this reason and also because the boundaries of community-types may arbitrarily exclude intermediate stands, the figure by itself does not provide any evidence for either continuous floristic gradients, or for discontinuities'. The development of electronic computers has made possible the application of more sophisticated statistical techniques of multi-variate and factor analysis to the investigation of

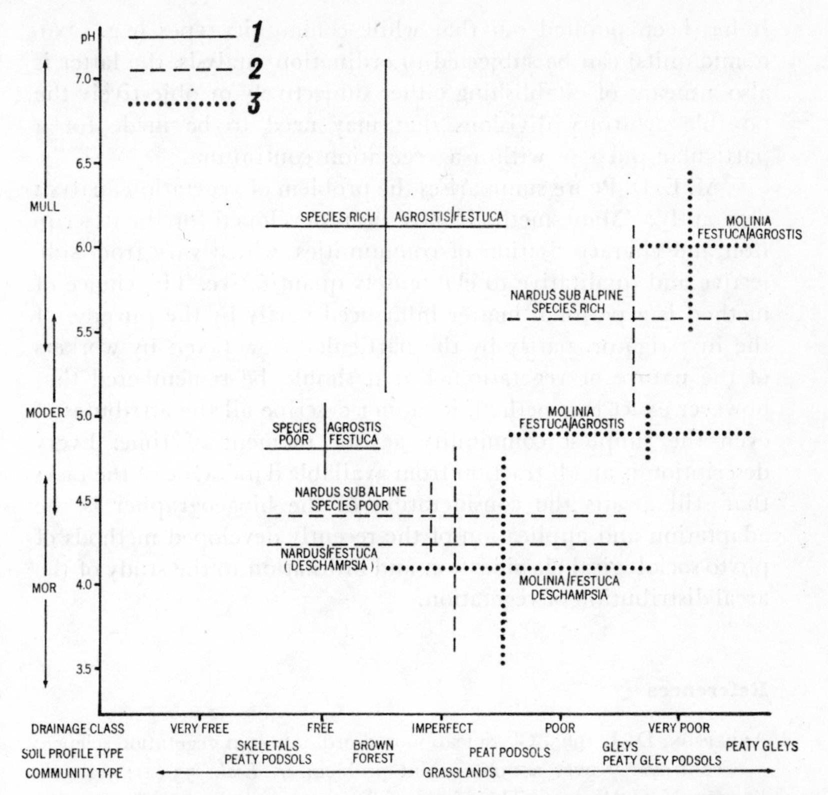

Fig. 34 Interrelationships of grass moorland community-types in terms pH and soil drainage class. (adapted from: Burnett, J. H. 1964) Dominants: 1. *Agrostis/Festuca;* 2. *Nardus Strieta;* 3. *Molinia Caerulea.*

an even greater number of ecological and phyto-sociological variables. Such methods, still at a relatively early stage of application to vegetation study, are however beyond the scope of this book.

Vegetation, because of the multiplicity of its attributes, is among the most difficult of natural phenomena to describe and analyse. Classification and ordination are two means to this end. However, as R. P. McIntosh is at some pains to emphasise, the community-type (or association) and continuum concepts are not mutually exclusive; they represent two different but complementary approaches to the same problems. He maintains that the purpose of both the traditional hierarchical classification systems and ordination techniques is 'to arrange or organise a set of observations and to seek meaningful patterns of relationships'; both are explanatory devices 'to elucidate a large mass of observations whether qualitative and subjective or quantitative and objective'.

It has been pointed out that while community-types (e.g. taxonomic units) can be subjected to ordination analysis, the latter is also a means of establishing either subjectively or objectively the possible arbitrary divisions that may need to be made for a particular purpose within a vegetation continuum.

M. E. D. Poore summarises the problem of vegetation analysis very aptly: 'Many methods have been developed for the description and characterisation of communities, which vary from subjective and qualitative to elaborately quantitative. The choice of method is a personal matter influenced partly by the purpose of the investigator, partly by the particular view taken by workers of the nature of vegetation. But it should be remembered that however exact the method, it cannot describe all the attributes of even the simplest community at one moment of time. Every description is an abstraction from available data'. One of the tasks that still awaits the consideration of the biogeographer is the adaptation and application of the recently developed methods of phyto-sociological classification and ordination to the study of the areal distribution of vegetation.

References

ANDERSON, D. J. 1965. Classification and ordination in vegetation science: controversy over a non-existent problem. *J. Ecol.*, **53** (2): 521–526.

BEADLE, N. C. W. 1951. The misuse of climate as an indicator of vegetation and soils. *Ecology*, **32** (2): 343–345.

BIROT, P. 1965. *Les formations végétales du globe.* S.E.D.E.S., Paris.

BRAUN-BLANQUET, J. 1932. *Plant sociology* (Translation revised and edited by George D. Fuller and Henry S. Conrad). London.

BRAY, R. J. and CURTIS, J. T. 1957. An ordination of the upland forest communities of southern Wisconsin. *Ecol. Monogr.*, **27** (4): 325–349.

BURNETT, J. D. (Ed.) 1964. *The vegetation of Scotland.* Oliver & Boyd, Edinburgh.

CAIN, S. A. 1950. Lifeforms and phytoclimate. *Bot. Rev.* **16** (1): 1–32.

CAIN, S. A. and CASTRO, G, M, DE O. 1956. *Manual of vegetation analysis.* Harper Bros., New York.

CURTIS, J. T. and MCINTOSH, R. P. 1951. An upland forest continuum in the prairie-forest border region of Wisconsin. *Ecology*, **32** (3): 476–496.

CURTIS, J. T. 1955. A prairie continuum in Wisconsin. *Ecology*, **36** (4): 558–566.

CURTIS, J. T. 1957. *The vegetation of Wisconsin: an ordination of plant communities.* Univ. of Wisconsin Press, Wisconsin.

DANSEREAV, P. 1951. Description and recording of vegetation on a structural basis. *Ecology*, **32** (2): 172–229.

Dansereau, P. 1957. *Biogeography: an ecological perspective*. Ronald Press, New York.

Daubenmire, R. F. 1960. Some major problems in vegetation classification. *Silva fenn.*, **105**: 22–25.

Daubenmire, R. F. 1966. Vegetation: identification of typal communities. *Science*, **151**: 291–298.

de Laubenfels, D. J. 1968. The variation of vegetation from place to place. *Prof. Geogr.* **20** (2)

Dice, L. R. 1952. *Natural communities*. Univ. of Michigan Press, Ann Arbor.

Egler, F. E. 1942. *Vegetation as an object of study. Philosophy Sci.*, **9**: 245–60.

Evans, F. S. and Dahl, E. 1955. The vegetational structure of an abandoned field in South eastern Michigan in relation to environmental factors. *Ecology*, **36** (4): 685–706.

Eyre, S. R. 1963. *Vegetation and soil: a world picture*. Arnold, London.

Gaussen, H. 1954. *Géographie des plantes*. 2nd Ed. Collection Armand Colin, Paris.

Gimmingham, C. H. 1951. The use of life form and growth form in the analysis of community structure as illustrated by a comparison of two dune communities. *J. Ecol.*, **39** (2): 396–406.

Gimmingham, C. H. 1961. North European heath communities: 'network of variation'. *J. Ecol.*, **49** (3): 655–94.

Greig-Smith, P. 1964. *Quantitative plant ecology*, 2nd Ed. Butterworths, London.

Hanson, H. C. 1958. Principles concerned in the formation and classification of communities. *Bot. Rev.*, **24** (2 and 3): 65–125.

Hanson, H. C. and Churchhill, E. O. 1961. *The plant community*. Chapman and Hall, London.

Hare, K. F. 1951. Climatic classification. In *London Essays in Geography*, Ed. by L. Dudley Stamp and S. W. Woolridge. London.

Hare, K. F. 1950. Climate and zonal divisions of the boreal forest formations in east Canada. *Geogr. Rev.*, **40** (4): 615–635.

Holridge, L. R. 1947. Determination of world plant formations from simple climatic data. *Science*, **105**: 367–368.

Köppen, W. 1918. Klassification der Klimate nach Temperatur Niederschlag und Jahreslauf. *Petermanns Mitt.*, **64**: 193–203; 243–248.

Küchler, A. W. 1949. A physiognomic classification of vegetation. *Ann. Ass. Am. Geogr.*, **39** (3): 201–10.

Langford. A. N. and Murray, F. B. 1969. Integration identity and stability in the plant association. *Adv. Ecol. Res.*, **6**: 83–135.

McIntosh, R. P. 1967. The continuum concept of vegetation. *Bot. Rev.*, **33** (4): 130–187.

McVean, D. N. and Ratcliffe, D. A. 1962. *Plant communities of the Scottish highlands*. Monographs of the Nature Conservancy, No. 1. H.M.S.O., London.

Mather, J. R. and Toshiola, G. A. 1968. The role of climate in the distribution of vegetation. *Ann. Ass. Am. Geogr.*, **58** (1): 29–41.

Miller, A. A. 1950. Climatic requirements of some major vegetation formations. *Advmt. Sci. Lond.*, **195**, 7 (23): 90–94. *Climatology*. Methuen.

Oosting, H. J. 1956. *The study of plant communities* (2nd Ed.). San Francisco.

Pears, N. V. 1968. Some recent trends in classification and description of vegetation. *Geogr. Annalr* (Stockholm), **50A** (3): 162–172.

Phillips, E. A. 1959. *Methods of vegetation study*. Henry Holt and Co.

Poore, M. E. D. 1955, 1956, The use of phyto–sociological methods in ecological investigations. *J. Ecol.*, **43** (1): 226–269; **43** (2): 606–651; **44** (1), 28–50.

Poore, M. E. D. 1962, The method of successive approximation in descriptive ecology. *Adv. Ecol. Res.*, **1**: 35–66.

Poore, M. E. D. 1964. Integration in the plant community. *British Ecological Soc. Jubilee Symposium*, 1963. Ed. Macfadyen A. and Newbould, P. J.

Poore, M. E. D. and McVean, D. N. 1957. A new approach to Scottish mountain vegetation. *J. Ecol.*, **45** (4): 401–439.

Raunkiaer, C. 1934. *The life forms of plants and statistical plant geography*. Clarendon Press, Oxford.

Smith, A. D. 1940. A discussion of the application of a climatological diagram, the hypergraph, to the distribution of natural vegetation types. *Ecology*, **21** (2): 184–191.

Tansley, A. G. 1939. *The British Islands and their vegetation*, Vols. I and II. Cambridge Univ. Press.

Thornthwaite, C. W. 1933. The climates of the earth. *Geogr. Rev.*, **23** (3): 433–440.

Thornthwaite, C. W. 1943. Problems in the classification of climates. *Geogr. Rev.*, **33** (2): 233–255.

Tosi, J. A. 1964. Climate control of terrestrial ecosystems: a report on the Holdridge model. *Econ. Geogr.*, **40** (2): 173–181.

Trewartha, G. T. 1954. *An introduction to climate*. McGraw-Hill, London.

Watt, A. S. 1964. The community and the individual. *British Ecological Society. Jubilee Symposium* (1963), Ed. by Macfadyen, A. and Newbould, P. J.

Webb, D. A. 1954. Is the classification of plant communities either possible or desirable? *Bot. Tidsskr.*, **51**: 362–370.

Whittaker, R. H. 1951. Criticism of plant association and climatic climax concepts. *N.W. Sci.*, **25** (1): 17–31.

Whittaker, R. H. 1962. Classification of natural communities. *Bot. Rev.*, **28** (1): 1–239.

9
Vegetation Change and Stability

One of the most fundamental characteristics of vegetation is its susceptibility to *change*. The plant communities of which it is composed vary in relation to variations in the complex of environmental factors that occur from place to place. But the composition and structure of any given community also reflects the interaction between its component members and between them and their habitat through time. Any environmental changes must inevitably result in correspondingly minor or major changes in the associated plant community. They will tend to make the habitat less favourable for some, more favourable for other species. Some may not be able to tolerate the new conditions, or to survive the competition of those better adapted to them. Some species will tend to disappear from the habitat, some will decrease, others increase in number, while yet others, formerly not present, will be able to establish themselves in the community. The result of changes in a particular habitat, then, will be that one assemblage of plant species will be replaced or succeeded by another of different structure and composition. This process whereby one community replaces another on a given site, as a result of habitat change, is called plant or vegetation *succession*.

There are two ways in which a particular habitat can be altered or modified. It can, on the one hand, be by reason of a change in one or other of those environmental factors independent of the plants themselves, such as climate, soil, land-form or the direct or indirect activities of man. Pollen grains preserved in peat deposits provide—as has already been noted (p. 120)—striking evidence of the succession from one type of vegetation to another following the climatic changes in the post-glacial period in North America and in Europe. The deliberate drainage of peat bogs by man can effect, within less than a generation, a replacement of communities formerly dominated by bog moss

(*Sphagnum*) and other species tolerant of acid water-logged conditions, by those dominated by heather (*Calluna vulgaris*) or even birch (*Betula spp.*) scrub. Furthermore continuous change in composition of vegetation is characteristic of areas subjected to regular grazing or burning of varying intensities. In contrast to these essentially 'extrinsic' habitat changes, there are those 'intrinsic' to all plant communities. It has already been stressed that, plants by their very presence modify the physical conditions of the sites on which they grow. They contribute to the production and stabilisation of weathered mineral material. Through the addition of organic matter they effect changes in the moisture content, temperature, aeration and nutrient-status of the soil; they are, as we know, one of the important factors in soil formation. Above ground, the vegetation so modifies light, temperature and humidity conditions as to create a micro-climate significantly different from that which would prevail in the absence of a vegetation cover.

From the moment, therefore, that a particular site is first colonised by plants the resultant modification of the habitat initiates changes which tend to bring about the replacement of one collection of plants by another: colonisation initiates succession. The addition of organic matter permits the growth of other plants which are more demanding of water, nutrients or anchorage and are more vigorous, taller and more aggressive than the preceding species. The shelter and protection afforded by one group of plants allows shade-tolerant seedlings, incapable of germinating in open exposed sites to establish themselves. The preceding light-demanding species and those which cannot compete with more aggressive newcomers will tend to die out. As Eugene Odum has so aptly remarked, 'the action of the community on the habitat tends to make the area less favourable for itself and more favourable for other sets of organisms'; each community has within itself, so to speak, a potentiality for self-destruction! Succession is, in fact, a universal biological process—a 'law of nature'—characteristic of plant (indeed all organic) communities. It is the process whereby plant communities, and the vegetation cover they form, *develop* in a particular area through time. It results in the progressive development from simple to more complex, from 'youthful' to 'mature', from unstable to more stable plant communities.

Characteristic of succession is a change in the species of plants (and animals), an increase in the number of species and in the variety of growth-forms occupying a particular site (see Table 23).

	Embryo Dunes	Yellow Dunes	Fixed Dunes	Late-Fixed Dunes	Dune Scrub
Annuals	9 (50%)	44 (60%)	60 (41·3%)	17 (10%)	—
Biennials	2 (11·1%)	5 (7%)	18 (12·4%)	4 (2·5%)	—
Perennials (herbaceous)	7 (38·9%)	24 (33%)	65 (44·8%)	124 (77·5%)	68 (44·7%)
Total (herbaceous spp.)	18	73	143	145	68
Shrubs	—	—	2 (1·3%)	14 (8·7%)	29 (43·2%)
Trees	—	—	—	2 (1·2%)	8 (12%)
Total herbs, shrubs, trees	18	73	145	161	105

Table 23 Number of species in phases of dune vegetation· (From: Salisbury, Sir E. 1952. Summary of totals obtained from Salisbury's examination of the most important dune systems in Great Britain together with published records).

There is a sequence of communities—of differing assemblages of species—each of which, through soil-formation and micro-climate modification, prepares the way for a succeeding more diverse and complex one. The culmination of this process is the final establishment of the *climax* community or type of vegetation. This, according to A. G. Tansley, will be 'dominated usually by the largest and particularly by the tallest plants which can arrive on the area and can flourish under the particular conditions which it presents'. It will be the most diverse and complex that can be achieved; as such, it will represent the most complete expression possible of the prevailing environmental conditions by plant growth. The climax community will represent the maximum use possible of a given habitat by organisms. At this stage the habitat will have reached 'saturation point', so to speak, in terms of the plant (and hence animal) life it can support and the invasion and establishment of other species will, as a result, be inhibited. The community is 'closed', all potential niches are occupied and progressive succession is at an end. *Theoretically,* provided environmental conditions do not alter drastically, the climax community will be relatively stable, capable of reproducing itself and maintaining a constant composition, structure and biomass. It will then represent a stage in which a balance has been achieved in the cycling of nutrients, between the uptake of nutrients by plants and their eventual return to the physical habitat and in the flow of energy, between the accumulation by photosynthesis and loss by respiration (see Figs. 35 and 36).

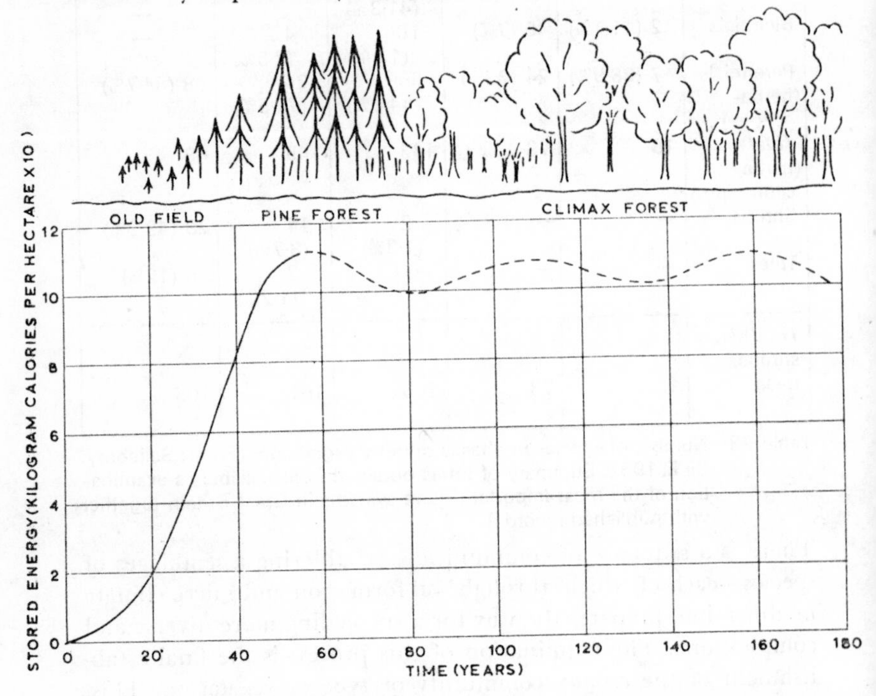

Fig. 35 Plant succession and accumulation of energy typical of terrestrial ecosystems in eastern North America (From: Woodwell, George M. 1963)

The sequence of communities by which the climax is attained is termed a *sere*, of which those communities preceding the climax are the *seral* stages. Primary successions (or *priseres*) are those initiated on new, fresh, biologically unmodified sites. Although the species-composition and the number of seral communities involved may vary from place to place, the principal stages, in areas where forest is the potential climax, are basically similar. They are characterised by the dominance of mosses and lichens and/or herbs, shrubs and trees respectively. Initial plant colonisation in a new area will obviously depend on the prevailing climatic conditions, the nature of the site, and the ability of available seeds to migrate into the area. However, the pioneer community which establishes itself will tend to be composed of species tolerant of

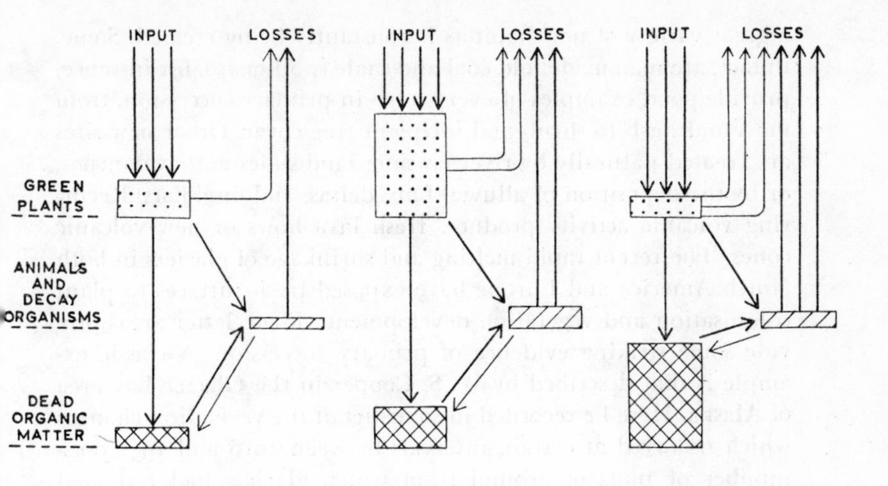

Fig. 36 Diagrammatic representation of energy flow in ecosystem with stage of development. Input is solar energy fixed by photosynthesis : losses are by respiration, death and decay and export ; from left to right are represented :

I *Successional stages* when input exceeds losses, energy accumulates and ecosystem develops ;

II *Climax or stable stage* when input equals losses ;

III *Disturbed or degraded stages* when because of flood, storm, overgrazing or overburning losses exceed input and amount of energy in system declines.

(From : Woodwell, George M, 1963)

open, exposed and often unstable conditions, as on sand or scree, and frequently, as on little-weathered bare rock-surfaces, deficient in available plant nutrients and of limited rooting depth. The first colonists may include lichens, mosses and/or hardy annual herbs. These begin the process of soil-formation and stabilisation and produce conditions in which mat-forming perennial herbs and low-growing shrubby plants can form a complete cover and accelerate the process of soil enrichment. This, together with the increased protection and shelter provided by a denser ground-cover of vegetation, paves the way for the establishment of quick-growing shrubs and, later, for the slower-maturing, taller trees. In the early stages soil development tends to be the most important factor controlling succession; in the later stages, the increasing modification of the micro-climate by the taller shrubs and trees, which oust those plants unable to tolerate their shade, plays an increasingly important part.

Although areas where primary succession is in progress at present are, in relation to the total land-surface, relatively limited a great variety of new habitats is constantly being created. Some, in fact, are man-made; old coal and shale spoil heaps, for instance, provide good examples of every stage in primary succession, from moss and herb to shrub and incipient tree cover. Other new sites are created naturally by river erosion, land-slides and avalanches, or by the deposition of alluvial fans, deltas, and mud-flats. Recurring volcanic activity produces fresh lava flows or new volcanic cones. The recent rapid melting and shrinkage of glaciers in both North America and Europe have exposed fresh surfaces to plant colonisation and vegetation development. These latter areas provide some striking evidence of primary succession. A classic example is that described by W. S. Cooper in the Glacier Bay area of Alaska. Here he recorded in some detail the vegetation changes which occurred at certain intervals between 1916 and 1931 on a number of plots of ground from which glaciers had retreated within a known period of time. His observations revealed a succession in which three principal stages could be recognised: pioneer, thicket and forest. The *pioneer stage* was characterised by the initial colonisation of hardy *Rhacomitrium* mosses (*R. lanuginosum* and *canescens*), a gradual increase in perennial herbs, particularly the broad-leaved willow-herb (*Epilobium latifolium*) and the horsetail (*Equisetum variegatum*), and finally the establishment of *Dryas drummondii*. This latter species is a low-growing evergreen under-shrub whose mat-forming habit plays an important rôle in soil stabilisation and erosion control. The transition from the pioneer (moss-herb) to the *thicket* (shrub) stage was marked by the appearance of dwarf, creeping willows which together with the *Dryas* increased in abundance and suppressed the shade-intolerant mosses and herbs. Then followed the establishment of shrubs such as willow (*Salix spp.*) and alder (*Alnus spp.*), with the latter in particular becoming the principal and sometimes the sole dominant of the thicket stage. The alder, a characteristic shrub of many other recently deglaciated areas, is an important element in soil-nitrogen enrichment. In the Glacier Bay area it prepares the way for the final development of the *forest* stage, dominated by Sitka spruce (*Picea sitchensis*) and, later, a mixture of spruce and hemlock (*Tsuga spp.*).

The gradual recession of a glacier will result in the continuous exposure of a new land-surface and its progressive colonisation by plants. Hence there will also be a spatial sequence or zonation of plant communities each representing a different stage in the primary succession from the pioneer on the newest, most-recently exposed, to climax forest on the older, earliest exposed sites. Such a spatial zonation of the various stages in plant succession is particularly characteristic of those areas where a new surface is continually being built up by the progressive deposition of mud, silt or sand. It is well illustrated in sheltered bays and tidal estuaries where the gradual accumulation of sediments in the inter-tidal zone is reflected in the various stages in the development of salt-marsh vegetation. Here there is a progressive change in plant communities from the lowest zone where mud-banks or flats are exposed for short periods twice daily, to the highest where silting has built up a surface out of range of all but the highest tides. The lowest outer edge of the salt marsh, just above low-tide level, is where initial colonisation begins. The pioneer stage is characterised by a few highly specialised plants capable of tolerating an extremely mobile, frequently inundated and highly saline habitat. As they become established, at first sporadically and then in increasing numbers, they tend to check the rate of water flow, stabilise the mobile substratum and accelerate the process of silting. As the surface level increases in height the length of time during which the salt marsh is subjected to inundation, and consequently the degree of salinity, decreases. The pioneer species of the low marsh are gradually replaced in the high or middle marsh (submerged for a varying period only at spring-tides) by an increasing number of species less tolerant of high salinity and a mobile substratum. The highest levels covered only occasionally, during exceptionally high water, can be colonised by non-halophytic rushes and grasses. If the surface level is built up sufficiently above the water table this higher and drier surface can then be colonised by scrub and woodland vegetation. A somewhat comparable zonation of communities (though composed of different species) accompanies the progressive silting of fresh-water lakes. As the depth of water decreases, submerged and floating-plants which anchor themselves on the lake floor and help to trap and accumulate silt are gradually replaced by reeds and rushes. The continued deposition of silt and, in such water-logged conditions, an increasing accumulation of organic matter may eventually raise the rooting medium high enough to allow colonisation by willow and alder scrub, and eventually oakwood. In both these cases the zones of vegetation reflect a gradual and continuous spatial change in habitat conditions; but each zone will tend to be replaced in time by the next stage in the succession as silting continues to extend the fresh or salt marsh progressively further from the shore.

In much the same way the gradual accumulation of wind-blown sand piling up around coasts and lake shores reveals a successive development of communities from those of the youngest mobile to the older fixed and stabilised dunes (see Table 23, p. 295 and Fig. 37). Under these circumstances the accumulation

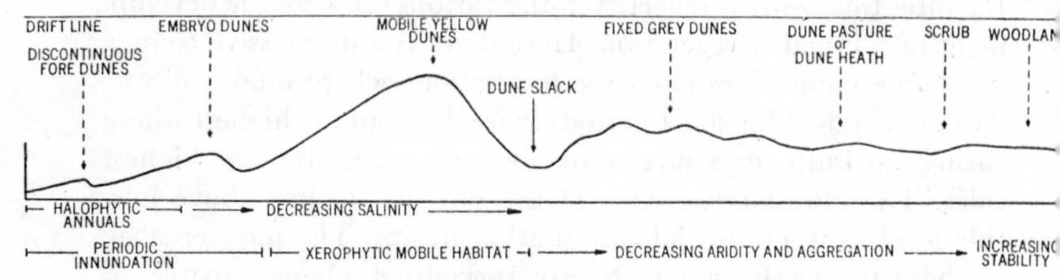

Fig. 37 Diagrammatic representation of habitat zonation characteristic of a coastal sand-dune complex.

and stabilisation of drifting sand is accomplished by plants which can tolerate not only a mobile but a very dry habitat. The pioneers are commonly such drought-resistant grasses as lyme grass (*Elymus arenaria*) and marram (*Ammophila arenaria*). They check wind-force and trap sand. Their upward growth can keep pace with the continued accumulation of sand, they have in addition extensive and rapidly spreading rhizomes and a ramifying net-work of very fine roots which are constantly being renewed at higher levels. The growth of these 'dune-building' plants not only stabilises the sand but ameliorates its aridity by the addition of water-retentive organic matter. Eventually (and particularly if shelter is provided by the formation seaward of a new line of dunes) plants less tolerant of mobility and aridity can establish themselves; lichens and mosses and a variety of herbaceous plants provide a complete ground-cover and increase further the humus and water-content of the dune soils. The formerly mobile dune becomes 'fixed', the pioneer species gradually die out in face of competition and, where not disturbed by man, shrubs such as hawthorn, gorse, broom and buckthorn can establish themselves on the improved habitat preliminary to the final development of woodlands. In this respect it is interesting to note (see Table 23, p. 195) that the maximum number of species is attained before that of the climax. With the establishment of a shrub and/or tree stratum there is an accompanying reduction in, particularly, heliophytic herbs.

An important feature common to all primary successions is the gradual amelioration of the initial habitat conditions. The progression from the pioneer and early seral stages to the climax is marked by a trend from more to less extreme, from more severe to more equable conditions. Primary seres initiated on sites which, as on bare rock or sand, tend under similar climatic conditions to be excessively dry, are called *xeroseres,* those on sites submerged or waterlogged, *hydroseres.* The stabilisation, protection and accumulation of mineral and/or organic matter promoted by plant growth will result in habitat conditions and a climax vegetation less xerophytic or hydrophytic (more mesophytic) in character than that of the pioneer stage. Under similar climatic

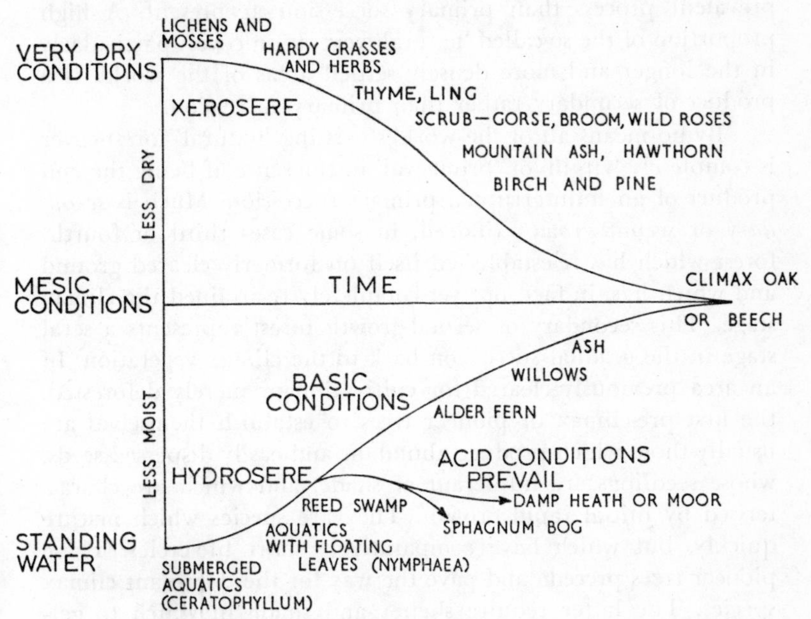

Fig. 38 Diagram summarising relationships between primary successions. (From: Ashby, Maurice. 1961)

conditions a variety of pioneer communities in different local habitats tend to 'converge' towards a similar climax type of vegetation or formation (see Fig. 38).

If the existing vegetation cover is partially or completely removed, provided no environmental change occurs and no inhibiting factor is present, a successional development of plant communities towards the climax will be re-initiated. Any succession which takes place on a site which was formerly vegetated is called a *secondary succession* (or *subsere*). Such are characteristic of land once cultivated or grazed and since abandoned, of former forest areas clear-felled or devastated by catastrophic fires or wind storms. Secondary successions differ from primary ones in that they occur on sites that have already been biologically modified, with soils formed and an amount of organic matter already present, and possibly part of the original plant community still surviving. In the absence of continued disturbance or alteration, succession tends to be more rapid than on completely 'new' sites. The early pioneer stages characteristic of primary succession may be non-existent or fleeting; they may tend to become 'telescoped' and, in humid areas, the development of shrub and tree stages can be relatively rapid. Indeed, secondary succession is a more prevalent process than primary succession at present. A high proportion of the so-called 'natural' vegetation cover, particularly in the longer and more densely settled areas of the world, is a product of secondary rather than primary succession.

By no means all of the world's existing 'natural' forest-cover is completely 'virgin' or 'primaeval' in the sense of being the end product of an uninterrupted primary succession. Much is *secondary* or *second-growth* (indeed, in some cases third or fourth) forest which has re-established itself on formerly cleared ground and which has, in fact, not yet completely re-attained the climax stage. This secondary or second-growth forest represents a seral stage in the gradual succession back to the climax vegetation. In an area previously cleared for cultivation or merely deforested, the first pre-climax or pioneer trees to establish themselves are usually those which produce abundant and easily dispersed seeds, whose seedlings are intolerant of shade, and which are characterised by initial rapid growth. They are species which mature quickly, but which have comparatively short life-cycles. These pioneer trees precede and pave the way for the dominant climax species. The latter require shelter and shade in which to germinate; they are dispersed, grow and reproduce more slowly, but they are longer-lived. Growing to greater heights, they eventually over-top and suppress the earlier established pioneer trees. We have already noted that the alder is the percursor of the climax spruce in primary successions in deglaciated areas of Alaska. In Britain one of the commonest colonisers of cleared woodland is the birch (*Betula spp.*), which can produce, after only fifteen to twenty years, large quantities of small wind-borne seeds annually. Its growth is rapid. Birchwoods which have established themselves on cut-over woods and plantations, or have recolonised moorlands cleared of sheep, are one of the most prevalent types of woodland in the Highland areas of Scotland. Birch, however, cannot hold its own in competition with taller, shade-casting and longer-lived trees, and in undisturbed conditions would normally be replaced by pine or oak in areas where these species are capable of growing successfully.

Of the deciduous, mixed deciduous and coniferous forests which still cover a high proportion of the eastern states of the U.S.A. (seventy-five per cent. of New England and forty to fifty per cent. of many of the other states east of the Mississippi) relatively little, if indeed any, is primary growth. It is rather the product of secondary succession on land which may have been cut over or devastated by fire more than once since or even before the first white settlement of the area. In this area much poor or hilly land, once cultivated, has been and continues to be abandoned. On the old, idle fields secondary succession is re-establishing a forest cover. These second-growth forests are dominated by such pioneer trees as the paper birch, aspen and white pine; they are often even-aged, the result of widespread destruction or removal followed by rapid regeneration in a limited time. Similarly in the coniferous forests of the North-Western States second-growth forests dominated by lodge-pole pine (*P. contorta*) or aspen have followed the destruction by fire of the formerly more diverse climax of spruce and fir. In Mediterranean Europe prolific seed production and rapid root development of the Aleppo pine (*P. halepensis*) make it a very successful and now widespread pioneer species on land formerly cultivated or grazed. In many of the more densely populated areas of the humid tropics the primary 'virgin' rain forest has been replaced by secondary growth, as a result of the widespread practice of shifting agriculture. Patches of forest are felled. After one or two years the plot is abandoned, and the secondary forest which precedes the re-establishment of the climax has many characteristics in common with *second-growth* temperate forests. It is composed of easily and rapidly dispersed, light-demanding and extremely quick-growing trees which are smaller and shorter-lived than the dominant climax species. In contrast to the latter, which are predominantly hardwood, trees of the secondary tropical forest have light soft-textured wood.

The rate of both primary and secondary succession, and the length of time necessary for the final establishment of the climax community is extremely variable. It would appear that succession is more rapid the more severe are climatic conditions; the sere is shorter and the number of seral stages limited. In the Glacier Bay area of Alaska it was noted that the time interval between the disappearance of the ice and the arrival of the first trees was ten to twelve years in some areas; in others they appeared almost on the heels of the retreating ice-front. Indeed, trees growing on terminal moraines adjacent to or on ablation moraines on top of the glacier edge are not an uncommon sight in many parts of the world. It has also been suggested that in cold and hot deserts succession is so limited as to be non-existent. The severity of the environmental conditions so restricts the development of vegetation that the extent to which it can modify its habitat is very limited in comparison to other more favourable areas. Under comparable environmental conditions secondary succession will tend, particularly in its initial stages, to be more rapid than primary succession. The rate of both will be dependent on the nature of the initial site. Primary succession tends to be most rapid on moist, unconsolidated or easily weathered sediments and in hydroseres where the process is accelerated by the rapid accumulation of silt and organic matter. It will be less rapid on bare, highly resistant rock-surfaces. Secondary succession will be particularly rapid on formerly cultivated land. In eastern North America the natural re-establishment of a near-climax deciduous forest on old fields has been estimated to take about a hundred and fifty to two hundred years. Succession may be delayed by the presence of a long-established grass cover under which a well-developed closely-knit turf can very effectively inhibit the penetration of tree-seedling roots. Such evidence as is available suggests that secondary successions in the humid tropics are more protracted than in temperate regions. It has been noted, for instance, that the forest which occupies the site of Ankor Vat (Cambodia), destroyed by fire some five to six hundred years ago, now resembles virgin forest but still preserves some differences. Pierre Gourou notes that after a lapse of fourteen hundred years the forests on the sites of ancient Maya settlements in Central America are still composed of pre-climax trees. The speed and facility with which secondary succession takes place will, however, depend on the length of time which has elapsed since the removal or disturbance of the climax and the extent to which the habitat has been modified in the interim. But even on comparable sites the rate of both primary and secondary successions may vary according to the availability of colonists and the rate of seed dispersal.

Today man is the major and most powerful instrument of vegetation change, either directly or indirectly, through his increasing modification of the physical environment. In many areas he has completely removed the pre-existing vegetation cover and replaced it with his 'unnatural' and highly specialised communities of field crops or forest plantations. In others, where he has destroyed both the vegetation and the soil cover, he has left in his wake man-made deserts on which plant succession has not been able to recommence. Relatively few parts of the remaining uncultivated vegetation cover have completely escaped his activities, and regions where primary (virgin) climax communities still exist intact are fast dwindling. Over considerable areas man has removed or, through grazing and firing, prevented the regeneration of forest and scrub growth. The continued operation of these factors has effectively checked the natural succession back to the climax, and has maintained the vegetation cover at what is referred to as a sub-climax or 'semi-natural' stage. Grassland, heath or scrub which exist in areas which, were it not for the activities of man, could support a forest vegetation are sometimes called *biotic* or *anthropogenic* climaxes, since the maximum vegetation development is controlled primarily by these rather than climatic or edaphic factors. Much of the savanna grasslands of tropical regions is now considered to be a biotic (or more explicitly 'fire') climax which has replaced formerly existing tropical deciduous or evergreen rain forest, and which is maintained in this condition by periodic and deliberate burning, as well as by grazing.

Man's activities, however, tend not only to check succession and the re-establishment of the climax but to give rise to a type of vegetation that might not otherwise have developed. Where a protective tree-cover is removed and its regeneration inhibited, the resulting change in micro-climatic conditions can have considerable repercussions on the nature of the soil. Less water is lost by transpiration and the surface is, in addition, unprotected from the direct onslaught of rainfall. In cool, humid climates increased leaching and soil saturation may result. Where temperatures are higher, the elimination of shade opens up the surface to full sunlight, and increased temperatures near the ground accelerate evaporation from the soil surface and the rate of organic decomposition. The cumulative effect of such changes may initiate changes in the composition of the vegetation that might not have taken place had natural succession been able to recommence immediately. In the former the dominance of more hydrophytic, in the latter more xerophytic species than existed before may be favoured. Similarly, continuous burning or grazing tends to give advantage, at the expense of others, to species tolerant of, and whose increased growth may in fact be promoted by these activities. They may be species which, though present in the climax vegetation, were previously relatively minor or even inconspicuous elements; they may be plants formerly excluded by competition from the climatic climax community. In Britain, for instance, heather-dominated communities now form one of the commonest types of vegetation on once forested uplands. This species (*Calluna vulgaris*) would have occurred naturally, though much less prolifically along with other shade-tolerant and calcifuge species in the field-layer of former birch and pine woods. Its present dominance has been promoted and maintained by burning which has tended to eliminate less fire-resistant species and which, together with sheep grazing, has prevented the regeneration of climax forest. In other places heather has invaded new areas where the removal of oak and/or birch wood has been followed by increased soil leaching and acidity. In yet other instances it has become prolific as the result of the deliberate drainage, burning and, consequently, surface drying of former *Sphagnum* bogs.

Such changes, or replacements of one type of community by another as a result of the direct or indirect activities of man, are called *biotic* successions. Since these changes tend to alter or *deflect* plant succession from its 'natural' course and in doing so produce a type of vegetation that might otherwise not have arisen, the term *deflected succession* or more impressively though no more precisely, *plagioseres* (i.e. Gr. *plagio* = oblique) is commonly employed. In contrast to 'natural' successions which, under constant environmental conditions, are progressive (effecting a development of the most diverse and complex community that can be attained) biotic successions or plagioseres, of their very nature, tend to be *retrogressive*. Continuous grazing and burning result in continuous and cumulative changes in the habitat. Domestic grazing animals inevitably remove more plant nutrients in the herbage they eat than is returned in their dung; relatively few of these animals die or decay on their grazing grounds. While burning releases nutrients in plant-ash for re-use by the growing vegetation much is removed by wind erosion and leaching, and hence lost from the soil. Similarly any changes which promote increased leaching or organic decomposition accelerate the loss of plant nutrients. As a result of all these factors the depletion of nutrients will tend to be greater than can be made good by the slow weathering of soil parent material. The disruption of the nutrient cycle (see Fig. 39), of the equilibrium between the community and its environment, leads to declining fertility and decreased organic productivity. Communities more tolerant of the impoverished conditions will replace those less able to grow and compete successfully. The result will be a 'degradation' or 'degeneration' of the vegetation from more to less complex and diverse communities and consequently a decrease in the plant and animal biomass. Degradation and the concomitant changes in the habitat may be such that even in the event of the cessation of burning or grazing, the re-establishment of the pre-existing

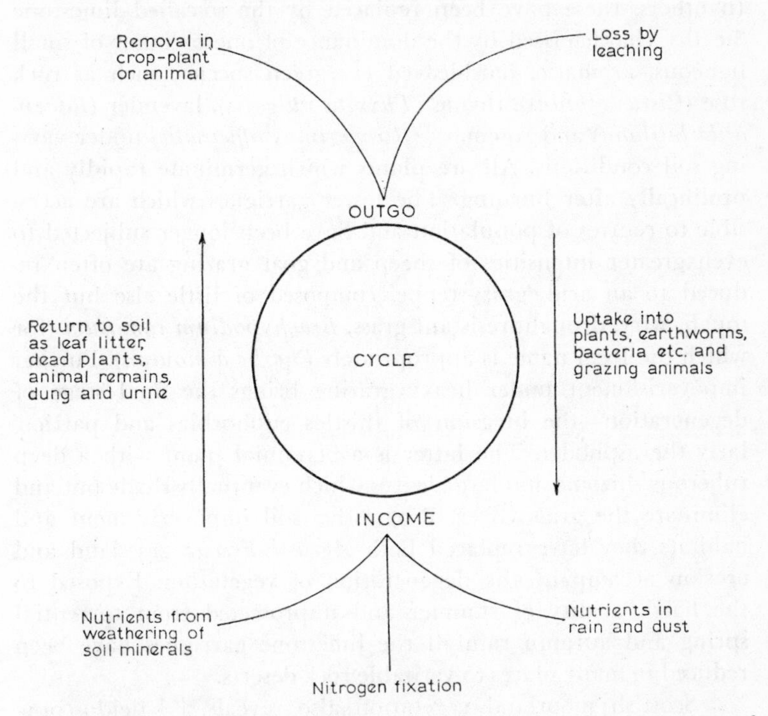

Fig. 39 The nutrient cycle. Diagrammatic representation of the movement of mineral nutrients through the ecosystem. (From : Report of the Nature Conservancy, 1960)

climax may be difficult, exceedingly slow or in some cases impossible.

Biotic successions and the degradation of vegetation is all too common and characteristic a feature of those areas of the world which have long been subjected to intensive grazing. The vegetation of the sub-alpine moorlands of Scotland and of the *garrigues* of the limestone hills and plateaus of southern France alike is a product of the progressive modification and deflection of the climax vegetation by fire and grazing over a long period of time. On the latter, little now remains of the former climax forest of evergreen holly-oak (*Quercus ilex*). Much was deliberately cut; fire and grazing have prevented regeneration. The existing vegetation of the French garrigues is composed of a patchwork of communities representing various stages in the progressive degeneration of the vegetation cover. Some areas carry a low, open scrub of deep-rooted drought-resistant dwarf oak (*Q. coccifera*), box (*Buxus sempervirens*) and juniper (*Juniperus oxycedrus* and *communis*). In others, these have been replaced by the so-called limestone 'heaths' characterised by the dominance of one or other of small ligneous, aromatic, hard-leaved evergreen species, such as rock rose (*Cistus albidus*), thyme (*Thymus vulgaris*), lavender (*Lavandula latifolia*) and rosemary (*Rosmerimus officinalis*) under varying soil conditions. All are plants which germinate rapidly and prolifically after burning. The lower garrigues which are accessible to centres of population and have been longer subjected to even greater intensities of sheep and goat grazing are often reduced to an arid 'grass-steppe' composed of little else but the tough, wiry, drought-resistant grass, *Brachypodium ramosum* (for which the local name is appropriately *l'herbe à mouton*). Further impoverishment under heavy grazing brings the final stage of degeneration—the invasion of thistles euphorbias and particularly the asphodel. The latter is a perennial plant with a deep tuberous rhizome and large leaves which eventually shade out and eliminate the grass cover. Progressive soil impoverishment and habitats they have replaced both *Agrostis-Fescue* grassland and erosion accompany the degeneration of vegetation. Exposed to the harsh aridity of summer and unprotected from torrential spring and autumn rainfall the limestone garrigues have been reduced in many places to veritable rock-deserts.

Scottish moorland vegetation also reveals a kaleidoscopic pattern of vegetation communities which represent different stages in biotic succession resulting from varying intensities of grazing and burning. Although the destruction of the pre-existing climax forest was later than in the longer-settled Mediterranean region, the combined effects of grazing and burning have also been accompanied by a progressive deterioration of soil and vegetation. The concentrated accumulation of poor, acid, heather-litter has undoubtedly promoted the formation of 'mor' humus and intensified soil leaching, with a resulting decline in the availability of plant nutrients. Intense and selective sheep grazing combined with over-burning has resulted in the replacement of heather by grass moorlands—particularly in the Southern Uplands which have, for a much longer period, been subjected to a higher intensity of grazing than the Highlands. The spread of the less nutritious moor mat grass (*Nardus stricta*) in the drier and of the purple moor grass (*Molinia caerulea*) in the wetter habitats has also been promoted by these activities; both tend to be avoided by sheep in favour of the more nutritious and demanding species. In varying habitats they have replaced both *Agrostis-Fescue* grassland and heather moorland. In addition biotic factors have been responsible for the rapid spread of bracken (*Pteridium aquilinum*), now the most virulent and intractable weed of hill land. Formerly an associate of deciduous woodland, it grows with greater vigour in full sunlight and it can propagate rapidly by means of underground rhizomes. It is unpalatable to grazing animals. Within a relatively short time it has, in many areas, invaded and replaced heavily grazed grassland or severely burnt heather-moorland. Finally the progressive deterioration of vegetation as a result of over-grazing and burning has culminated, in many parts of upland Scotland, in accelerated soil erosion which, though not so spectacular as in the more arid Mediterranean region, has reduced steep slopes to unstable rock screes.

In other parts of the world on land laid bare by cultivation the depletion of the organic content by long-continued cropping reduces the stability and water-holding capacity of the soil. As a result it becomes increasingly susceptible to removal by wind or rain. Soil destruction consequent upon either the complete removal or the gradual degeneration of the natural vegetation cover results in the production of 'biological deserts'. Where slopes are steep or where rainfall is normally low, the susceptibility to erosion and 'desertification' is particularly great. Around the margins of the more arid regions in North Africa and Asia man has brought about, as a result of vegetation and soil destruction, an extension of desert conditions. In any area, however, complete soil destruction results in the creation of a 'new', raw, biologically unmodified, and extreme habitat. Under these circumstances, unless artificially rehabilitated by man, the natural re-establishment of a vegetation cover will have to be by the slow process of primary succession. Where erosion, particularly on steep slopes or in arid areas, once initiated continues with increasing intensity, the re-establishment of vegetation may be delayed or interrupted for a very long period. In some cases it would appear that biotic activities can effect irreversible soil changes. This is probably nowhere more vividly illustrated than in tropical regions where the destruction of forest or savanna vegetation can result in the exposure of hard lateritic crusts: composed of hydrated oxides of iron and aluminium, impermeable, resistant to weathering and erosion, and containing no plant nutrients they are in extreme cases completely sterile.

The Climax Concept

Plant succession is the *process* of vegetation change from less to more, or from more to less complex and stable plant communities. It is basic to the concept of climax vegetation. The latter is, theoretically, the stage at which a plant community attains the maximum development possible under the prevailing environmental conditions. Further development or progressive succession ceases and the climax community can then maintain itself in a 'state of equilibrium' so long as environmental conditions remain unchanged. In contrast to seral communities the climax is, therefore, relatively stable. The concept of climax vegetation evolved as a result of the recognition and intensive study of plant successions by, particularly, American ecologists at the end of the last and the beginning of this century. Outstanding among these workers was the botanist, Frederick Clements, to whom we owe the elaboration of the idea of the climax community as the final or terminal stage in the progressive development of vegetation in a given area.

Clements maintained that the final stage in the development of vegetation was determined or controlled by climatic conditions and that the climax community was the most complex that could be attained, *and* maintained, in a state of equilibruim with the prevailing climate. Within a particular climatic region, given time and freedom from environmental disturbance, plant succession would eventually result in the development of the same type of vegetation and soil irrespective of original variations in rock type, drainage or landform. The climax community would then be characterised by the dominance of plants of a form best adapted to compete successfully in a given type of climate. This *mono-climax* concept of Clements conceived of only one type of climax, the *climatic* climax, and of only one type of climax community, the climatic climax *formation*. Indeed, in his terminology, climax is synonymous with 'climatic climax' and the latter with the largest community possible, the *formation* (formation-type was designated pan-formation), which occupies a particular climatic region. In his own words, 'a formation, in short, is the final stage of vegetational development in a climatic unit. It is the climax community of a succession that terminates in the highest life-form possible in the climate concerned'. Communities which differed from the climax because of particular soil or landform characteristics were regarded as sub-climax. They represented stages in the development of the climax on sites where succession had been arrested or retarded because of more extreme habitat conditions. Clements assumed, however, that these sub-climax communities could develop eventually, though more slowly, to the climatic climax. In time, all the vegetation in a given climatic region would be of a similar form, and the climax vegetation would be associated with a 'mature soil'—the zonal soil—which reflected the combined effect of vegetation and climate irrespective of the parent rock type. It is, however, doubtful if climatic or relief conditions have ever remained constant long enough to allow this theoretically ideal stage of complete correlation between climate, vegetation, and soil to be attained. Those communities which resulted from the modification of the 'true climax' by biotic factors, particularly the effects of man, Clements designated *dis-* (or *disturbed*) climaxes; he considered that they also had the potential to resume succession to the climax when the 'disturbance' ceased. *Associations* were regarded as the major sub-divisions of the climax formation, characterised by the dominance of one or a few species of similar form, reflecting, particularly, inter-regional variations of climate.

While accepting the concept of the climax as a relatively stable community in dynamic equilibruim with the prevailing environmental conditions other ecologists have since rejected in part, or in whole, the Clementsian mono-climax theory and its associated implications of 'climatic determinism'. Advocates of what, in contrast, has been called the *poly-climax* concept maintain that the climax is not necessarily determined by climate alone but by the interaction of *all* the factors which influence plant growth and distribution. Within any climatic region there will be a number of climax communities (all of which may or may not be characterised by the same life-form dominants) corresponding to a number of habitats dependent on varying combinations of local climate, relief, parent material and biotic factors. Other ecologists, like A. G. Tansley, have defined the climax in terms of the principal factor limiting or determining the maximum development of vegetation, whether climate, soil, relief, biotic or anthropogenic activities. They, however, still regarded the climatic climax as the 'natural' or 'potential' natural vegetation on 'normal', freely drained, 'mesic' sites. Communities differing from the climatic climax because of extreme soil conditions—excessively dry or wet, extremely acid or alkaline for instance—or because of the continued operation of biotic activities can, if they have reached a stage of equilibruim with the major controlling factor, be regarded as *edaphic* or *biotic* (*anthropogenic*) climaxes respectively. There is no implication that they have the potential to progress eventually towards the climatic climax; in many cases there is no evidence that they can or will. It has been pointed out by man. Indeed, one of the main objections to the climatic climax would have been regarded as sub-climax are not only often widespread but are also persistent, particularly when they have for several thousand years been subject to disturbance or modification by man. Indeed, one of the main objections to the climatic climax concept must be the tendency to regard disturbance as 'unnatural' or 'abnormal', and to search for or to reconstruct a theoretical climax which may or may not have existed in the past and which, in any case, is unlikely in many areas ever to return.

More recently, other workers, including R. H. Wittaker, have suggested that there is no *absolute* type of climax determined by one or a few controlling factors. Any particular climax community will be composed of an assemblage or combination of species (a 'population' of plants and animals) the composition of which is determined by the sum total of all the interacting habitat factors. There will be, therefore, as many climax communities as there are possible combinations of environmental and biotic conditions. A *type* of climax community will, then, be composed of similar, though not necessarily absolutely identical, 'populations' associated with similar sites. The main criterion distinguishing a climax community, according to R. H. Whittaker, is that it should be relatively permanent and self-maintaining so long as the particular combination of environmental conditions does not alter.

The climax concept, however interpreted, implies a degree of stability, a dynamic equilibrium, in the sense that a balance is achieved between not only the plant community and its physical habitat, but all the organisms which are part of the same ecosystem. But the climax is by no means static. In the first place the members of any community will be continually renewing themselves, some rapidly, some slowly, as old individuals die and young ones replace them. It has been shown that many communities are subject to cyclic changes in composition unrelated, as far as is known, to external environmental changes, but due rather to the continuing effect of the community on its own habitat. For instance, it has been noted that in the presumed climax tropical forest, seedlings and saplings of the dominant trees are often scarce or absent. P. W. Richards suggests that although the climax forest persists, its composition may vary not only from place to place but during time, as one assemblage of dominant species is replaced by another, different group. A. S. Watt has noted a comparable feature in beechwoods in southern England. In certain circumstances the dominant 'climax' species, beech, may create light and soil conditions under its canopy unfavourable for the regeneration of its own seedlings. When mature beeches die they may be replaced by more tolerant trees such as birch or ash which, in time, provide more suitable conditions for the establishment and successful growth of beech seedlings; the latter will eventually suppress the intermediate stage. Hence the climax in such a case will be composed of alternating phases or stages rather than one persistent type of community. The 'equilibrium' of a climax community may be disrupted by the invasion by a more successful and aggressive species, as a result of migration, introduction or the slow process of evolution. Had the sycamore tree managed to reach Britain earlier and without the aid of man, the natural deciduous woodlands might well have been dominated by this tree or a mixture of it with oak, ash or beech. It has the potential to compete successfully with and even displace these species. The stability of a climax plant community can be upset by changes in the balance of animal populations which may result in the destruction of seedlings or the decimation of the dominant trees by pathogenic organisms. While such events have undoubtedly been increased by the activities of man, they are inherent in all communities. Furthermore, Eugene Odum points out 'it is not known if any community can be completely self-perpetuating and permanent even assuming no change in regional climate'.

Finally the concept of community equilibrium or stability can, in the long term, only be relative since environmental conditions are themselves subject to recurring changes, some of which may be sudden and catastrophic, some of which are gradual and continuous over a long period of time. One of the most variable of environmental factors is that of climate. The latter is the average condition from which the actual atmospheric conditions, or weather, fluctuate to a lesser or greater extent daily, seasonally or annually. Such short-term fluctuations may not affect mature long-lived perennial plants; they do however affect seed-production and germination, and the number of seedlings and annual plants (as well as the number of animals dependent upon them for food) present in any community from one year to another. But longer-term fluctuations involving secular climatic change towards wetter or drier, warmer or colder average conditions have been a constant, indeed a normal rather than abnormal, feature of the earth's environmental history, particularly in high latitudes. The climatic changes preceding, during and since the Pleistocene era were such as to initiate quite radical changes in vegetation and to cause the replacement of one type of 'climax' by another. The rapid recession of many present day glaciers already referred to is indicative of continuing change.

There must, however, inevitably be a time-lag between short and long-term fluctuations of climate and the resulting re-adjustment in the balance of community populations and structure. The effects of a particularly severe winter, or a cool or dry growing-season, on seedling mortality in temperate regions will not become manifest until the following season and may then persist for some time after the event. Detailed studies revealed that the recovery of the North American plains grassland vegetation from the effect of the severe and protracted droughts of the nineteen-thirties was much slower than the subsequent climatic change to more humid conditions. In the case of forests with long-lived dominant trees the time-lag must inevitably be even greater. Climatic change might effect fairly rapidly the regeneration of the climax trees but it would be some time before the mature trees died and new climax dominants replaced them. Because of this time-lag between climatic change and vegetation readjustment, a state of equilibrium may never be attained since the climate may well change before this is effected. Because of this it has been suggested that the climax community is a concept only, never existing in reality either because of the catastrophic initiation of fresh seres or because of an ever changing environment. S. A. Graham likens the climax to 'a phantom always moving ahead into the future and becoming visible for only relatively brief periods in small areas' and wonders whether the assumption that culmination in an ultimate climax is either the necessary outcome of, or must mark the end of, succession.

References

Allred, B. W. and Clements, E. S. (Eds.) 1949. *The dynamics of vegetation*: Selections from the writings of Frederick E. Clements. The H. W. Wilson Co., New York.

Ashby, M. 1961. *Introduction to plant ecology*. Macmillan. London.

Burnett, J. H. (Ed.) 1964. *The vegetation of Scotland*. Oliver and Boyd, Edinburgh.

Cain, S. A. 1939. The climax and its complexities. *Am., Midl. Nat.*, **21**: 146–181.

Clements, F. E. 1936. The nature and structure of the climax. *Ecol.*, **24** (1): 253–284.

Churchill, E. D. and Hanson, H. C. 1958. The concept of climax in Arctic and Alpine vegetation. *Bot. Rev.*, **24**: 127–191.

Cooper, W. S. 1926. The fundamentals of vegetational change. *Ecology*, **7** (4): 391–413.

Cooper, W. S. 1931. A third expedition to Glacier Bay, Alaska. *Ecology*, **12** (1): 61–95.

Crocker, R. L. and Major, J. 1955. Soil development in relation to vegetation and surface age at Glacier Bay, Alaska. *Ecol.* **43** (2): 427–448.

Dansereau, P. 1954. Climax vegetation and the regional shift of controls. *Ecology*, **35** (4): 575–579.

Ellison, L. 1960. Influence of grazing on plant succession of rangelands. *Bot. Rev.*, **26**: 1–78.

Fridriksson, S. 1969. Life arrives on Surtsey. *New Scient.*, **37** (590): 684–687.

Godwin, H. 1929. The sub-climax and deflected succession. *J. Ecol.* **17** (1): 144–47.

Gorham, E. 1954. An early view of the relation between plant distribution and environmental factors. *Ecology*, **35** (1): 97–98.

Gorham, E. 1955. Vegetation and the alignment of environmental factors. *Ecology*, **36** (3): 514–15.

Gourou, P. 1961. *The tropical world*, 3rd ed. Longmans, London.

Graham, S. A. Climax forests of the Upper Peninsula of Michigan. *Ecology*, **22**: 355–362.

Hanson, H. C. and Churchill, E. O. 1961. *The plant community*. Reinhold, New York.

Hewetson, C. F. 1955. A discussion on the 'climax' concept in relation to tropical rain and deciduous forest. *Emp. For. Rev.*, **35**: 274–291.

Jones, E. W. 1945. Structure and reproduction of the virgin forest of the north temperate zone. *New Phytol.*, **44** (2): 130–148.

Kitteridge, J. 1934. Evidence of the rate of forest succession on Star Island, Minnesota. *Ecology*, **15** (1): 24–35.

Odum, E. and Odum, H. T. 1959. *Fundamentals of ecology*, 2nd ed. W. B. Saunders Co., Philadelphia and London.

Olsen, J. S. 1958. Rates of succession and soil changes on southern Lake Michigan sand dunes. *Bot. Gaz.*, **119** (3): 125–170.

Oosting, H. J. 1956. *The study of plant communities*, 2nd ed. W. H. Freeman and Company, San Francisco.

Phillips, J. 1934. Succession, development, the climax, the complex organism: an analysis of concepts. *J. Ecol.*, Pt. I, **22** (2): 554–571; Pt. II **23** (1): 210–246; 488–508.

Richards, P. W. 1952. *The tropical rain forest*. Cambridge University Press.

Richards, P. W. 1958. The concept of the climax as applied to tropical vegetation. In *Study of tropical vegetation*. UNESCO, Paris.

Salisbury, Sir E. 1952. *Downs and dunes: their plant life and its environment*. Bell, London.

Selleck, G. W. 1960. The climax concept. *Bot. Rev.*, **26** (4): 537–545.

Tansley, A. G. 1935. The use and abuse of vegetational concepts and terms. *Ecology*, **16** (3): 284–307.

Tansley, A. G. 1949. *The British Islands and their vegetation.* Cambridge Univ. Press.

Watt, A. S. 1947. Pattern and process in the plant community. *J. Ecol.*, **35** (1 and 2): 1–22.

Whittaker, R. H. 1951. A criticism of the plant association and climatic climax concept. *NW. Sci.*, **25**: 17–31.

Whittaker, R. H. 1953. A consideration of climax theory: the climax as a population and pattern. *Ecol. Monogr.*, **23** (1): 41–78.

Woodwell, G. M. 1963. The ecological effects of radiation. *Scient. Am.*, June, **208** (6): 40–49.

10
The Marine Ecosystem

No greater or more fundamental contrast exists in the biosphere than between the environmental conditions and associated forms of life of the land and those of the sea. The oceans cover over two-thirds of the earth's surface and, habitable throughout the whole of their depths, they support a volume of living organisms—a biomass—at least five and probably as much as ten times as great as that on the land. The marine environment is, in fact, a major source of food, and of mineral resources, whose full potential man has yet to develop. However, in spite of the distinctive and particular nature of this environment and of the organisms which inhabit it, the whole cycle of life, the basic functioning of the marine ecosystem, is the same as that of terrestrial ecosystems. Requirements for life are the same; and organic production and distribution in the sea are determined by factors similar to those on land.

The sea, however, is in many respects a more favourable and certainly a more equable medium for life than is the land. The problems of desiccation are unknown, except in the inter-tidal zone—transitional between marine and land habitats—where organisms must be adapted to exist both in and out of sea water. Further, organisms adapted to live in the sea are immersed in a solution that contains all the nutrient elements necessary for the maintenance of life. The two most essential gases—oxygen and carbon dioxide—are readily soluble and available in sea-water. Relatively easily absorbed from the atmosphere, their replenishment and distribution from the surface downwards is effected by the mixing action of an ever-mobile medium, as well as by the biological processes of photosynthesis, respiration and decomposition. Of the two, carbon dioxide is the more readily soluble; it reacts with water to form carbonic acid (H_2CO_3) and can be easily fixed in carbonate ($-CO_3$) or bicarbonate ($-HCO_3$) compounds. The concentration of carbon dioxide in sea-water is about fifty times higher than in the atmosphere. That of oxygen, however, is less attaining a maximum of only 0·9 per cent. in the sea as compared with twenty per cent. in the atmosphere. It is also less evenly distributed than carbon dioxide. The oxygen content is higher in surface than deeper water, as a result on the one hand of direct contact with the atmosphere and, on the other of the concentration of photosynthetic activity in the upper illuminated layers. The latter contain the maximum amount possible under given temperature conditions, this being greater in cooler than in warmer waters. However, except in localised basins, such as those of some deep fjords or the Mediterranean and Black Seas, where deep, stagnant oxygen-deficient water is cut off from contact with adjoining water-bodies, continual oceanic circulation between bottom and surface waters ensures a replenishment of oxygen sufficient to maintain life even at the greatest depths.

In addition, dissolved in the sea are varying amounts of many of the minerals found in the earth's crust. This salt content, or *salinity,* is probably the most distinctive characteristic of the marine environment. Successful adaptation to life in a saline solution is one of the principal features which distinguishes the types of life found in the sea from those on land or in freshwater; and adaptation to varying degrees of salinity is an important factor determining the distribution of organisms within the sea itself. It has been estimated that in a cubic kilometre of sea water there are on average one hundred tonnes of dissolved mineral salts, and that if evaporated the total salt content of the oceans would be sufficient to cover all land surfaces to a depth of about one hundred metres. In relation to the immense volume of the oceans, however, this does not give an excessive concentration. The most saline conditions occur where temperatures, and hence evaporation, are greatest. Among the highest values recorded are those for the Red Sea whose average salinity is of the order of forty grams of salt per thousand grams of sea water. (40–41°/oo). Minimum salinity, on the other hand, is found where the in-flow of large volumes of fresh water from adjacent land areas or from melting ice, causes dilution. In the Baltic Sea, for instance, where a large volume of fresh river-water pours into a relatively small and partially enclosed basin, salinity during the summer months is barely 5°/oo. In the surface waters of the open seas, however, the range of salinity—from 37·5°/oo in tropical seas to 33°/oo in polar regions—is much less. Nevertheless, many marine organisms are very delicately adapted to an existence in a particular salt-concentration; they are, hence, extremely sensitive to quite slight variations of salinity which may, as a result, rigidly limit their distribution to particular areas or depths in the sea.

Some forty-five elements are known to contribute to the total salt content of the sea; many more, as yet undetected, may be present. Some occur in larger quantities than others (see Table 24). The major constituents are present in remarkably constant

Element	Concentration in mg/kg (parts per million)	Element	Concentration in mg/tonne (parts per thousand million)
Chlorine	18 980	Arsenic	10–20
Sodium	10 561	Iron	0–20
Magnesium	1 272	Manganese	0–10
Sulphur	884	Copper	0–10
Calcium	400	Zinc	5
Potassium	380	Lead	4
Bromine	65	Selenium	4
Carbon	28	Caesium	2
Strontium	13	Uranium	1·5
Boron	4·6	Molybdenum	0·5
Silicon	0–4·0	Thorium	0·5
Fluorine	1·4	Cerium	0·4
Nitrogen	0–0·7	Silver	0·3
Aluminium	0·5	Vanadium	0·3
Rubidium	0·2	Lanthanum	0·3
Lithium	0·1	Yttrium	0·3
Phosphorus	0–0·1	Nickel	0·1
Barium	0·05	Mercury	0·03
		Gold	0·006
		Radium	0·0000002

Table 24 Concentration of elements (in order of abundance) in sea-water of salinity 34·33 parts per thousand. (From: Cox, Roland A. 1959)

proportions, irrespective of the degree of salinity. Of these the most abundant are sodium (30·9 per cent.) and chlorine (55·3 per cent.), the main determinants of the 'saltiness' of sea water; together with smaller quantities of magnesium, sulphur, calcium, potassium and bromine they account for a little over ninety-nine per cent. of the total salt content. The remainder occurring in minute, and often variable, quantities contribute less than one per cent. Among the latter, however, are those nutritive elements, such as nitrogen and phosphorus, essential for plant growth and upon whose availability the potential fertility of the marine environment depends. Sea water is not just a saline medium; it is a nutritive solution whose composition is remarkably similar to the body fluids of the organisms which inhabit it, and indeed is not dissimilar to those of all living organisms (see Table 25).

Physically as well as chemically the sea provides a particularly suitable medium for life, and one which is in many respects more uniform than that provided by the land. The specific heat of water is higher than that of solids; the former absorbs heat and

	Na	K	Ca	Mg	Cl	SO_4
Sea water	100	3·6	3·8	12·1	18·0	25·2
King Crab	100	5·6	4·1	11·2	18·7	13·4
Lobster	100	3·7	4·8	1·7	17·1	6·7
Cod	100	9·5	3·9	1·4	149·7	—
Pollock	100	4·3	3·10	1·4	137·8	—
Frog	100	11·8	3·17	·79	135·6	—
Dog	100	6·6	2·8	·76	139·5	—

Table 25 Comparison between sea-water and body fluids of various organisms. (From: Sverdup, Johnson and Fleming, 1942). Nutrients in each organism expressed as a percentage of its Na content.

loses it more slowly. Spatial and seasonal variations of temperature are, as a result, much less in the ocean than on land masses. The difference between the surface-temperature of the warmest (about 32°C in the Persian Gulf) and the coldest (about −2°C in polar areas) seas is of the order of 34°C, as compared with a range of 87–92°C on land. Also, the annual range of temperature over the oceans rarely exceeds 10°C, and is usually much less. Around the coasts of Britain, for instance, surface-water temperatures vary from about 8°C in winter to 17°C in summer. Most of the heat derived from solar radiation is rapidly absorbed in the surface layers of the sea and so, with increasing depth, seasonal and latitudinal variations experienced at the surface, decrease. Below about a hundred fathoms there is relatively little seasonal fluctuation and comparably low temperatures persist over extensive areas. On the floors of the deep oceans temperatures vary only a few degrees above or below freezing point.

Life in the sea does not have to contend with such extreme or rapid variations of temperature as on land. The uniformity of environmental conditions over extensive areas is facilitated by the continuous and unbroken extent of the water-area, and by the vertical and horizontal movements of ocean currents. Vertical movements are engendered by variations in density resulting from differences in temperature and salinity; colder or more saline water being denser than that which is warmer or fresher will tend to sink below the latter. Also, under the influence of the rotation of the earth and the friction of prevailing winds, surface waters are kept in constant circulation. Water heated in tropical regions

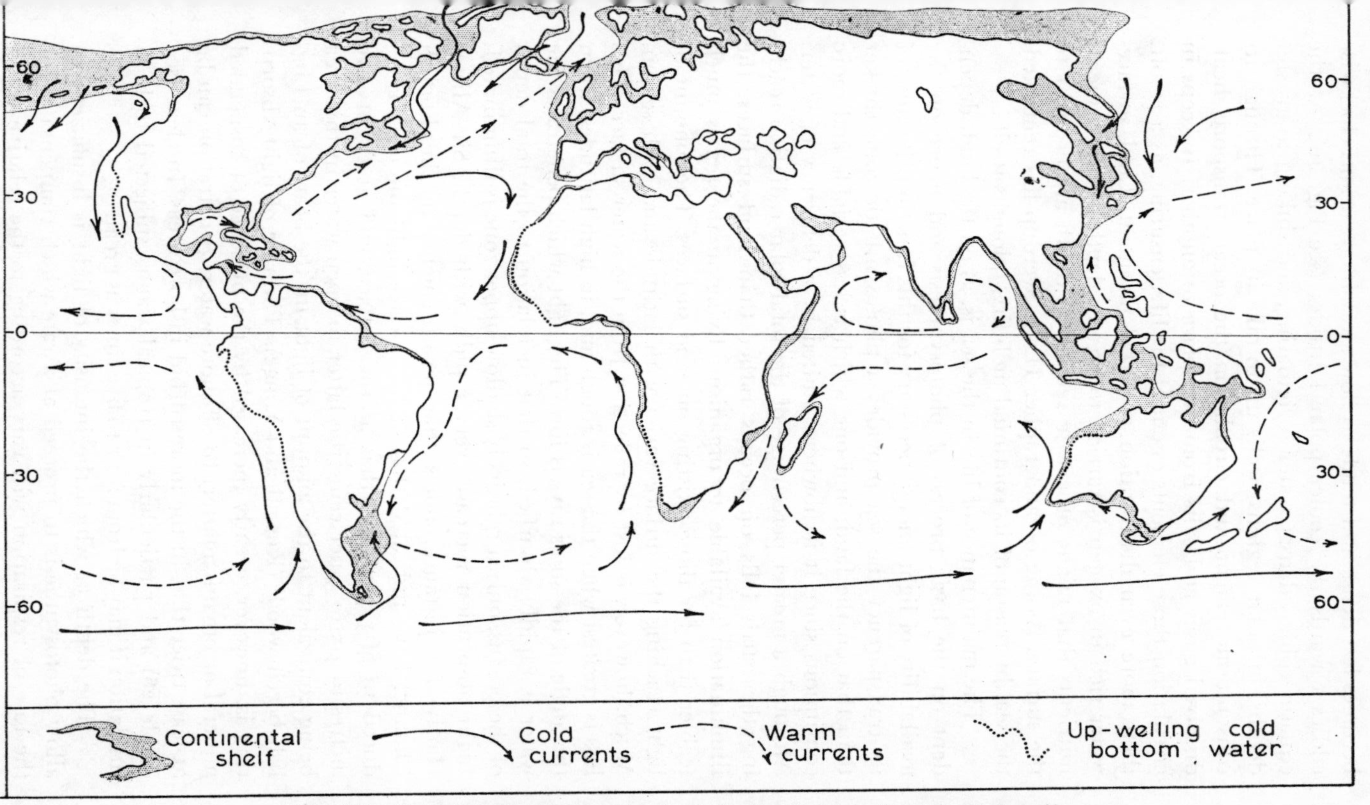

Fig. 40. Major features of oceanic circulation and extent of continental shelf.

becomes less dense and tends to flow away in great anti-clockwise eddies towards the cooler polar latitudes (see Fig. 40). At the equator colder, denser water (originating, and sinking to greater depths, in polar regions) wells up to take its place. Through its constant movement and circulation the ocean transports heat, dissolved gases and salts from one place to another. It keeps in circulation those elements essential for life, ensuring, as a result, their more even distribution and replenishment. Cold surface-water rich in oxygen is carried to greater depths, while mineral nutrients that might otherwise remain at depth are brought to the surface. As the oceanographer H. U. Sverdup has remarked, the sea, by reason of its continual mobility, 'ploughs itself'.

The maintenance of life in the sea is, as on the land, dependent on the basic process of photosynthesis and hence on the availability of light energy necessary for this function. Because of its transparency the sea provides a photosynthetic zone thicker than that on the land, and one which is less variable and more continuous since it is nowhere limited by a deficiency of water. Although a greater percentage of the total solar radiation reaching the earth falls on oceans rather than land surfaces, the illumination available to organisms living in the sea is much feebler than for those existing on land surfaces. The amount of light reaching the surface varies with latitude and with season. Much, however, is lost by reflection from the water surface. This loss is greatest when the sea is rough and, in high latitudes, when the angle of the sun's rays is low. The light which penetrates the water is rapidly absorbed, so that in relation to the total depth of the oceans, only a relatively shallow upper zone is illuminated. Light absorption increases very rapidly with depth. Sir Alister Hardy, for instance, notes that measurements in the English Channel have indicated that at five metres light intensity is reduced to fifty per cent. that at the surface, and at twenty-five metres to 1·5 to 3 per cent; the latter in open water in this area being equivalent to the amount of light in 'the heart of an English beech wood' (Russell and Yonge). The rate of light absorption is, however, greatly increased by the presence of suspended mineral or organic matter. In off-shore waters penetration can be at least twice that in the more turbid inshore areas. In clear (see Table 26) and particularly in tropical waters subjected to high insolation it may be from four to five times as great.

The depth at which the intensity of light is insufficient to allow photosynthesis to proceed at a rate which compensates for the rate of respiration will vary according to the relative transparency of the ocean, and with diurnal and seasonal variations in the altitude of the sun. In inshore waters it may be as little as ten to fifteen metres, in the open ocean as much as a hundred metres. The maximum depth of what is referred to as the

% light absorbed	Depth in inshore waters	Depth in water 16 kilometres off-shore
50	2 metres	4·5 metres
75	4·5 ,,	9·0 ,,
90	8·0 ,,	17·0 ,,

Table 26 Light absorption with depth in English Channel. (From: Hardy, Alister C. Sir. 1956, p.54)

euphotic zone (within which there is sufficient light for photosynthesis) is probably about a hundred metres, which corresponds to the *average* depth of the continental shelf. The maximum limit of penetration (about two hundred metres) coincides with the average depth of the outer edge of this shallow submarine bench (see Fig. 40, p. 221). But this deeper *disphotic zone,*between a hundred and two hundred metres, is so dimly lit that the possibility of photosynthesis taking place within it must be discounted. Below this level the seas are dark and sunless. Hence it is only around the continental margins on the gently shelving, shallow floor of the continental shelf that light can penetrate from the surface to the bottom of the sea.

As on land the basic process of photosynthesis is dependent on chlorophyll-containing plant life, the distribution of which is limited to the euphotic zone. In comparison with land plants, however, those of the sea are much less diverse. The latter are dominated practically entirely by that most primitive, least evolved and specialised group—the algae. Other than these only about thirty species of higher plants (all of which are Angiosperms) are represented in the marine flora and their contribution to the total mass of marine vegetation is relatively insignificant. The most obvious and clearly visible types of marine algae are the 'seaweeds'. But apart from a few rootless, drifting species, such as those which comprise the marooned and floating vegetation of the Sargasso Sea, they are confined to a narrow and localised zone around the coasts. Requiring anchorage, they can exist only where water is shallow enough to allow sufficient light to penetrate to the sea floor. The maximum possible depth to which bottom-rooted plants are thought to be able to grow is no more, and probably less, than fifty metres. About ninety-nine per cent.

of the marine vegetation is composed of much less conspicuous types of plants—of microscopic (average diameter twenty-five microns) one-celled organisms containing chlorophyll which float or drift, suspended in the upper layers of the sea, a little below the surface. They can occur in such quantities—estimates have indicated densities in some areas of the order of 36 million per cubic metre—as to impart a distinctive colour to the sea. They form part of the life which floats in the upper layers of the ocean and is known as the *plankton*.

The most important and abundant members of the *plant* or *phyto-plankton* belong to the large family of the *diatoms*. Unicellular organisms, distinguished by delicate and intricate 'skeletons' made of silica, they are particularly prolific in the waters of cold and cool-temperate latitudes. Second only in abundance to the diatoms are the *dinoflagellates*, a group in which the distinction between plant and animal becomes blurred. Some species can undertake photosynthesis, others gain nutrition by ingesting organic matter, while some obtain their food by both processes. Together with the even more minute *coccoliths*, they are more characteristic of the warmer waters of low latitudes. In size and form the phyto-plankton is particularly well adapted to its particular environment. Not only can these plants remain suspended in the euphotic zone but their minute size provides the maximum area possible in relation to total volume, for the absorption of gases and nutrients from the surrounding water. Indeed the efficiency of such a form for photosynthesis may well explain the overwhelming dominance of this relatively primitive type of plant life in the sea. In addition, the life-cycles of the phyto-plankton are short and uncomplicated. Given favourable conditions for growth they are capable of rapid proliferation by repeated simple cell-division, and can as a result produce in a short space of time a tremendous volume of vegetable matter. The whole food economy of the oceans is based on the phyto-plankton—often popularly described as the 'grassland' or 'pasture' of the sea. The phyto-plankton are responsible for all but a very small, indeed insignificant, fraction of the primary biological productivity of the marine ecosystem.

It has been estimated that the mean annual primary productivity of the seas is twenty-five to fifty per cent. less than that on land (see Fig. 41.) Two factors contribute to this difference: firstly the relative paucity of such nutrients as nitrogen, phosphorus and potassium available for plant growth in the euphotic zone; and secondly, the higher rate of respiration and consequently lower

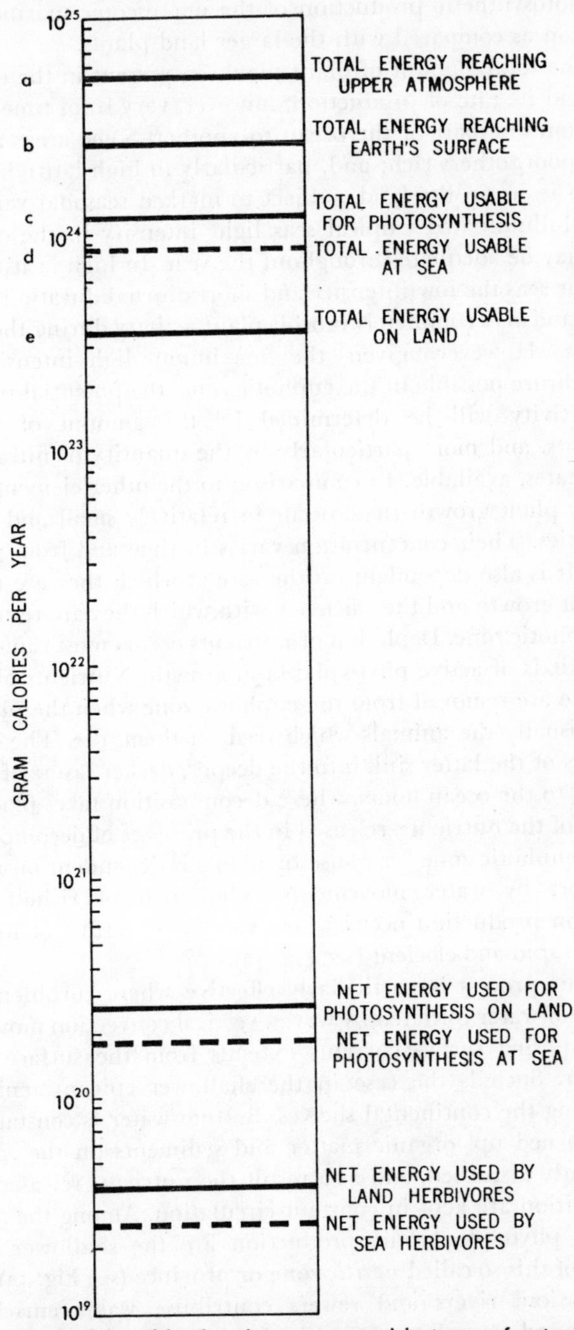

Fig. 41 Productivity of land and sea compared in terms of net amount of energy assimilated by green plants (From : Isaacs, John, D. 1969)

net photosynthetic production of the microscopic marine phyto-plankton as compared with the larger land plants.

The total amount of plant plankton present in the euphotic zone and its rate of production, however, vary from time to time and from one part of the ocean to another. Some areas are relatively poor, others rich; and, particularly in high latitudes, plant growth is, as on the land, subject to marked seasonal variations. In well-illuminated tropical seas light intensity in the euphotic zone may be adequate throughout the year. In high latitudes and in polar seas the low intensity and short diurnal duration of light checks and may completely inhibit plant activity during the winter months. However, given the maximum light-intensity and temperature possible in the euphotic zone, the potential plankton productivity will be determined by the amount of mineral nutrients, and more particularly by the quantity of nitrates and phosphates, available. In comparison to the other elements essential for plant growth these occur in relatively small and limited quantities. Their concentration varies in time and from place to place. It is also dependent on the rate at which they are used up in plant growth and the efficiency with which they are returned to the euphotic zone. Depletion of nutrients occurs most rapidly during periods of active phyto-plankton growth. Nutrients absorbed by these are removed from the euphotic zone when the plants, or more usually the animals which feed on them, die. The organic remains of the latter sink into the deeper, darker layers of the sea and on to the ocean floors, where decomposition takes place. The return of the nutrients, released in the processes of decomposition, to the euphotic zone for re-use by plants is dependent on upward transport by water movements. The areas of richest phyto-plankton production occur where this cyclic return of nutrients is most rapid and efficient (see Fig. 42).

This process is particularly effective where turbulence and mixing of water initiated by waves, vertical convection movements and opposing surface-currents extends from the surface to the sea-floor. Such is the case in the shallower epicontinental seas over-lying the continental shelves. Bottom water is constantly being churned up, organic matter and sediments on the sea floor constantly disturbed, and as a result the nutrients released in decomposition are kept in constant circulation. Among the areas of highest phyto-planktonic production are the shallower coastal waters of this so-called *neritic* zone or province (see Fig. 43). Here, in addition, rivers and sewers contribute water enriched in nitrates and phosphates from the land. Despite the more limited

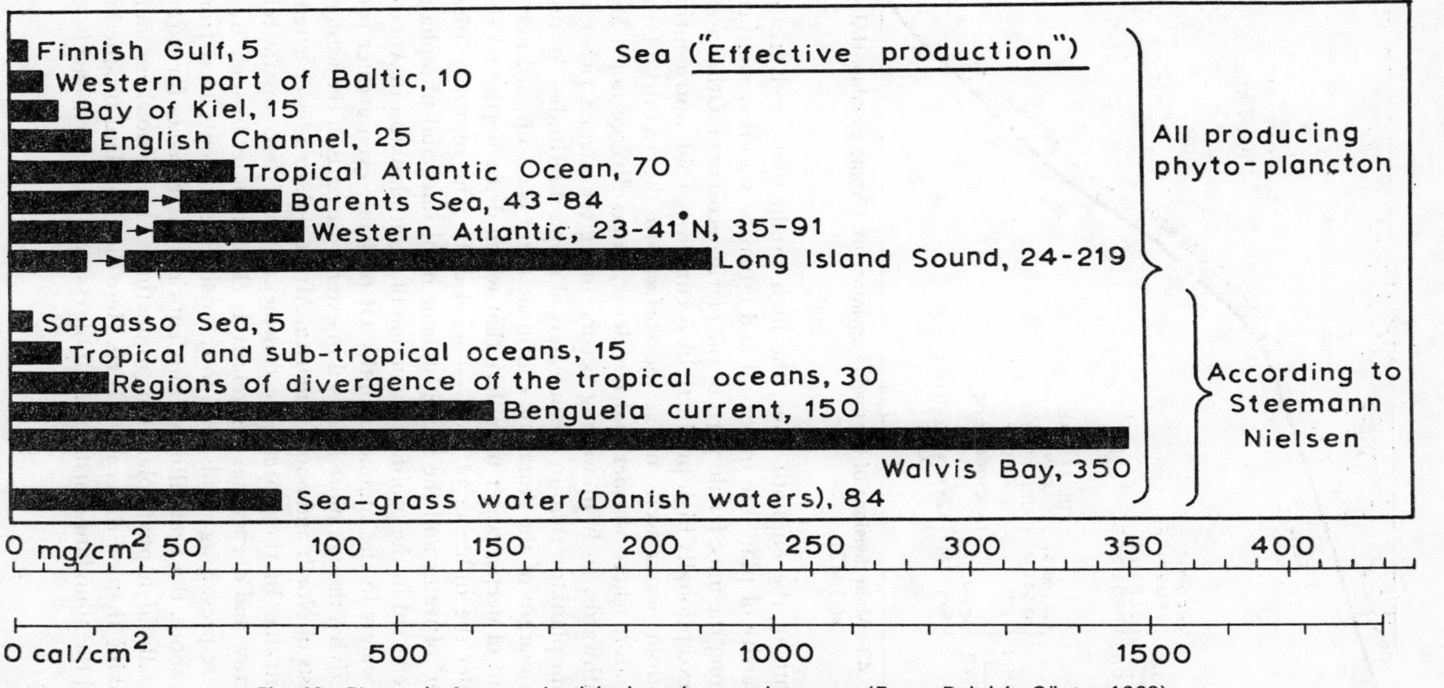

Fig. 42 Phyto-plankton productivity in various marine areas. (From: Deitrich, Günter, 1963)

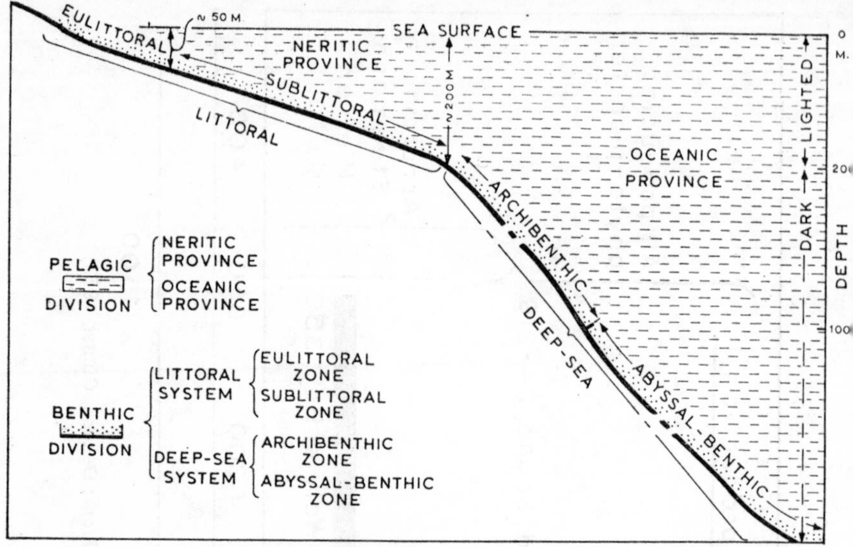

Fig. 43 Main divisions of the marine environment. (From: Sverdup, H.U. *et al.* 1942)

depth of the euphotic zone, the high productivity and greater diversity of plant-life in coastal and estuarine waters make them among the most fertile parts of the marine ecosystem. Other areas of exceptionally high productivity occur where cold, nutrient-rich 'bottom' water wells up at the surface. In the vicinity of the equator, where warmer and less dense waters diverge north and southwards, such up-welling occurs and gives a zone richer in plant plankton than in ocean waters of tropical latitudes. Around the margin of the Antarctic continent there is an upward movement of deeper water to replace that which drifts north and east under the influence of the west wind drift. The resulting 'turn-over' of water and the replenishment in nitrates and phosphates is reflected in the prolific production of plant-plankton in Antarctic waters in the summer months. Of probably even greater productivity, though more localised in extent, are parts of the western coasts of North and South America, Africa and, to a lesser extent, Australia. In these areas the continental shelf is comparatively narrow and deep water lies relatively close inshore. In such areas where prevailing winds tend to carry surface currents away from the shore, the up-welling of nutrient-rich water makes for exceptionally high phyto-plankton production. This is nowhere more striking than off the coast of Peru. Here the rich plant production and the abundant marine animal life it supports are reflected in an

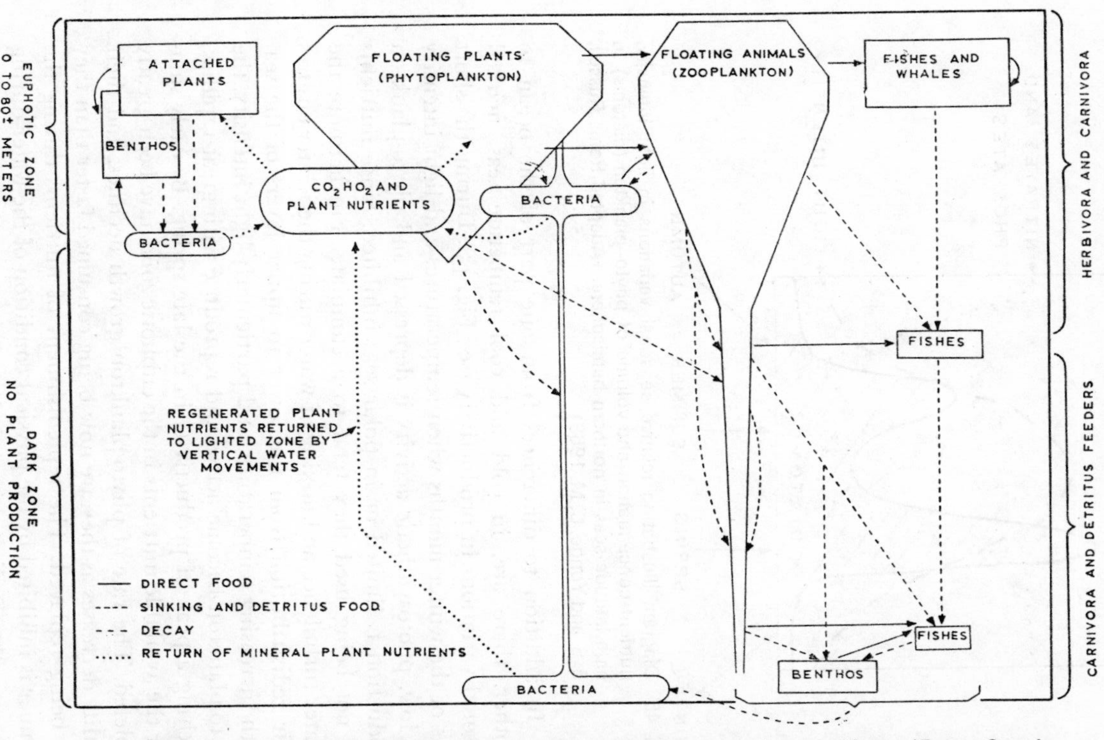

Fig. 44 The main features of the feeding inter relationships between marine organisms. (From: Sverdup H. U. *et. al.*, 1942)

exceptionally high density of birds. Their droppings have built up phosphatic 'guano' deposits as much as thirty metres thick along the shore and on the off-shore islands of this region.

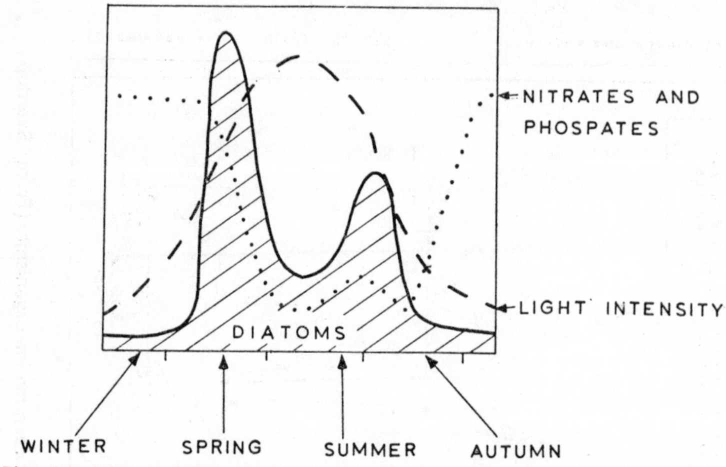

Fig. 45 Diagram illustrating relative seasonal variations in light intensity, nutrient-concentration and volume of phyto–plankton (diatoms) in high latitude seas in northern hemisphere. (Adapted from: Russell, E. S. and Yonge, C. M. 1963)

In addition to differences from one part of the ocean to another, there are, in cold and cool temperate seas, marked seasonal variations in productivity (see Fig. 45). During the short days of the winter months when temperatures and light intensity are low, photosynthetic activity is depressed and phyto-plankton production is limited or in polar seas inhibited. Since nutrients are not being used they tend to accumulate. Furthermore the greater turbulence and mixing of water during the winter ensures their redistribution from the lower to upper layers of the sea. With increasing temperature, and particularly light intensity, the phyto-plankton become active and reproduce with great rapidity, reaching a peak of productivity in the late spring. By early summer the available nutrients in the euphotic zone have been greatly depleted. The rate of phyto-plankton growth declines, and their quality decreases as they are now being consumed faster than they are being replaced. The replenishment of nutrients during the summer is inhibited by the physical condition of the water during these months. During this period surface water is heated and its decrease in density tends to inhibit vertical movement. Since water is a poor conductor of heat the warmer surface water becomes, in the absence of turbulence, sharply separated from the underlying cooler layers. At the junction between the warmer and cooler water there may be a marked and rapid change of temperature—of the order of 4°C. This junction, between the warmer surface and cooler deeper water is called the *discontinuity layer* or *thermocline*; around the coasts of Britain this may occur, on average (dependent on seasonal conditions), at a depth of about fifteen metres. The establishment of the thermocline, together with calmer surface water conditions in summer, effectively prevents or slows down interchange of surface- and bottom-water, and hence the return of nutrients to the euphotic zone. Not until the autumn, when surface waters are cooled, become denser and tend to sink while equinoxial gales promote further disturbance and mixing, is the thermocline disrupted and temperatures and nutrients are again more evenly distributed. In temperate latitudes of the northern hemisphere this is accompanied by a renewed autumn outburst of planktonic activity. It is, however, of lesser intensity and duration than that in spring because the period of favourable light and temperature is relatively short-lived. In high latitudes of the southern hemisphere, however, only one peak period of plant production occurs, coincident with the favourable light and temperature conditions of late spring and early summer. The high frequency of gales and the continual disturbance of the surface waters inhibits the establishment of the summer temperature-discontinuity or thermocline characteristic of northern waters. In tropical latitudes, on the other hand, exposed to continually high insolation the thermocline tends to be a more permanent feature. Light and temperature conditions are such that plant growth can continue unchecked throughout the year. But the level of nutrients is low and their replenishment in the euphotic zone slow. Hence, although temperature conditions are higher and the rate of reproduction and decomposition is rapid, the average annual productivity is low. The open oceans of the tropics (particularly in the becalmed Doldrums where the admixture of fresh deep water is negligible) are comparable in this respect to the desert areas of the terrestrial ecosystem (see Fig. 46).

The phyto-plankton provides all but a very small proportion of the primary food on which the animal life of the sea and its relative abundance ultimately depend. But the nature of this plant life, particularly its minute size and rapidity of reproduction, gives rise to plant-animal food relationships somewhat different from those on the land. Firstly because of the microscopic proportions of the individual members of the phyto-plankton, their direct use by large 'grazing-animals' is restricted. Large

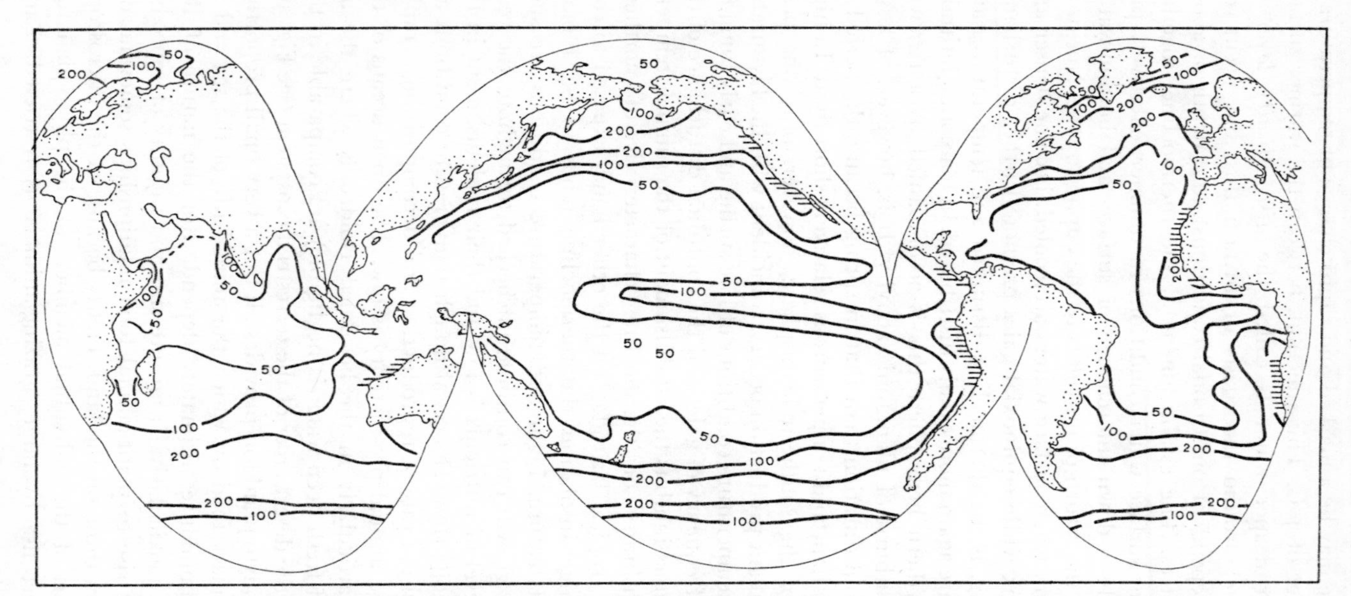

Fig. 46 Estimation of the production of organic matter in the oceans (gms/m³/annum. Horizontal shading indicates areas characterised by up-swelling of bottom waters. (From : Sverdup, H. U. 1952)

herbivores such as those which feed directly on the bulkier land plants do not exist in the sea. The majority of the phyto-planktonic 'grazers' are among the smallest of marine animals. They are of a size capable of utilising the available source of plant food most efficiently and thereby converting it into proportions more manageable by larger carnivores. For this reason food-chains in the ocean tend to be longer than on the land, and the standing-crop (or biomass) of larger marine flesh-producers represents a smaller proportion of the primary plant productivity than do land animals. Secondly most, if not all, of the phyto-plankton is consumed by the grazing animals of the sea. Although there is still some doubt on this matter it would appear that probably only a small proportion of the plant plankton exists in a dead or decaying 'detrital' form. This is in striking contrast to the situation in the terrestrial ecosystem, where a relatively small proportion of the living plant material is consumed by animals and a much greater proportion passes as dead and decaying organic matter on and into the surface layers of the soil, there to support a large number of saprophytic organisms which are responsible for its decomposition. To put it another way, the grazing-link in the food-chains in the sea plays a more important part than it does on the land, where the saprophytic or detrital link assumes a greater significance. Also on the land the total plant biomass (the standing crop of plants) tends to be much greater than that of the animal biomass. In the sea, on the other hand, the situation is reversed, the total quantity of animal matter actually exceeding that of the vegetation. Consumption, in fact, tends to keep pace with production: the life-span of the minute phyto-plankton, which may under favourable conditions effect cell-division every ten to thirty-six hours, is very much shorter than that of most of the animals either directly or indirectly dependent upon them. Not only is the total volume of animal tissue greater than that of plants, but they contribute the bulk of the decaying, detrital organic matter present in sea water.

Those animals (the main herbivorous grazers of the sea) which consume the greatest bulk of the phyto-plankton, belong to the animal or *zoo-plankton*. Like the phyto-plankton this comprises a group of microscopic or small animals which also form part of the drifting population of the sea. Members of the zoo-plankton are, however, not only generally larger in size but also much more diverse in form than those of the phyto-plankton. The number of different species which make up the zoo-plankton are legion; they range from minute one-celled organisms to such larger and more readily visible species as the jelly-fish. Some depend entirely on the phyto-plankton for food, others prey on smaller members of the group itself. Some, the permanent plankton, spend the whole of their lives as drifters; others, the temporary plankton, include the young or larval stages of larger bottom-living (*benthos*) or free-swimming (*nekton*) animals. The temporary zoo-plankton are most abundant in shallower inshore waters where the density of the benthos is greatest and where, in addition, the spawning grounds of many species of fish are located. Among the most important of the myriad of species which comprise the zoo-plankton are such small crustaceans as the *copepods* (of which the genus *Calanus* is one of the major sources of fish-food in cooler, northern waters); and the shrimp-like euphausiids or 'krill' which provide the main diet of the Antarctic whalebone whales. The dominance of a small number of species in cooler waters as compared with a much greater diversity of species in warmer and tropical sea is a feature which characterises not only the zoo-plankton but all other forms of marine life.

While the distribution and productivity of the zoo-plankton are dependent on the availability of plant food, they are not, as in the latter instance, directly dependent on light. According to F. S. Russell and C. M. Yonge, there is apparently 'no depth to which some planktonic animals do not exist'. The different species adapted to varying conditions of temperature and salinity are characteristic of particular types and depths of water. Many undergo marked vertical mirgations upwards towards the surface during hours of darkness, and downwards to deeper levels with increasing light-intensity during the day. Few exist in the brightly lit euphotic zone. Many of the deeper-living zoo-plankton must therefore derive their food from material, much of it probably in detrital form, which sinks down from the upper levels of the sea. Although the productivity and amount of zoo-plankton decrease with depth, there is apparently sufficient either in living or decaying form to support life even in the deepest zones of the ocean.

Apart from the zoo-plankton, the only other direct plant-feeders are some of the bottom-living fauna of shallow seas which have feeding mechanisms capable of trapping small particles of organic matter. In areas where the euphotic zone extends from the surface to the sea floor, as over considerable areas of the continental shelf, the phyto-plankton are readily accessible to these animals. It has, for instance, been suggested that in the English Channel about half the daily production of phyto-plankton may be consumed by bottom-living animals, the other half by the zoo-plankton. Outwith these shallow zones, however, the zoo-plankton forms the major link in the marine food-chains between the primary plant production and other forms of marine life. It is the principal food of such surface-water pelagic fish as herring, mackerel and sardines in cooler, and tunny and basking shark in warmer, waters. It is also the main diet of the largest of the marine animals, the whalebone whale, and its production constitutes the simplest, shortest and most efficient food-chain in the sea. Finally, the zoo-plankton also provides an important source of food for many of the bottom-living invertebrates as well. The latter are, in turn, preyed upon by such demersal fish as the cod, plaice and haddock, which live and obtain their food as carnivores or scavengers on or near the sea floor. But the benthos also includes among its bewildering variety of invertebrate life many species that are carnivorous predators or scavengers, which may prey on or compete with demersal fish for food; while among both the demersal and pelagic fish are those which prey on each other. The animals which inhabit the various parts of the marine environment are, in fact, interrelated by an intricate series of food links the main outlines of which are summarised in Fig. 44 (see p. 229). Their relative abundance is dependent, in the first instance, on that of the primary plant-food supplied by the phyto-plankton. The density and productivity of marine animals closely parallels that of the phyto-plankton being greater in shallow, inshore seas than in the open ocean, and in the upper layers of the sea than at greater depths.

As has already been stressed, the total biomass of the sea far exceeds that of the land. In comparison with the latter it is a resource whose full potential as a source of food for man has yet to be fully developed. At present only some one per cent. of what man eats comes from the sea. All but a negligible amount of this comes primarily from the *end* of the marine food chains in the form of animal food, the fish and so-called 'shell fish'. These are secondary or tertiary consumers and as such they represent a relatively small proportion of the primary food produced in photosynthesis by marine plants. Of the latter, only the seaweeds provide a direct but infinitesimal amount of human food. It is likely, as far as the foreseeable future is concerned, that the seas will remain a source of animal protein rather than plant-food for man. The phyto-planktonic biomass existing at any one time is less than that of the animals. In addition, the minute size of its individual members would make its direct harvesting from the sea extremely difficult and expensive. One estimate, made in a particularly fertile area off the coast of north-east America, showed that something over seven thousand cubic metres of water would need to be filtered in order to supply one man's daily food ration of three thousand calories. This *might* in the future become feasible with the installation of barrages and stations designed to exploit the power potential of tidal flow.

In his exploitation of marine resources for food man has focussed on those animals whose size and concentration of numbers in particular areas make for greatest ease of catching, and which because of custom or prejudice he has tended to favour. In fact, man has been fairly selective of the types of marine life that he has been prepared to eat. This tendency is, however, more pronounced in the cold and cool temperate than in warmer tropical seas which are characterised by a greater diversity of species. In the former the number of species is fewer but they more frequently occur concentrated in such numbers as to make the catching of one particular type more economic. The bulk of the commercial fish-catch of the world is taken from northern waters where it is composed of a relatively small number of species for which there is the greatest demand.

Fish constitute not only the one important source of protein food that is still hunted but also the only important food-animal that is not domesticated! While the techniques for catching fish have become increasingly efficient, the exploitation of this marine resource is still, with a few exceptions, an extractive 'robber' process. Under natural conditions, the total stock or biomass of a particular species will depend on the amount of food available, the relative balance from year to year between reproduction and growth and mortality from natural causes. When, therefore fish are extracted from the biomass at a rate greater than can be replaced by reproduction and growth, it will obviously decline. Over-fishing would result which, if continued, could lead not only to the depletion but to the eventual elimination of a particular stock. The symptoms of over-fishing reveal themselves when, although the intensity and efficiency of fishing remain constant, there is (1) a diminution in the average size of fish caught; (2) an ever increasing proportion of the weight of the catch composed of smaller and younger fish than of larger and more mature individuals; (3) a decline in the actual yield in terms of the weight of fish caught. Over-fishing and the depletion of fish stocks have already affected those areas and species which have been subjected to the greatest intensity of commercial fishing. Among the species which have suffered most severely are the whales. Their reproduction rate is slow; they may take between two to three years to reach maturity and thereafter only produce one calf every two years. The great demand for whale products during the course of the late-eighteenth and nineteenth centuries resulted in the virtual extermination of the 'Right' Whales of Arctic waters. Among the fishes the most susceptible to over-fishing have been the demersal species such as the plaice, haddock and cod. Firstly, the demersal fish are less abundant than the pelagic, plankton-feeders, secondly the use of trawl nets dragged along the sea-floor has greatly increased mortality rates among the young stock of demersal fish. Some are swept out to sea and fall prey to other animals, many more are killed by the rough handling to which they are subjected in the trawl nets. Demersal fish have been particularly severely depleted in the North Sea, probably the most intensively exploited of all the commercial fishing grounds.

The danger of over-fishing, together with the problem of maintaining and, if possible, increasing food production from the sea are matters of increasing urgency in view of the ever-increasing world population. The aim must be to exploit the potential stock of marine food in such a way as to ensure the maximum sustained yield of those products which are of most use to man. Up to a certain level, the extraction of fish can help to increase productivity; the removal of larger, older fish whose growth-rate is negligible makes more food available for the more rapid growth of the younger individuals of the species. The optimum intensity of fishing is that which promotes the maximum yield in terms of weight caught, without depleting the weight of the standing stock, i.e. the breeding stock. It is that which maintains the maximum rate of production and growth of fish stocks. Efforts as have been made, through the agency of International Fishing Councils, to impose closed seasons, to regulate the mesh-size of nets and to restrict the size of fish landed are among the first steps that have been taken towards the rational exploitation of marine resources. A further possibility, which has been the subject of much research, is that of re-stocking depleted fishing grounds with young fry reared in large quantities in special hatcheries. Attempts to re-stock the sea in this way—although initiated at the end of the nineteenth century in both Europe and America—have only been made on a limited scale and with varying degrees of success. One of the most outstanding achievements was the establishment of a successful shad fishing industry off the Californian Coast as a result of the transference to this area of fry from hatcheries established on the east coast of North America. The re-stocking of sea-fisheries will undoubtedly increase in importance as the problems of ensuring the survival of fry of a particular species from excessive loss when transferred from hatcheries to open water are overcome, and when the necessity for the measure becomes even more urgent.

The problems of conserving the stocks of certain fish, and other forms of marine life, are immediate. Those of increasing man's food resources are now of more than merely academic interest. The actual amount of food—in the form of the larger fishes and crustaceans—most valuable to man and most easily obtained from the sea by him is, as has already been noted, limited. It represents a relatively small proportion of the total marine organic resource. To increase productivity would require an increase in fish food-supplies in the sea. 'Fish-farming', in fresh-water ponds and lakes, is an ancient tradition which dates back at least two thousand years in China and was a common practice in Medieval Europe. Today, particularly in southern China and South-East Asia it still provides a valuable means of supplementing the meagre protein diet in these highly-populated agricultural areas. In fish ponds a high production of selected types of fish can be maintained by liberal supplies of fish food or by the application of organic or inorganic fertilizers to promote planktonic growth. At the end of the Second World War experiments conducted in Loch Fyne on the west coast of Scotland showed that the addition of nitrates and phosphates resulted in a greatly increased production of plankton and, consequently, of those organisms directly or indirectly dependent upon it for food. But while there was an increase in growth of fish, growth of seaweeds and bottom-living scavengers and predators of little value to man was in fact proportionately much greater. In any case, the possibility of increasing marine productivity in the open sea by the direct addition of 'mineral nutrients' would hardly be feasible or economic in view of the volume and mobility of the great mass of water involved.

It has, however, been pointed out by marine biologists that the potential *fish food* in the sea is considerable. A large proportion is consumed by animals, other than fish, which are of little use to man as food. The greatest competition for this food exists in the shallow seas overlying the continental shelf. In such areas where the productivity of both the phyto- and zoo-plankton is exceptionally high, only a small percentage (somewhere between one and two per cent at most) is consumed by fish. The bulk, in fact, sustains a large volume of invertebrate animals. It has been calculated that if only a quarter of such 'pests' as sea-urchins, star-fish and crabs, which compete with demersal fish for food were eliminated, at least ten times the existing weight of fish could be supported. Considering this situation, Sir Alister Hardy, invokes a Wellsian vision of the perhaps not too distant future when man might 'harrow' and 'weed' the floor of the continental shelf and thereby increase fish production. He speculates even further on the possibility of man pursuing some form of sub-marine 'ranching' or 'fish-herding' whereby he will exert a degree of control and management of his marine stocks comparable to that which he already exercises over his domestic land animals. But the vastness of the sea, the extreme mobility and the relative invisibility of its life together with the particular nature of its 'pastures' are problems peculiar to the rational exploitation of the marine ecosystem which man has yet to solve. The ultimate solution and the possible domestication of marine fish would further require a degree of international co-operation at present difficult to envisage.

References

BATES, M. 1963. *The forest and the sea: a look at the economy, nature and ecology of man.* Random House, New York.

BOWERS, A. B. 1960. Farming marine fish. *K.C.S. Sci., J.*, June.

CARRINGTON, A. 1960. *A biography of the sea.* Chatto & Windus, London.

CARSON, A. 1951. *The sea around us.* Staples Press, London.

CLARKE, G. L. 1946. Dynamics of production in a marine area. *Ecol. Monogr.* **16** (4): 321–335.

COX, R. A. 1959. The chemistry of sea water. *New Scient.*, 24 September, **6** (149): 518–521.

CRISP, J. D. (Ed.) 1964. *Grazing in terrestrial and marine environments.* British Ecological Society Symposium Number 4. Blackwell, London.

CURRIE, R. 1959. Organic production in the Sea. *New Scient.*, 1 October. **6** (150): 584–587.

CUSHING, D. H. 1959. On the nature of production in the sea. Min. of Agric. and Food. *Fisheries Investigations* (2), **18** (7).

DEITRICH, G. 1963. (transl. by F. Ostapoff). *General oceanography.* Wiley London.

FLEMING, L. H. and LAEVASTU, T. 1956. The influence of hydrographic conditions on the behaviour of fish. *Fish. Bull., F.A.O.*, **9** (4): 181–196.

HARDY, Sir A. 1956. *The open sea; Pt.I. World of the plankton.* Collins, London.

HARDY, Sir A. 1959. *The Open Sea; Pt. II. Fish and fisheries*, Collins, London.

HARDY, Sir A. 1962–3. Man and the beneficent sea. *Advm. Sci.* **19** (82): 533–544.

HARVEY, H. W. 1942. Production of life in the sea. *Biol. Rev.*, **17**: 221–246.

Holt, S. J. 1969. The food resources of the ocean. *Scient. Am.*, September **221** (3): 178–197.

Issacs, John D. The nature of oceanic life. *Scient. Am.* September, **221** (3): 146–165.

King, C. A. M. 1962. *Oceanography for geographers.* Arnold, London.

McKee, A. 1967. *Farming the sea: the first steps into inner space.* Souvenir Press, London.

Mann, K. H. 1969. The dynamics of aquatic ecosystems. *Adv. Ecol. Res.* **6**: 1–81.

Murphy, R. C. 1962. The oceanic life of the Antarctic. *Scient. Am.*, September, **207** (3): 186–211.

Neilson, S. E. 1964. Recent advances in measuring and understanding marine primary production. *British Ecological Society Jubilee Symposium* 1963. Eds. Macfadyen, A. and Newbould, P. J., 119–130.

Ommaney, F. D. 1949. *The oceans.* Oxford Univ. Press.

Orr, A. P. and Marshall, S. M. 1969. *The fertile sea.* Fishing News (Books) Ltd., London.

Pequegant, W. E. 1958. Whales, plankton and man. *Scient. Am.*, January, **198**, (1): 84–90.

Raymont, J. E. G. 1954. Food from the seas. *The Listener*, 9 December.

Raymont, J. E. G. 1963. *Plankton and productivity in the oceans.* Pergamon Press, London.

Rhyther, J. 1959. Potential productivity of the sea. *Science, N.Y.*, **130** (3376): 602–608.

Riley, G. A. 1949. Food from the sea. *Scient. Am.*, October, **181** (4): 16–19.

Russell, E. S. 1942, *The over-fishing problem.* Cambridge Univ. Press.

Russell, F. S. and Yonge, C. M. *The seas*, New revised edition. Frederick Warne, London.

Scientific American (The Ocean) September, 1969, **221** (3).

Scientific American (The Antarctic). September, 1962. **207** (3).

Sears. M. (Ed.) 1961, *Oceanography.* Publ. No. 67 of the American Association of Science, Washington, D.C.

Shelbourne, J. E. 1959. Could sea fish be farmed? *New Scient.* **5** (118): 413–415.

Sverdup, H. U. *et al.* 1942. *The oceans, their physics, chemistry and general biology.* Prentice-Hall, New York.

Sverdup, H. U. 1952. Some aspects of the primary productivity of the sea. *Fish. Bull. F.A.O.*, **5** (6): 215–223.

Wimpenny, R. S. 1966. *The plankton of the sea.* Faber and Faber. London.

Wooster, W. S. 1969. The ocean and man. *Scient. Am.*, September, **221** (3): 218–234.

11
Trees and the Forest Ecosystem

Forest and woodland are the most extensive, complex and biologically productive of the terrestrial ecosystems. It has been estimated that formerly they covered at least two-thirds of the earth's surface and, although now reduced to almost half their original extent, they still occupy an area greater than that of the world's agricultural land. They occur over a wider range of ecological conditions than any other type of vegetation. Producing the largest biomass per acre, their impact on atmosphere and soil is correspondingly great. The biological productivity of forests is high. At the peak of production their annual rate of photosynthesis probably approaches the maximum that can be achieved by plant growth for a particular site; and in their efficiency of utilisation of solar energy it has been said that they are comparable to the highest yielding crops. Forests constitute one of the basic primary resources with an ever-increasing diversity of uses and products for which demand is still growing.

The ecological dominance and the economic significance of trees is to a great extent a function of their size and longevity. These two characteristics, which distinguish them from other types of plants, have ensured their success in competition with smaller annual and perennial forms. Size (and more particularly height) is an important if somewhat arbitrary criterion whereby a distinction is made between trees and other woody forms such as shrubs. The former have been defined as woody plants that attain a height of at least ten feet, have a single main stem or trunk, and a distinct crown; in contrast, shrubs are less than three metres, have several basal stems and no distinct crown. However, while botanists such as Christian Raunkiaer, for example (see p. 167), have put the emphasis on height of perennating buds above the ground-surface, the possession of a single trunk and a well-defined crown, irrespective of height, are often regarded as more important distinguishing characteristics by the forester. The superior size of trees is a result of their continued increase in height and girth. Growth in height takes place primarily from buds which develop at the tip of branches. Many species, however, possess 'epicormic buds' which are formed and remain quiescent along the trunk and lower branches. Should the terminal buds be accidentally or deliberately damaged, growth of the epicormic buds is stimulated and new lateral shoots, or 'suckers', are produced. The pollarded form of trees such as willows or the multi-stemmed coppice of hazel are the result of the deliberate suppression of terminal buds and leading shoots by cutting back the crown and the trunk.

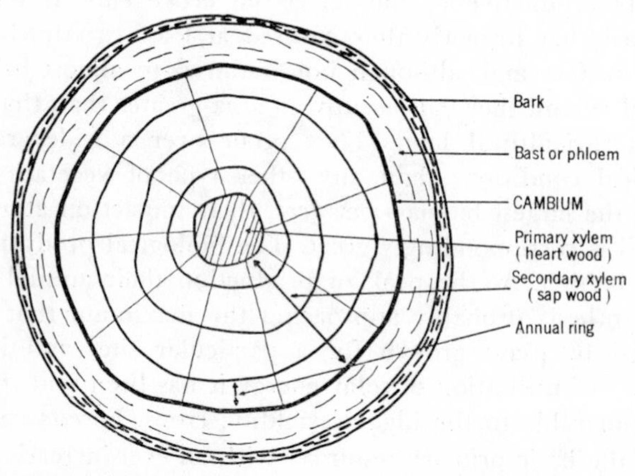

Fig. 47 Diagrammatic cross-section of tree trunk showing relationship of main anatomical features.

In most perennial woody plants, elongation involving the addition of new primary tissues to those already formed is accompanied by secondary growth which results in increased girth. It is dependent on the presence of the *cambium*—a layer of 'growth' cells lying between the inner *xylem* or wood and outer *phloem* (or bast)—in trunks and stems (see Fig. 47). The xylem is composed of cells and fibres whose walls are thickened and reinforced by a mixture of cellulose and lignin. It forms the bulk of the so-called 'wood' and in the living tree it provides support and the means whereby water and other solutes are transported from roots to leaves. The hardness of wood is dependent basically on the amount of lignification and the proportion of thick-walled fibres. However, while the phloem may contain lignified fibres, its principal components are living cells whose function is the storage and transport of the complex organic products of photosynthesis. Most of the activity in secondary growth is involved in the production of new xylem; when growth is seasonal this is reflected in the characteristic annual growth-rings whose width and cell-size are dependent upon the environmental conditions at the time of their development. With increasing age and growth in girth the original xylem cells, often filled with resinous, fatty or other substances, become inactive; they form the core of what is known as the *'heartwood'* in contrast to the more recently formed active sap-wood. Increase in girth tends to rupture the outer superficial layers of epidermal and cortical cells which form the bark. As these crack, die and frequently peel off they are replaced by the growth of new epidermal cells. If the bark is removed (either deliberately for such products as tannin or cork, or accidentally by fire or animal 'rubbing') at a rate greater than it can be renewed, exposure and destruction of the underlying phloem and cambium layers will result in the eventual death of the tree. Indeed a long-recognised method of killing trees is to make an annular incision, or ring-cut, from the bark into and across the cambial layer or growth-zone.

It is possible that wind stress combined with the problem of raising water from the soil sets a limit to the maximum height that can be attained by trees. While potential growth-height is genetically controlled, the actual maximum height reached by a particular species will depend on the extent to which the site approaches optimum growth conditions. In humid temperate and tropical regions the average maximum heights of dominant trees range from forty to sixty metres. The tallest species—of heights between a hundred and a hundred and twenty metres—include the eucalypts of Australia; the conifers of the Pacific coast of North America (the greatest height ever recorded for a tree was just over a hundred and thirty metres for a Douglas fir (*Pseudotsuga taxifolia*) in British Columbia); and the sequoias of northern California where one tree of *S. gigantea*, with a height of over a hundred and twenty metres and a girth of thirty, contained a volume of wood equivalent to ten times that in half a hectare of mature spruce in Britain.

The life span of the majority of tree species far exceeds that of man and indeed, of most living organisms. While the average or maximum life-expectancy of trees under natural conditions is not known with certainty, the age of some existing oaks has been estimated at fifteen hundred, of yews and junipers at over two thousand and of certain Californian redwoods at over three thousand years. Although there would appear to be no absolute limit, there is an age beyond which the rate of growth and increase in volume begin to decline and eventually to become negligible. The tree then becomes increasingly susceptible to disease and wind-blow; but even after the trunk has rotted and decayed, growth may be continued by 'suckers' from the base or side branches. The maximum annual amount of growth (or in silvicultural terms the current annual increment, C.A.I.) varies with species and site conditions. The age at which the peak annual increment of the bole is attained for species commonly grown in Britain may range from fifteen to forty-five years. From the economic point of view, however, 'maturity' or maximum economic age is attained when the increase in volume drops below a level at which the continued maintenance of the tree is no longer economically justified. For the production of timber maturity would be coincident with the maximum mean annual increment (M.A.I.). In Britain on good quality (Class I) sites this might vary from eighty years for the principal coniferous species to a hundred and twenty to a hundred and fifty years for such hardwoods as beech and oak. The maximum economic age however also depends on the proposed use to which the timber is to be put, as well as on environmental conditions. Demand for pulpwood combined with the susceptibility to wind-blow in many Forestry Commission plantations in Scotland has recently initiated a re-evaluation of the 'economic age' of particular trees in certain areas; liability to wind damage with increasing height has reduced this to as low as twenty-five years for Sitka spruce in some particularly exposed areas.

In contrast to their size and longevity, the regeneration and establishment of trees is relatively precarious in comparison to other plant forms. Early development is slow and most trees take several years before they become reproductive. The age at which flowers develop varies with species. For some exotic pines it is as soon as two or three years after germination, but in general it is much longer. For most temperate species it is rarely less than ten years after germination, and ranges from fifteen to twenty for birch, pine and larch to as long as forty to fifty for beech and oak. Furthermore, in some, such as pine and oak, seeds are set only in the year after pollination. Seed production can be very variable; some species, as for example birch and sycamore, produce a good crop annually, while others as in the case of oak and beech do so only at long and variable intervals. In the latter species favourable 'mast' years are usually determined by the weather in the preceding season.

Not only does the protracted period before sexual maturity make regeneration and survival precarious, but the perennial habit together with the persistence of stem and branches increases the difficulties of seedling establishment and survival. This is dependent on the successful development of a root system which will provide adequate anchorage and will tap a source of moisture and nutrients sufficient to maintain growth. Tree roots usually grow more rapidly than the shoot in the seedling stage. Most species initially have a deep tap-root which is maintained in trees such as the ash, elm, oak, sycamore, pine and larch. Others, however, do not have a persistent tap-root but develop a deep and widely branching system such as is characteristic of the beech, or a shallow lateral system as in the spruce, birch, willow and poplar. While the root potential varies with species, the actual depth and extent of development is dependent on soil, and particularly soil-water conditions. In general trees are capable of exploiting a greater depth and volume of soil than herbaceous and smaller perennial woody plants.

According to Edlin trees are 'simply large land plants with persistent woody and more or less individual stems'. And this form is shared by a vast number of different species which vary in habit, ecological requirements and distribution. Apart from the 'tree-ferns' which were such important constituents of the Carboniferous forests and are now localised in certain New Zealand forests, the tree form is confined to the seed-bearing plants (*Phanergamae*). The Gymnosperms are all woody perennials; and of these the conifers are the largest and most prolific group extant. They range in size from dwarf shrubs to the tallest trees in existence. The majority are evergreen and many, particularly those native to northern latitudes, are characterised by small, hard, narrow or scale-like leaves. While it should not be forgotten that broad-leaved conifers do occur, such as *Auraucaria spp.* (Chilean pine or monkey-puzzle tree), the former are more abundant and attain their greatest extent in cool to cold regions in the northern hemisphere. The conifers are commercially the most important group of trees. They occur in relatively pure stands and, together with the so-called 'white woods' of birch and aspen, comprise the softwoods for which there is the greatest demand today. The lightness of their wood is combined with good physical and mechanical qualities, as well as ease of working; and the characteristic high resin content of many species contributes to its durability. The flowering plants (Angiosperms), while not exclusively woody, contain a greater number of woody species. This group includes the dicotyledons and the monocotyledons. Of the latter the principal woody representatives are the bamboos and the palms. The latter are characterised by growth in height but not in girth, a simple non-branching trunk terminating in a dense crown of large leaves and a predominantly tropical distribution. They make, however, a very limited contribution to the world's forest-cover and their economic value is related more to their fruit than their wood. The dicotyledons include the greatest number of tree species and variety of tree forms. Predominantly, though not exclusively, hardwoods, they range from sub-arctic to equatorial regions and dominate a greater area than the two preceding groups.

Limits of Tree Growth and Distribution

The factors controlling the actual and potential limits of tree growth have long been an important and hotly debated subject in biogeographical studies. Those characteristics which contribute most to the success and dominance of trees in fact pose very considerable physiological problems for their growth and survival. Their size is dependent upon their ability to increase in height and girth each year. Trees, therefore, require a period during which photosynthesis is sufficient to effect the production (and protection) of new woody tissues and perennating buds. Unlike herbaceous plants all of whose aerial parts are capable of photosynthesis, a very large proportion of a tree's biomass is contained in the trunk and branches which are incapable of assimilation but which require energy for their production and metabolism. Hence for tree growth the excess of material produced in photosynthesis over that expended in the provision of energy for metabolic activities must be correspondingly greater than in non-arboreal plants. In the absence of an adequate excess the amount of growth will be small and woody forms may not be able to exceed the height of shrubs. In addition, because of their height, water must be raised a considerable distance from the soil, and the taller the tree the greater the amount of energy required to overcome the friction of the conducting vessels.

Among the most important factors limiting photosynthesis, and hence the growth and distribution of trees, are an insufficiency of either water or heat. The former is characteristic particularly of semi-arid and arid areas of the world where low rainfall, high evapo-transpiration and protracted periods of drought create soil-water deficiencies for part or the whole of the year. With decreasing water availability trees tend to become smaller in stature and deformed, and they may assume a spreading habit as well as becoming more widely spaced. Even in humid regions, seasonal water deficiencies during particularly dry summers are reflected in the checking of the growth of upper branches. In comparison, lower branches will tend, under these conditions, to receive most if not all of the available moisture. However, attempts to determine the minimum amount of rainfall necessary for tree growth have been complicated, indeed thwarted, by the number of factors other than precipitation which determine the availability of soil-moisture. On the one hand, the amount of 'effective' rainfall which reaches the soil will be dependent on interception by trees themselves and the amount lost by evaporation. On the other hand, the amount of effective moisture may be increased by soil or relief conditions. The latter may, as along permanent water-courses or in local depressions, provide an easily accessible and permanent water-table. In the former, coarse-textured porous soil not only facilitates rapid percolation of rainfall but permits the penetration of deep roots to an adequate source of ground-water. However conditions resulting in a permanently high water-table, soil saturation and a deficiency of oxygen may, in areas otherwise climatically suitable, preclude tree growth.

Insufficient heat is assumed to be the main factor accounting for the Arctic and altitudinal limits of tree growth. In the latter area there is a general though by no means exact coincidence of the Boreal forest/tundra transition with the following temperature thresholds: (a) mean daily temperature of 10°C for the warmest month; (b) a growing season of three months; (c) 10°C of accumulated month degrees over the threshold value of 6°C. In addition, with decreasing summer warmth, the depth, temperature and persistence of the thawed soil layer above the permanently frozen ground (*permafrost*) decreases. Where this is less than 0·5 metre root development is restricted, and water-logging combined with low soil temperatures, which impair the efficiency of water and mineral (particularly nitrogen) absorption, are detrimental to tree growth. This can, however, be maintained further north on warmer well-drained soils and along the less-exposed valley floors where northward-flowing rivers transport warmth and nutrients from further south. However recent work has emphasised the fact that the Arctic tree limit is determined basically by a decline in seed production and by the ability of such viable seeds as are produced to germinate and establish themselves successfully.

Analyses of the relative importance of the factors involved have been further complicated by somewhat conflicting and at times contradictory evidence that the Arctic tree-line is still in a condition of post-glacial instability. In some areas in the Eurasian Arctic, plant geographers regard the presence of dead and dying trees on the northern limits of the Boreal forest as indicative of a retreat still in progress—the result of climatic deterioration since the post-glacial thermal maximum. In contrast, in western Alaska the small but young trees along the forest edge suggest that a northward extension of the forest zone is still taking place following either a general post-glacial climatic amelioration or local and short-term fluctuations in climate.

Because of their greater height, trees are more exposed to the effects of high wind-force than low growing plants. On particularly exposed sites such as coastal areas, plains and plateaux, and hill summits increased evapo-transpiration concomitant with increased wind speed may check or inhibit growth. Trees tend to become stunted with flattened 'umbrella-shaped' or markedly asymmetrical crowns planed in the direction of the dominant wind force. Exposure is an important factor reinforcing the effect of low temperatures and (as has been demonstrated in the Cairngorms) such biotic factors as grazing in determining the altitudinal limits to which trees can grow. While their form makes trees particularly susceptible to adverse conditions, their slow initial growth can make their regeneration precarious even in climatically favourable areas. The tree seedling is particularly vulnerable to competition, grazing and burning which may destroy it before it can attain maturity. By reason of his direct or indirect effects on regeneration man is undoubtedly the most important agent determining the actual limits of tree growth at present. Apart from the replacement of former forest and woodland by agricultural land, his intensification of grazing and browsing, and his use of fire to promote the growth of forage plants has effectively checked regeneration in many areas. Anthropogenic factors cannot be ignored even at the Arctic limits, where reindeer herding and the use of fire have tended to give the tundra an advantage over the forest. In certain areas it has been calculated that seventy-five to ninety-two per cent. of the dead trees along the northern limit of tree growth had been damaged by reindeer and/or fire. Even where such anthropogenic activities have ceased forest re-establishment may be either difficult or impossible. Seed material may not be available. The successful establishment of young seedlings may be inhibited or prevented by their inability to penetrate tough swards or to maintain themselves in face of competition with a well-established vegetation cover. In the extreme cases the absence of necessary symbiotic root-fungi (mycorrhiza), the presence of hard pans or particularly severe soil erosion may retard, if not prevent, natural reforestation.

Whatever the origins of the frontier between arboreal and non-arboreal vegetation, once established it can remain relatively stable long after the initial causal factors have ceased to operate. On the one hand trees and the communities they dominate possess a high degree of *inertia*. Because of their longevity they can persist despite the fact that environmental conditions may no longer permit effective reproduction. On the other hand, the vulnerability and difficulty of establishment of seedlings may delay the spread of forest into climatically favourable areas where an already well-established non-arboreal vegetation cover exists.

The Forest Ecosystem

Because of their size and persistence trees form the most massive and complex of terrestrial ecosystems. They are characterised by a height which ranges from an *average* of thirty metres in temperate to fifty-five metres in tropical regions and a multi-layered, highly stratified structure of both aerial shoots and subterranean stems and roots. This is a result of the ability of plants of progressively smaller life forms to occupy the space below the tree canopy and to tap varying soil levels. In addition the trees and shrubs provide an anchorage for a wide variety of epiphytic plants from mosses and liverworts to woody lianas. Forest and woodland ecosystems are also distinguished by the volume of the living plant biomass they contain and the high proportion of this accounted for by the tree layer. It has been suggested that on average, in a mature well-grown stand three-quarters of the plant biomass is contained in the living trees with that in the trunk far exceeding that in the roots and canopy combined. (see Fig. 48). In a comparative study of contiguous prairie, savanna, and oakwood communities found under similar site conditions in east-central Minnesota, J. D. Ovington measured the amount and distribution of plant matter present in the three ecosystems (See Table 27). This indicated that at a period of maximum productivity the oakwood

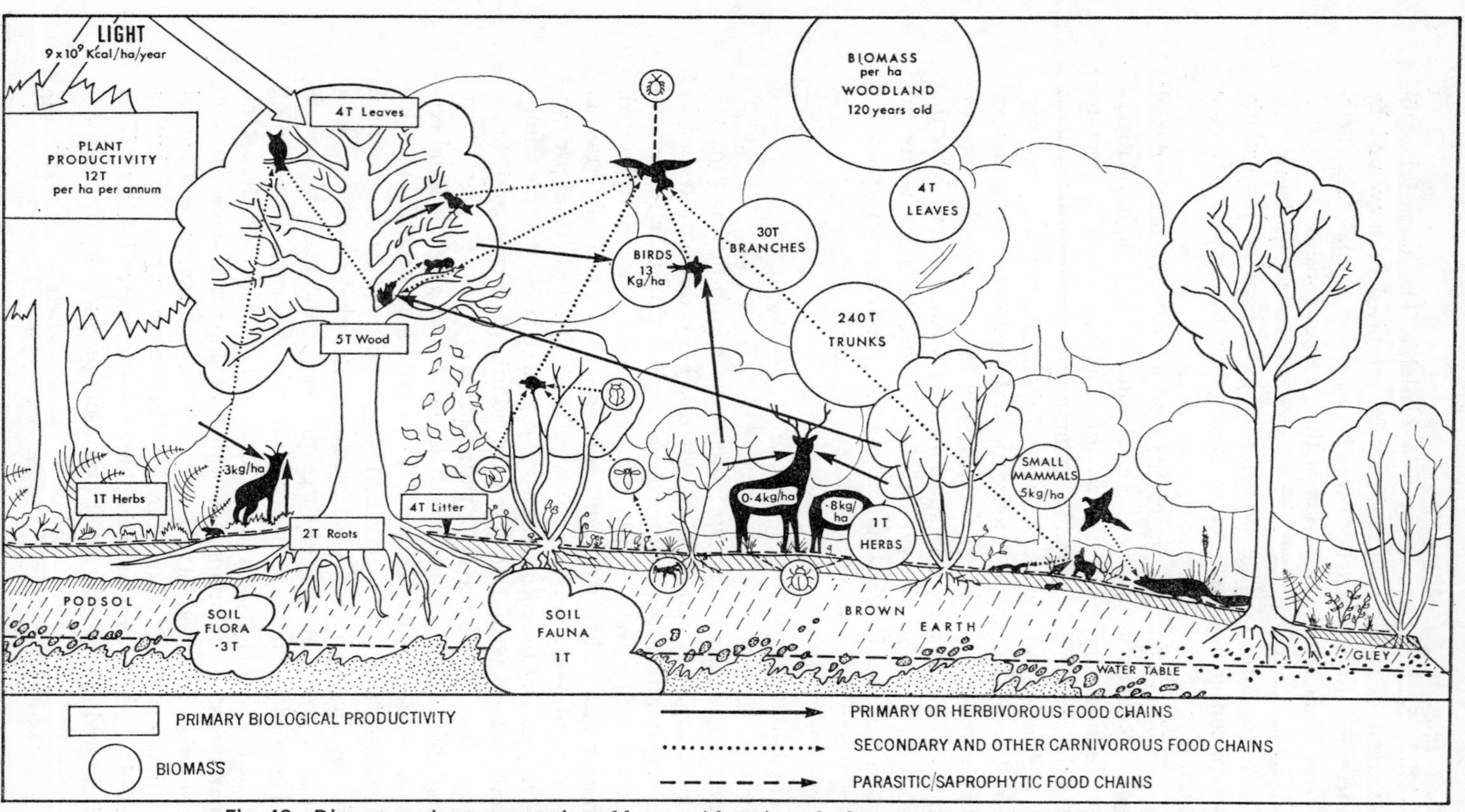

Fig. 48 Diagrammatic representation of form and function of a forest ecosystem in Belgium. (Simplified from Ministerie van Nationale Opvoeding en Cultuur. *Ecosystem en Biosfer:* Documentatie 23.

Sampling months in 1959	April	May	June	July	Aug	Sept	Oct	Nov
Prairie								
Living plant material above ground	0·1	0·1	0·5	0·7	1·0	0·9	0·4	0·2
Dead plant material above ground	2·9	2·0	2·5	3·0	2·7	2·4	3·0	3·8
Roots and subterranean stems	6·7	4·0	4·7	6·0	3·5	5·4	3·8	4·4
Total plant biomass	9·7	6·1	7·7	9·7	7·2	8·7	7·2	8·4
Savanna								
Living plant material above ground	30·2	30·3	32·5	32·8	35·0	33·5	31·0	30·9
Dead plant material above ground	11·1	12·2	10·9	12·9	14·3	14·8	16·6	16·5
Roots and subterranean stems	13·1	13·0	12·9	11·9	8·1	14·9	8·8	11·7
Total plant biomass	54·4	55·5	56·3	57·6	57·4	63·2	56·4	59·1
Oakwood								
Living plant material above ground	161·2	162·1	164·0	164·4	164·8	165·8	163·9	163·0
Dead plant material above ground	56·1	59·0	73·8	60·5	45·8	53·3	63·1	57·1
Roots and subterranean stems	13·0	15·5	19·3	20·7	13·6	15·9	11·9	10·1
Total plant biomass	230·3	236·6	257·1	245·6	224·2	235·0	238·9	230·2

Table 27 Comparison of biomass in prairie, savanna and oakwood ecosystems in central Minnesota during period April to November. Savanna used in morphological sense to describe a type of vegetation with a well-developed ecologically dominant herbaceous stratum. Oven dry weight plant material in Kg × $10^3 Ha^{-1}$. (From: Ovington, J. D. 1964)

Trees	*Pinus nigra*	*Pinus sylvestris*	*Betula verrucosa*	*Quercus borealis*	*Picea abies*	*Nothofagus truncata*	*Pseudotsuga taxifolia*	Evergreen gallery forest
Location	N. E. Scotland	E. England	Moscow, USSR	Minnesota U.S.A.	Sweden	New Zealand	Washington State, USA.	Thailand
Status	Plantation	Plantation	Natural	Natural	Natural	Natural	Natural	Natural
Age (Years)	48	55	67	57	58	110	52	—
Height (m)	14	16	26	17	17	21	17	19
Number trees per ha	1 112	760	—	800	924	490	1 157	16 200
BIOMASS								
Tree leaves	5·6	7·2	2·8	3·5	9·1	2·7	12·0	19·0
Tree branches	11·2	12·3	11·3	49·5	14·3	42·0	17·9	50·0
Tree trunks	95·1	96·7	156·7	111·9	85·2	224·8	174·8	225·2
Shrubs and herbs	7·0	2·6	2·0*	0·6	1·0*	0*	0·1	0·2
Roots	34·0	34·1	43·1	15·0	60·0*	39·2	12·3	88·5
Dead branches on trees	10·0	10·0	2·0*	21·9	2·6	1·1	11·2	—
Organic matter on ground	22·0	45·0	3·0	36·7	78·0	16·7	117·3	3·0

Table 28 Plant biomass in selected woodlands. Oven dry weight 10^3kg/ha. *Estimated from other woodlands. (Ovington, J. D. 1965)

contained twenty-eight times as much living plant matter as in the prairie, and nearly eight that in the savanna. Seventy-five per cent. of the total plant biomass in the oakwood was living material, of which sixty-five per cent. was contained in the tree layers.

The amount and distribution of living and dead plant material is, however, dependent on a number of interacting variables, the most important of which are site conditions, species, age and not least the type and intensity of use and/or management of the forest in question. A comparison of data on the amount and distribution of plant biomass in selected forest stands of roughly comparable age (although incomplete and of varying reliability) illustrates the magnitude of difference with latitude (see Table 28). Under comparable climates, however, forest biomass can vary with soil—particularly water and nutrient—conditions. Given similar sites it can further vary as a result of the growth potential of the species and/or the density of the tree canopy. Also the volume of plant material may be a direct result of use in so far as this controls the composition of the forest community and the balance between that harvested and that produced over a period of time. Selective management in plantations may give the maximum biomass possible of trees but the deliberate suppression of an under-storey may result in a volume of living matter less than in a community with a more complex stratification. Finally the amount and distribution of biomass varies with age and hence the stage of woodland development. In youth the proportion of photosynthetic material to woody tissues is much greater than later when the trunks and branches increase in volume and the canopy begins to open out (see Fig. 49).

The composition and structure of forest and woodland is dependent on a similar complex of environmental factors as is the plant biomass and the resulting diversity of communities is correspondingly great. The classification and description, together with suggested explanations of the ecological significance and status of the global range of forest types, have formed the bulk of much biogeographical literature up-to-date. Forest communities have been identified on the basis of the life-form and/or floral composition of their dominant tree strata. In the former case there is, as has been noted already, a striking but less than perfect correlation between the general appearance or physiognomy of forest types and zonal climatic conditions. Of the species available, those whose form allows them to compete most successfully under the prevailing growing conditions still tend to be dominant. This is reflected in such similarities as the evergreen, needle-leaved form of the Boreal forest and the deciduousness of temperate and tropical forests in areas with a marked seasonal alternation of either rainfall or temperature. Such a 'convergence' of form of the dominant trees in similar climatic regions is most striking in the Taiga and Selva forests. In other areas the occurrence of a particular

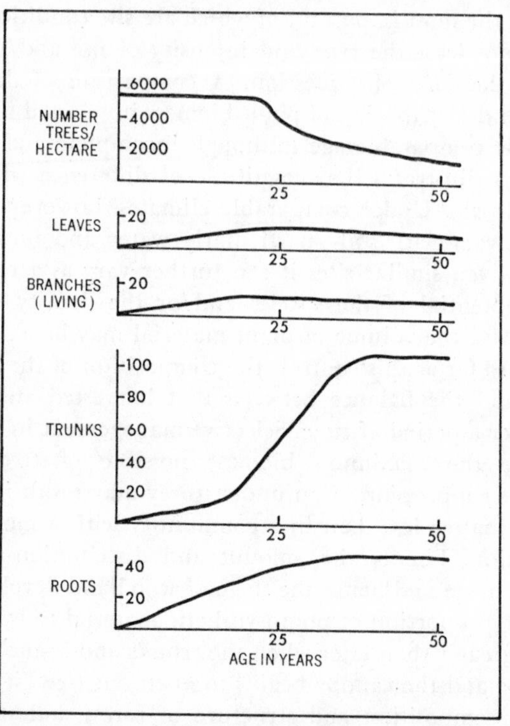

Fig. 49 Oven dry weights (10^3Kg/ha) of the tree stock (biomass) of Scots pine plantations of different ages growing under similar environmental conditions. (Ovington, J. D. 1965)

type of woodland or forest formation may be due to habitat factors, such as soil or man, which have given a particular form a competitive advantage over other climatically more demanding species.

A somewhat comparable relationship can be demonstrated between climatic conditions and forest floras. On a world scale increasing severity of environmental, and particularly climatic conditions is accompanied by decreasing diversity. It is reflected in the contrast between the overwhelming richness of the tropical rain forest and the relative poverty of the Boreal forest. In the former over forty, and sometimes as many as a hundred, species per hectare is not uncommon and, particularly in areas little disturbed by man, dominance of one or two species is rare. In the latter the tree flora is much less diverse and vast areas are dominated by one or two species. Under comparable physical conditions on a continental or regional scale, however, other ecological factors may result in variation in floral diversity. It has already been noted, in another context, that the Pleistocene glaciation in northwest Europe caused a much greater depletion of pre-glacial flora than in comparable latitudes and climatic regions in North America and Asia. Others have suggested that the single-species dominance characteristic of the northern Boreal forest is as much a function of the relatively short time since the area was deglaciated as of the harsh environmental conditions. It is now a well-established fact, particularly in temperate and tropical deciduous forests, as well as the coniferous forests of western North America, that the direct or indirect activities of man have reduced floral diversity very considerably. Recurring fire over a long period of time is probably the principal factor which gave a competitive advantage and extended the range of such species as the lodgepole (*P. contorta*) and ponderosa (*P. ponderosa*) pines in western North America and of the long-leaved, loblolly and slash pines in the south-eastern areas of the U.S.A. The relatively pure stands of the fire-resistant teak (*Tectona grandis*) and sal (*Shorea robusta*) in areas of tropical deciduous or monsoon forest have been attributed to the same cause. The dangers of reconstructing (and hence interpreting) the 'natural' or 'climatic climax' forest from existing remnants have been amply demonstrated in Britain. The dominance of one or two species in semi-natural woodlands is in many cases a legacy of either planting policy and/or selective felling in the past. This has been very clearly proved in the oakwoods on the islands and west shore of Loch Lomond. Formerly interpreted as modified remnants of 'natural' woodlands, recently discovered documentary evidence has established their origin as the result of deliberate planting and management by the Duke of Montrose in the mid-eighteenth century. Where, as on the Nature Reserve island of Inchailloch, they have been protected from cutting, and more particularly from grazing, for nearly a hundred years the incursion of other tree species has given rise to a diversity which contrasts with the relative floristic poverty of the woodlands on the adjacent mainland.

The degree of structural complexity also reflects the effect of the nature of the tree canopy on environmental conditions. The density of the canopy may be such as to inhibit or prevent the existence of non-arboreal strata. In the continuously hot, humid equatorial regions where climatic conditions are optimal for growth, the tree biomass is greater and the tree stratum more highly developed than in any other type of forest. The dense multi-storied, evergreen canopy, composed of three to four strata of trees of varying light requirements and height potential, can attain a depth of twenty-five to thirty metres. The consequent reduction in light limits the development of shrub and field strata. The latter is sparse and open, containing a large proportion of saprophytic or parasitic plants. The number of lianas and other epiphytic species is high. Under less favourable climatic conditions the volume and complexity of the tree stratum is less; but in the case of dense stands of evergreen species the development of undergrowth may be similarly restricted. This is particularly noticeable in young coniferous plantations where the density and depth of the canopy may be such as to reduce light intensity to less than one per cent of that in the open, and where even a ground-layer is absent.

Structural development is greater where spatial or seasonal variations occur in the density of the tree canopy. The former are characteristic where physical environmental conditions are sub-optimal for tree growth. Trees are more widely spaced and form parkland, savanna or open woodland, in which there may be a field and ground layer whose biomass is equal to or greater than that of the trees. On the other hands in deciduous temperate or tropical woodland the canopy may be continuous and deep but seasonal variations in density, combined with phenological synchronisation allows the development of the maximum number of strata. The number, form and composition of the component strata are conditioned to a large extent by the effect of the dominant tree strata on micro-climatic and soil conditions. The composition of the non-arboreal strata, however, often provides a more sensitive indication of local variations in site conditions than that of the dominant trees. Oakwood in Britain can exist under a variety of site conditions—the accompanying species will however vary with soil depth, drainage and available nutrients etc.

The extent of anthropogenic factors on forest or woodland structure throughout the world is difficult to assess, and has complicated the problems of classification as well as of understanding the interrelationship between forest types and environmental factors. Of existing woodland, particularly in heavily-populated areas there is little primary or 'virgin' growth; a high proportion is the result of second or third growth after felling. Little has escaped some degree of man-promoted burning or grazing; even in the two most extensive forest areas of the world—in circum-equatorial and circum-boreal latitudes respectively—there is no unanimity about their status. Much forest is the product of relatively recent selective planting. In the planted tree crop, the forest community attains its simplest structure and most uniform composition.

The forest community is distinguished by its large volume, and, as a result of a complex structure, large surface area. In contrast, the proportion of available primary plant-food at a particular point in time is relatively small. The effect of these basic characteristics on animal populations in the forest ecosystem is four-fold. First, the animal biomass is small in comparison with that of plants. Second, the diversity of fauna is great as a result of the multiplicity of available micro-habitats. Third, because of the diversity of micro-habitats and fauna the interrelationships between various trophic levels are more complex than in any other ecosystem. Fourthly, the density and variety of animal life is greater on the margins than in the interior of dense forests since at the zone of contact with another type of vegetation the variety of habitats is increased, and the range and habit of particular animals allows them to take advantage of both. As is characteristic of terrestrial ecosystems the volume and variety of the soil fauna exceeds that living on or above the surface. The latter include saprophytic invertebrates which play a major rôle in the communition and decomposition of dead organic matter, others are burrowing animals which may breed and live underground but are surface herbivorous feeders. The forest floor is the main habitat for large mammals. Populations tend to be sparse and characterised by scattered individuals or small groups, with a density usually proportionate to size. Very large mammals are exceptional. Above the surface the most densely populated part of the forest community is the canopy. As well as mammals (whether for example squirrels, monkeys or tree sloths and reptiles anatomically adapted to an arboreal existence) it harbours a great variety of birds and insects. The latter, in fact, are the most prolific and important; in both the canopy and the soil they form the vital link in the complex food-chains between plants and animals both above and below ground. The variety, density and distribution of animal population is dependent on the composition and structure of the forest, and the proportionate development of the main strata within it.

As Ovington has remarked 'energy capture by fast-growing woodland ecosystems at their most productive period is probably approaching the maximum possible under natural conditions.' This efficiency and productivity of the forest is primarily a result of its structure. Its assimilating tissues occupy an area which in some cases exceeds the euphotic zone of the marine ecosystem with which it is often compared in terms of structure and productivity. The volume of leaves varies from one to ten tonnes dry weight per hectare and is equivalent to a total surface area which may be *at least* eight times that of the ground below. The assimilating surface is therefore very large. In addition, photosynthetic efficiency is increased by the uneven canopy-surface and the mutual shading of plants and leaves. In the case of the former the area exposed to direct radiation is much greater than that of the ground while the undulating crown surface gives rise to variations in light intensity. The latter feature is even more important within the canopy and the understoreys. Tree leaves exposed to the sun are usually smaller and lighter in colour, and as a result absorb less of the available light than those in the shade. It has been noted that at low light-intensities the productivity of the shade leaves of beech can be four to five times that of the sun leaves because of their greater photosynthetic efficiency under these conditions.

In comparison with many herbaceous and annual field crops, the seasonal duration of tree leaves is long. In addition the diversity of life forms and phenology of the stratified forest ecosystem permits a longer and more efficient use of the potential growing-season than is possible in herbaceous vegetation or annual crops. It has also been suggested that absorption of radiation by trees is greater than by agricultural crops and grassland because of the darker colour and the uneven surface, both of which reduce loss by reflection. Thus it has been maintained that 'a pine forest in northern Scotland may absorb more radiation than a pasture in south-east England although incoming solar radiation is less'. On comparable sites evergreen and particularly coniferous trees are more productive than deciduous species, because of their ability to continue assimilating, albeit at a slow rate, when cold or aridity may inhibit the activity of other plants. The comparative efficiency of conifers is further increased by the relatively rapid development of a closed canopy; a frequently dark leaf-colour; a high volume of leaves per unit area of land and, in many species, a conical form which increases the unevenness of the canopy surface.

Finally, the vertical development of the forest ecosystem is supplemented by the extent and volume of its root system. This allows the community to tap a large volume of soil and to obtain the considerable amount of water and nutrients necessary to maintain the high level of primary production. It has been estimated that in British forest plantations, the mean maximum annual rate of energy fixation is about one per cent. of incoming solar radiation or about two per cent. that of a wave length suitable for photosynthesis. Average annual efficiencies of 2·5 per cent. for a fifty-year old Scots pine (*P. sylvestris*) and 2·7 per cent. for a twenty-one year old Siberian spruce (*Picea omricka*) stand have been recorded. And such estimates as are available indicate that values of three per cent. are not uncommon in humid tropical regions.

Productivity however varies not only with species and life forms, as has been indicated, but also with site and age. It reaches a maximum when the contribution of the canopy and roots—of total assimilating material—to total biomass is greatest (see Fig. 48). At this stage the excess of photosynthesis over respiration is greatest and hence the accumulation of organic matter highest. In coniferous plantations it has been demonstrated that maximum annual production of organic matter is achieved at what is referred to as the 'pole stage'—just before the increase in trunk-height raises the canopy above ground level. With increasing age the canopy opens (or is thinned) out, becomes reduced in depth and volume and what is even more important, the proportion of assimilating to non-assimilating woody material decreases.

Maximum potential productivity will be determined by climatic and soil conditions. Although the amount of leaf-fall is less than leaf production it provides a basis for comparison of net forest productivity. In a study of the forests of the world, R. J. Bray's results indicate that productivity in the tropical rain forest is at least three times that in the Boreal forest, and about ten times that in Arctic-Alpine habitats (see Table 29). However under comparable climatic conditions litter production on the best *sites* can be twice to three times that on the poorest.

The detailed pattern of energy fixation, flow, storage and release in a forest ecosystem is extremely complicated. Quantitative data are still relatively meagre and the process incompletely understood. However it has been possible (on the basis of information gathered mainly from managed plantations) to prepare

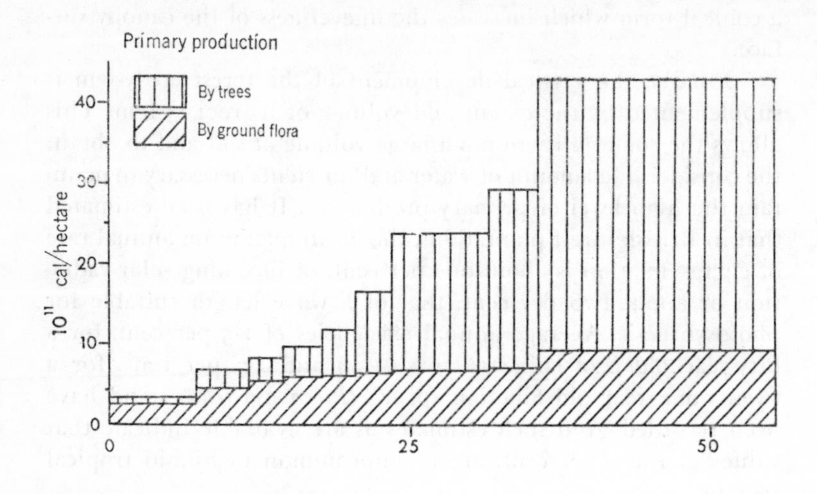

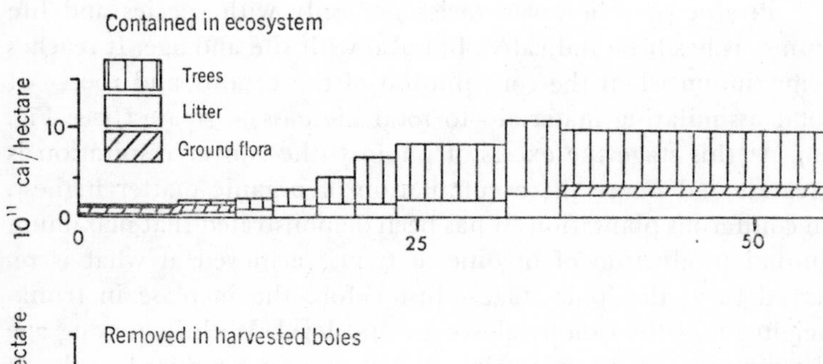

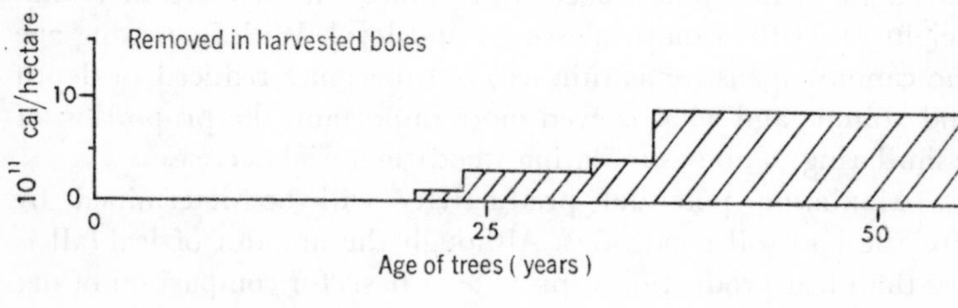

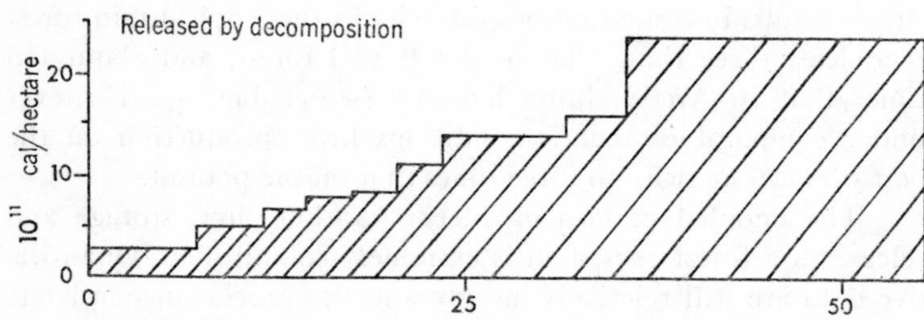

Fig. 50 Energy budget for plantations of *Pinus sylvestris* from time of planting. (From: Ovington, J. D. 1962)

Region	No regions averaged	Leaves	No regions averaged	Other	No regions averaged	Total
Arctic-Alpine	1	0·7	1	0·4	3	1·0
Cool Temperate	15	2·5	10	0·9	22	3·5
Warm Temperate	8	3·6	5	1·9	7	5·5
Equatorial	2	6·8	1	3·5	4	10·9

Table 29 Annual litter (metric tons/ha) production in four major climatic zones, (From: Bray, R. J. and Gorham, E. 1964)

	Energy content 10^{10} Cal/ha	Percentage utilisation total incident solar radiation
In gross primary production	340	2·46
Released by respiration of producer plants	160	1·16
In net primary production	180	1·30
Accumulation in living trees	61	0·44
,, ,, ground flora	<1	<0·01
,, ,, litter layers	6	0·04
Removed in boles of harvested trees	31	0·23
Left in roots of harvested trees	13	0·09
Released by litter decomposition	68	0·49

Table 30 Energy flow following photosynthesis of Scots pine plantations for the 18-year period from 17 to 35 years after planting. (From: Ovington, J. D. 1962)

energy budgets or balance sheets for standing crops. Table 30, for instance illustrates the absolute and proportionate distribution of energy in a particular example.

Although the absolute values will obviously vary according to age and habitat, characteristic of all forests is the large proportion of energy 'stored' in the system. It has been pointed out by many authors that the rapid release of this energy in forest fires results in very high temperatures, of at least 200°C and often exceeding 800°C at ground-level. For this reason forests, whether in fossil or living conditions, have provided man's primary and basic fuel resource.

(References for Chapters 11 and 12 are combined at the end of the latter)

12
Forest Environment and Resources

The forest ecosystem occupies a large volume of atmosphere and soil, the physical conditions of which are profoundly modified by the presence of this massive biomass. The dominant tree canopy insulates the space beneath it, as well as the ground surface, from the direct impact of solar radiation, precipitation and wind. As a result the character of the atmosphere in the 'trunk space' differs to a greater or lesser degree from that in the canopy above and from that over adjacent non-wooded areas. In addition, the circulation of large quantities of water through the system also affects the forest micro-climate and, together with the continual turnover of nutrients, influences the nature of the underlying soil. The mutual and dynamic interaction between the forest and its physical habitat creates not only a different, but a more equable and stable environment than would otherwise occur.

The nature of the forest environment however is dependent, on the one hand, on the size, structure and floristic composition of the particular community and, on the other, on the climatic, geological, and physiographic characteristics of the physical habitat. The canopy, in particular, plays a major rôle in the modification of atmospheric conditions and the creation of the forest micro-climate. A large proportion of the incident insolation is either reflected from or absorbed by it. In the former circumstances the *albedo* (or per cent. of incident light reflected from the canopy surface) can vary from ten to fourteen per cent. for darker-coloured species, such as some conifers, to sixteen to twenty-seven per cent. for deciduous trees in temperate latitudes. This compares favourably with that from a continuous cover of field crops or of short green grass (average twenty-six per cent.) and is much less than that recorded for desert scrub (thirty to thirty-eight per cent.). Absorption of light by the canopy results in a marked reduction in intensity from the surface downwards. The combined effect of reflection and absorption varies with species, their age and density as well as the seasonal duration of leaf cover.

Absorption of insolation during the day is accompanied by a decrease in air temperatures from the canopy downwards to the soil surface. However the presence of the canopy, together with a reduction of air movement under it, is thought to check heat loss by out-going radiation. The net effect is less extreme temperature conditions within a forest than in adjacent treeless areas. The exact type and intensity of micro-climatic modification, however, is dependent on the structure and composition of the forest as well as on the prevailing regional climatic conditions. Assessment of forest influences is handicapped by the paucity of quantitative data. In humid temperate latitudes (where the greatest number of measurements has been recorded) daily mean temperatures would appear to be somewhat lower in summer and higher in winter, with correspondingly lower mean annual temperatures and less pronounced diurnal and seasonal ranges than in adjacent non-forested areas. The reduction of summer temperatures is, usually, more marked than the amelioration of winter conditions. Available data indicates that the modification of temperature regimes in humid tropical forests (see Table 31) is similar to that in temperate regions.

The differences between temperatures in the open and under forest are greater in some circumstances than in others. A comparison of air temperatures beneath beech, Norway spruce and pine with those over open land in Germany has revealed that the reduction of summer temperatures is less under the coniferous than under the deciduous woodland (see Fig. 51). Except in

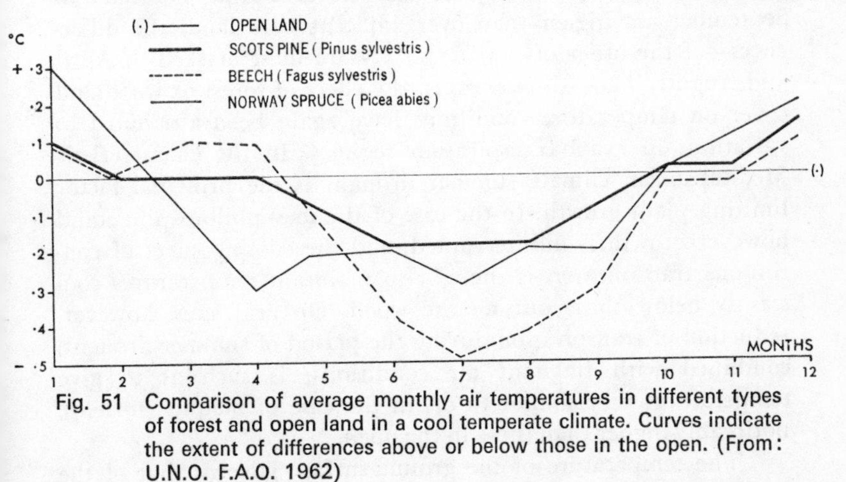

Fig. 51 Comparison of average monthly air temperatures in different types of forest and open land in a cool temperate climate. Curves indicate the extent of differences above or below those in the open. (From: U.N.O. F.A.O. 1962)

spring, increase in winter temperatures under the former tends to be more pronounced than under the latter. While variations in canopy density may account for the differences in winter temperatures, the more marked effect of beech in summer has been attributed to the greater transpiration rate of this species and the resulting loss of heat particularly during the growing season.

°Centrigrade	in open	under forest	difference
Annual mean maxima	32·1	28·4	−3·7
Annual mean minima	21·4	22·5	1·1
Difference	10·7	5·9	−4·8
Maximum diurnal range	18·5	13·5	−5·0
Minimum	1·0	6·3	−0·7
Difference between extreme maxima and extreme minimafor whole period	27·7	24·0	−3·7

Table 31 Comparison of air temperatures in open with those in experimental forest of Trang-Bom in South Vietnam during period 1933–37. (From: U.N.O./F.A.O. 1962)

Conversely, more recent recordings in different types of Mediterranean woodland indicate that, under certain conditions, the characteristic micro-climatic regime associated with humid temperate and tropical forests may be reversed (see Figs. 52A and 52B). For instance in the pine stands on damp mesophilous sites, air temperatures remain slightly lower (less than 1 °C) within than in the open; on sandy xerophilous soils, the reduction of temperature is less marked, particularly in summer. However, in the evergreen oak-coppice (*forteto*) air temperatures from February to September are higher than over adjacent bare land; the differences—of the order of +1 to +2 °C—are most marked in April and August. These differences in the effect of forest or woodland cover on temperature conditions have again been attributed to variations in evapo-transpiration regimes. In the characteristic Mediterranean climate summer drought is the principal factor limiting plant growth. In the case of the mesophilous pine-stand however growth is uninterrupted, and the cooling effect of continuous transpiration is thought to maintain temperatures constantly below those outside the wood. On arid sites, however, reduction of transpiration during the period of summer drought, combined with stagnant air conditions, is such as to give temperatures very similar to or, in the case of maquis (*forteto*), noticeably higher than those in the open.

The temperature of the ground-surface governs that of the

under-lying soil to at least a depth of twenty metres. Both the effect of the living and dead vegetation cover, as well as the soil-water regime, affect the amount of heat received and the amount, rate and depth of heat absorption and loss from the soil. Soil temperatures beneath a forest cover (as far as can be judged) are affected in a manner similar to those in the atmosphere, though to a more marked extent. Shade is greatest at ground level and its

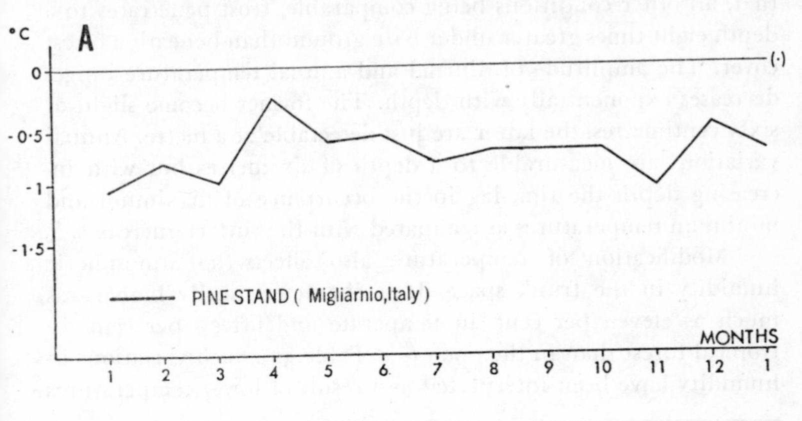

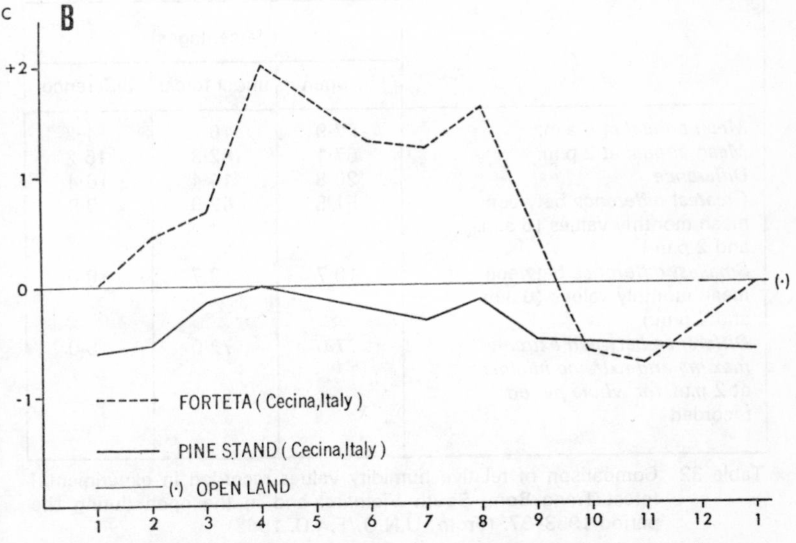

Fig. 52 Comparison of average monthly air temperatures in forest and open land in a Mediterranean climate. Curves indicate extent of differences above or below those in open.

A. *Pinus pinea* with understorey of *Quercus ilex* on damp mesophilous site.

B. *P. pinea* or *forteto* (dense maquis of *Q. ilex* coppice) on xerophilous site. (From: U.N.O./F.A.O. 1962)

effect is reinforced by an insulating litter cover. The mean daily maxima of the ground surface can be reduced by as much as 10°C below those experienced by adjacent bare soil; increases in mean mimima are, however, less pronounced. Soils under a forest-cover freeze later and less deeply than in the open, except when the presence of a snow-cover on open ground, and the absence of such a cover under trees, may reverse the situation. It has been estimated that, all other conditions being comparable, frost penetrates to a depth eight times greater under bare ground than beneath a forest cover. The amplitude of diurnal and annual temperature ranges decreases exponentially with depth. The former become slight at sixty centimetres, the latter are just detectable at a metre. Annual variations are measurable to a depth of six metres but with increasing depth the time-lag in the occurrence of maximum and minimum temperatures as compared with the surface increases.

Modification of temperature also affects the atmospheric humidity in the trunk space. It is characteristically higher—as much as eleven per cent. in temperate and fifteen per cent. in tropical forest than in the open (see Table 32). Such variations in humidity have been interpreted as a result of lower temperatures

	(percentages)		
	in open	under forest	difference
Mean annual at 6 a.m.	97·9	16·7	−1·2
Mean annual at 2 p.m.	67·1	82·3	15·2
Difference	30·8	14·4	−16·4
Greatest difference between mean monthly values (6 a.m. and 2 p.m.)	51·8	42·9	−8·9
Smallest difference between mean monthly values (6 a.m. and 2 p.m.)	10·7	2·7	−8·0
Difference between extreme maxima and extreme minima at 2 p.m. *for whole period* recorded	77·0	72·0	−5·0

Table 32 Comparison of relative humidity values recorded in experimental forest Trang-Bom, South Vietnam, and in the open during the period 1933–37. (From: U.N.O./F.A.O. 1962)

and/or reduced air-movement both of which lower the evaporation potential of the atmosphere.

Dependent on its size, structure and density a forest community forms an effective and often massive 'wind break'. Surface air-currents are deflected around or over it, while within the stand the speed of air, movement is very markedly checked. Wind velocity in the interior of a wood may range from as much as two-thirds to less than a tenth of that in the open. In an area of tropical rain forest in Panama wind speeds above the canopy of over two hundred times the velocity of those at ground-level below it have been recorded. Reduction of wind speed is a function of the density and extent of the forest cover. It has been estimated, in European forests of medium density, that at a distance of thirty metres in from the windward edge velocities are sixty to eighty per cent. of those in the open; at sixty metres fifty per cent., and at 120 metres seven per cent.

As well as modifying the conditions of the atmosphere it occupies, the forest also exerts a sensible effect on its immediate surroundings, both around and above the community. The extent of what has been called the *'forest-fringe'* or *'forest-edge'* effect varies according to the character of the forest and prevailing macro-climatic conditions. Firstly air temperatures immediately adjacent to the canopy surface or the vertical edge of the forest are usually characterised by more marked temperature ranges. This is a result of reflection and out-going radiation on the one hand and marked diurnal variations in the light and shade around the forest on the other. Secondly, the forest's influence on atmospheric conditions can extend to an appreciable distance above the canopy: more rapid up-draughts of warm air over large forested as compared to non-forested areas are familiar to glider pilots! Temperature differentials between the forest and free-atmosphere in the Mediterranean woods, previously mentioned, have been detected up to five hundred metres above them. Thirdly, the effect on the reduction of wind velocity extends beyond the leeside of the forest; and, in this respect, a considerable amount of experimental work concerned with the deliberate creation of *shelter-belts* has been undertaken. As Figure 53 illustrates, the amount and extent of wind-speed reduction varies with the density of the barriers. It is least immediately behind open, most behind dense barriers; in the case of medium-dense barriers, however, the shelter effect extends much further beyond the woodland edge.

The possible effect of particularly large areas of forest on the humidity of the surrounding atmosphere and on the amount of precipitation received by an area have been subjects of considerable speculation and debate. The idea that the presence of trees actually caused an appreciable increase in precipitation arose from the close association of forests with areas of abundant rainfall. It

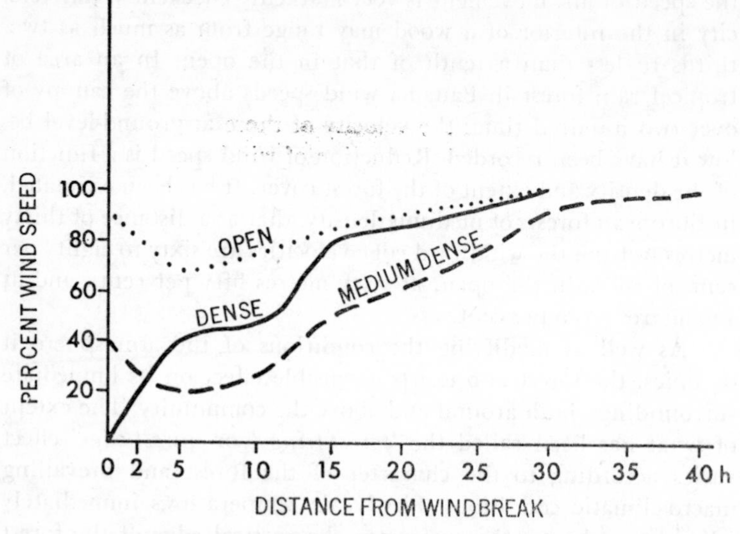

Fig. 53 Percentage wind-speed reduction downwind of shelter belts of varying density. 1. Open barrier: density of blockage ratio, 30 per cent; 2. Medium dense barrier: 50 per cent; 3. Dense barrier: 100 per cent. Distance from windbreak expressed in terms of height. (Data from: Gloyne, R. W. 1955)

was further strengthened by the assumption that trees transpired a much greater amount of water than any other type of vegetation. Indeed the passage of the Timber Culture Act of 1873, designed to promote tree planting in the Great Plains of the U.S.A. was based on the currently popular theory that trees were 'rain-makers'. While there is little direct evidence to prove or disprove this theory, it is now considered very doubtful that the presence of a forest or woodland, however extensive, can significantly influence the total amount of precipitation an area receives. It has, however, been suggested that slight local increases may result from: (1) the possible orographic effect of stands of very tall trees on the lifting and turbulence of unstable air masses; (2) condensation as a result of the lower air-temperatures and higher atmospheric humidity associated with many, though not all, forest micro-climates; and (3) the collection of water droplets from low-lying cloud or fog.

While the effect of the forest on the amount of precipitation received is negligible, the effect on its destination is considerable. The type of forest determines the routes and rates by which the water received in a given area finds its way back to the atmosphere.

Its effect on the hydrological cycle is complex and still not completely understood. In the first place, the amount of available precipitation reaching the surface of the ground is reduced, to a greater or lesser extent, as a result of interception by and evaporation from the tree canopy. The amount intercepted depends however on prevailing weather conditions, as well as the form and density of the cover. The percentage of rainfall intercepted decreases as the amount and heaviness of downfall increases; eighty per cent. of light showers but only ten per cent. of heavy downpours falling on a dry canopy may be lost. However, when the canopy leaves become completely wet they no longer check rainfall as effectively and losses by interception can become negligible. On the other hand, the percentage of snowfall intercepted invariably tends to be higher than that of rainfall. Losses will, for any type of precipitation be less in a light, open, leafless woodland than where a dense, closed evergreen canopy is present. Coniferous trees with their multitude of small leaves and extremely irregular canopy surface can trap a much higher percentage of incident rain or snowfall than can deciduous trees. The proportion of precipitation failing to reach the ground under British plantations has been estimated as varying from ten to fifty-five per cent.; and the loss by interception recorded for one year in a number of different forests was from thirty-two to thirty-six per cent. for deciduous and from thirty-seven to fifty-four per cent. for neighbouring coniferous stands. In deciduous woods seasonal variation can be considerable; in European beech forests summer losses exceeding forty per cent. contrast with less than nineteen per cent. in the winter when the trees are leafless. The magnitude and significance of interception by forests is not always fully appreciated. It is well to remember that in dense, evergreen, forests the amount of precipitation lost annually in this way may exceed that given off in transpiration.

Of the precipitation which reaches the forest floor a relatively high proportion penetrates into the soil, the absorptive capacity of which is often much greater than that of a bare mineral soil. Infiltration is facilitated by *'stem-flow'* (whereby water runs down the branch and trunk surfaces) which has the effect of directing and concentrating water along root channels, where the demand is greatest. The proportion of *'stem-flow'* to *'through-fall'* is to a marked extent a function of tree morphology. Slender trees with narrow, elongated crowns, as exemplified by the Lombardy poplar and the Lawson cypress have a greater channelling effect than those species with broad, widely spreading branches; the acute

Y-form is a well-known characteristic of many trees in arid and semi-arid regions. In addition, infiltration is further aided by the highly absorptive litter layer. Surface run-off and direct evaporation from the soil are as a result proportionately less than from a bare exposed soil.

Transpiration, however, continually removes a large amount of soil water. Because of the sheer magnitude of the plant biomass the forest contains a large volume of water at any one time. Most of this, however, is 'in transit'. In comparison the amount immobilised in the growing tissues is relatively small; it has been estimated that less than 0·1 per cent. of the annual precipitation becomes incorporated in the new organic matter formed each year. The effect of trees and forest-cover respectively on the amount and rate of evapo-transpiration and of the latter on the soil-water balance are still highly controversial subjects. Given an unlimited supply of water and a continuous vegetation cover with a uniform albedo, it has been maintained that the amount and rate of water loss would be entirely dependent on the atmospheric factors controlling evaporation rather than on the type of vegetation. However, when precipation is limited, trees, because of their more extensive root systems, tap a greater volume of soil and draw on less variable and deeper ground-water sources than other plants. They can, as a result, maintain high transpiration rates for longer periods and are less susceptible to drought than many shallow-rooted herbaceous plants. Also, under comparable climatic conditions the vegetative period, even of deciduous trees, is frequently longer than for many annual and perennial herbs. Further, it has been established that transpiration amounts can vary from one species to another. On *similar* sites in Britain daily losses per hectare during the summer of 24 000 kg for Scots pine, 38 000 kg for beech and 53 000 kg for Douglas fir have been recorded. Measurements in spruce plantations in Britain show output by evapo-transpiration from the soil to be as much as fifty per cent. of the precipitation received at ground level. On the sites indicated in Table 33 the percentage of water removed as a result of the combined effects of interception and evapo-transpiration is greater and surface plus ground-water run-off is smaller from forest covered than non-forested areas.

The net effect of these conflicting influences of the forest on soil-moisture conditions varies widely with relief, the physical characteristics of the soil and macro-climatic conditions, as well as with the type of forest community. In view of the relative paucity of measurements, generalisations can only be extremely

INSTALLATIONS

Location	Utah U.S.A.				Yorkshire England				Castricum Holland						San Dimas U.S.A.			
Vegetation Type	Aspen		Bare of vegetation		Sitka Spruce		Grassland		Black Pine		Hardwoods		Bare Sand		Pine		Bare Soil	
	Weight	%	Weight	%	Weight	%	Weight	%	Weight	%	Weight	%	Weight	%	Weight	%	Weight	%
Input (precipitation)	1 340	100	1 340	100	984	100	1 136	100	892	100	892	100	892	100	554	100	554	100
Output (evaporation + transpiration)	568	42	361	27	711	72	717	63	621	70	456	51	209	23	379	68	204	37
Output (surface + ground run-off)	772	58	979	73	273	28	419	37	271	30	436	49	683	77	175	32	350	63

CATCHMENT INSTALLATIONS

Location	Waggon Wheel U.S.A.				Coshocton U.S.A.				Coweeta U.S.A.				Sperbelgraben Switzerland		Rappengraben Switzerland	
Vegetation Type	Mixed Forest		Trees Felled		Mixed Forest		Grassland		Mixed Hardwood		Trees Felled		Mixed Forest		Grassland	
	Weight	%	Weight	%	Weight	%	Weight	%	Weight	%	Weight	%	Weight	%	Weight	%
Input (precipitation)	538	100	528	100	1 182	100	1 064	100	1 580	100	1 585	100	1 555	100	1 649	100
Output (evaporation + transpiration)	381	71	343	65	849	72	731	69	1 047	66	655	41	877	56	757	46
Output (surface + ground run-off)	157	29	185	35	333	28	333	31	533	34	930	59	678	44	892	54

Table 33 Annual water circulation on selected sites. Weight = 10^4kg/ha ; percentages—of precipitation. (From : Ovington, J. D. 1962)

tentative. On level, but well-drained sites under comparable climatic conditions, soils of similar texture might be expected to be drier under a forest cover than in the absence of woodland. However, due to reduction of surface evaporation beneath a tree canopy the situation may well be reversed in the upper A_0 and, to a lesser extent, A_1 soil horizons. On the other hand, where the liability to surface run-off is intensified by increased steepness of slope, forested areas would be expected to retain more moisture than otherwise, thereby retarding but at the same time regulating total discharge. The effect of a forest cover on soil-water content and drainage is probably most marked on heavy impermeable soils or those underlain at a relatively shallow depth by an impermeable substratum. Particularly in cool humid climates it is thought that the increase in soil-water following deforestation and the consequent reduction in evapo-transpiration may result in waterlogging, gleying and, in extreme cases, peat formation. The removal of a large percentage of precipitation by the forest reduces the surplus available to replenish ground water sources and stream flow. In controlled experiments in the U.S.A. the removal of part or all of a forest cover resulted in increased stream discharge. Stream flow accounted for thirty-four per cent. of precipitation in a densely wooded catchment area and fifty-nine per cent. in a comparable neighbouring area from which the forest cover had been cleared. It is now generally accepted that the yield of water available for use by man from forested areas falls far short of the potential in terms of 'actual' precipitation and is much less than from non-forested areas.

The forest biomass exerts a profound and distinctive influence on the physical and chemical properties of the soil as a result of the composition, amount and rate of decomposition of its contribution of organic material. Recent work using radio-active tracers together with detailed analyses of soil and plant nutrients in forests has provided interesting data on the magnitude and rate of nutrient cycling within forest ecosystems. Estimates of the mineral elements contained in plant tissues reveal that the forests' demands on the soil are apparently less than those made by agricultural crops on comparable sites (see Table 34). Rennie's work indicates that trees—largely because of the continual build-up of woody tissues—require a higher proportion of calcium (Ca) than of either potassium (K) or phosphorus (P); herbaceous plants, on the other hand, need more potassium. Uptake of nutrients can vary considerably with species and physical habitat. In Britain conifers, and more particularly pines are less

	Ca	K	P
Pines	203	91	21
Other conifers	438	234	41
Hardwoods	897	225	50
Agricultural crop	980	3 000	430

Table 34 Comparison of the amount of nutrients (kg/ac.) taken up from the soil and contained in tree organs (excluding litter) after 100 years growth and agricultural crop during 100 years cropping on the basis of a four-course rotation; oats, grass, potatoes, turnips. (From: Rennie, Peter J. 1956)

demanding than deciduous species. In this respect the calcium requirements of the latter are more nearly comparable to those of the main agricultural crops in this country. However, although absorption of minerals tends to be greater on nutrient-rich parent material, differences in the chemical composition of the biomass of the same species of equal age is not as great as might be expected from variation in the mineral content of contrasting sites. It has been suggested that the plant is capable of regulating the uptake and concentration of nutrients in its tissues.

While the average annual removal of nutrients from the soil by trees is less than by agricultural crops, the former retain and immobilise them in their boles for a much longer period than either annual or perennial herbaceous plants. Investigations of the percentage distribution of nutrients in different parts of the forest plant-biomass have shown that the annual uptake by tree leaves and small twigs together with the ground flora is greater than by the trunk and main branches. The annual return via litter therefore exceeds annual immobilisation in the wood. The percentage of mineral uptake 'fixed' annually however varies with age; while it can be of the order of thirty to fifty per cent. in young and hence rapidly growing birch trees, it can drop to ten

	Ca			K			P			Ca			K			P		
	H	O	P	H	O	P	H	O	P	H	O	P	H	O	P	H	O	P
Leaves	11	26	23	20	30	39	24	40	43	5	15	4	10	17	28	12	27	32
Brush	22	30	25	25	27	18	28	25	28	16	21	18	21	10	14	24	22	18
Stem bark	38	20	17	14	14	8	17	13	17	46	32	25	18	23	12	23	23	14
Stem wood	16	11	22	28	16	22	20	9	20	20	19	30	38	27	33	28	15	23
Roots	13	13	13	13	13	13	13	13	13	13	13	13	13	13	13	13	13	13

Table 35 Mean percentage distribution of nutrients in main parts of growing crop in H (hardwoods), O (other conifers), and P (pine) after 50 and a 100 years respectively. (After: Rennie, Peter J. 1956)

to fifteen per cent. in an old stand. Obviously the *cumulative* effect is such that in a mature forest seventy to ninety per cent. of the nutrient content of the plant biomass may be temporarily 'locked up' in its wood (see Table 35). While the forest ecosystem is relatively economical in its use of mineral nutrients, the amount

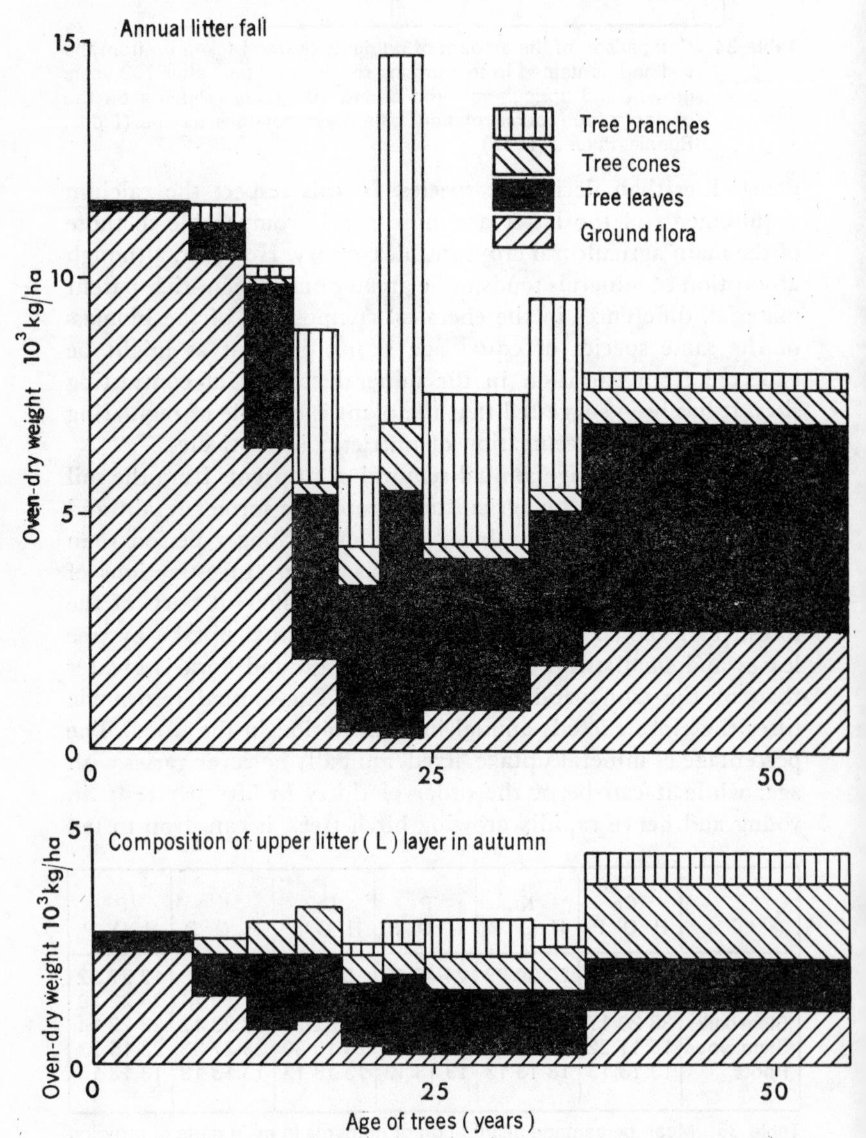

Fig. 54 Composition of annual litter fall and surface litter in plantations of *Pinus sylvestris* in Britain. (From: Ovington, J. D. 1962)

of organic material contributed to the soil, either in the form of living roots and stems, or dead and decaying plant and animal matter, is large. There are two features which distinguish the supply of organic matter to all forest soils. First the principal *source* is the tree stratum from which leaf-fall usually accounts for between fifty to eighty per cent. depending on the composition, structure and age of the forest (see Fig. 54). Second, because of the dominance of the arboreal contribution, the *supply* and *accumulation* of organic litter is concentrated on the surface of the ground. Forest soils are, as a result, characterised by the presence of a distinct surface organic layer of variable depth which is normally designated the A_0 horizon. This is composed of material in varying stages, which become more advanced with depth and age, of mechanical and chemical decomposition. In some cases it is possible to isolate these layers; a litter layer (L) composed of the most recent and often still-intact leaves; a fermenting layer (F) in which decomposition is further advanced, though the more resistant tissues usually preserve traces of the original form of the constituent material; and a humified layer (H) in which the organic material is reduced to a structureless, amorphous mass. The depth and composition of this A_0 horizon and that of the component L, F and H layers is dependent on the rate of decomposition. Total accumulation rarely exceeds a three to five years supply or, approximately, a depth of one metre.

Decomposition of organic matter and the resultant rate and magnitude of the nutrient cycle in the forest ecosystem is a function of several interacting variables. Of the latter the two most important are the composition of the litter and macro- and micro-climatic conditions. On comparable physical sites there is a close correlation between the nutrient status of the organic matter and the rate of its breakdown. Nutrient, and particularly base-rich, material with a low carbon: nitrogen ratio and a high proportion of the more easily degraded sugars, starches and proteins in relation to the less easily degraded cellulose and lignin can support a larger and more efficient soil population than can 'poorer' and 'tougher' material. In temperate regions the rate of decomposition of fresh, 'rich' deciduous litter may be such that there is little carry-over of undecomposed material from one year to the next (see Fig. 55). In autumn, under the fresh supply of fallen leaves, one may still detect recognisable remnants of the previous year's contribution. There is normally a rapid gradation into well-decomposed 'mould' completely humified and thoroughly mixed, as a result of intense biological activity, with the underlying mineral matter. Under these circumstances the A_0 horizon is rarely more than a few centimetres thick. The mild *mull* type of organic material contrasts with the *mor* produced by less demanding species such as pine or by hardwoods on nutrient deficient sites.

In the latter the litter, tougher and less rich in nutrients, provides a much less attractive substratum for living organisms than does *mull*. As a result of its acidity, the more efficient bacteria

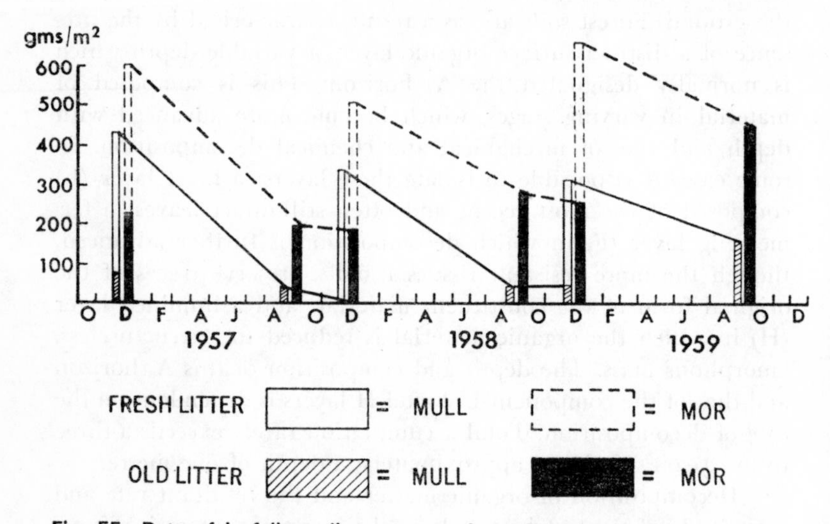

Fig. 55 Rate of leaf litter disappearance in mull and mor organic matter 1957–1959. (After: Witkamp, M. and J. van der Drift. 1961)

and earthworms—which play such an important rôle in the breakdown of *mull*—are sparse or absent. The less efficient fungi replace the bacteria as the principal agents of decomposition. The weight of soil organisms in *mor* is, in fact, usually less than half that in *mull*. Decomposition of organic matter is slower (see Fig. 55). It may take from three to five years and, as a result, a layer of only partially decomposed material accumulates and the L and F zones of the A_0 horizon tend to be well-developed in comparison to that of the H. Further, in the absence of burrowing animals, the organic matter forms a discrete A_0 horizon more clearly delimited from the underlying A_1 horizon than in *mull* soils.

The rate of decomposition is also affected by climate, particularly moisture and temperature conditions. This is strikingly illustrated by a comparison of litter production and accumulation under varying climatic regimes (see Table 36). In temperate forests the total volume of dead biomass far exceeds that of annual leaf-fall. In the humid tropics, on the other hand, the situation is reversed. Here a combination of high temperature and abundant soil-moisture promote optimum growth and decomposition. Litter fall in the tropical rain forest is continuous. Optimum

Climatic Zone	Litter production 10^3kg/ha/annum (1)	Biomass of litter, (L, F, H) 10^3kg/ha (2)
Cool Temperate	3·5	36–37
Warm Temperate	5·5	16–11
Equatorial	10·9	5.6–2.3

Table 36 Comparison of litter production and accumulation in forests in contrasting climatic zones ((1) based on Bray, R. J. and Gorham, E. 1964; (2) based on Ovington, J. D. 1965)

temperatures for organic-matter accumulation are between 20–25° C, for decomposition between 30–35°C. Above 25–30°C decomposition is so rapid that surface accumulation is negligible and for part of the year, at least, the soil may be virtually bare or have a thin discontinuous litter layer. More recent evidence, however, would suggest that in tropical latitudes, moisture is more important than temperature in determining the rate of decomposition. Microbial activity is apparently accelerated by alternating periods of wetting and drying and accumulation is even less than under conditions of constantly high temperature and rainfall. Also it has been assumed that because the skeletonisation and decomposition of litter in tropical forests is very rapid, the total humus content in the soil must be correspondingly low, if not negligible. However, it must be borne in mind that although decomposition is rapid so also is the formation of humus (the end-product of organic decay). Given the exceptionally high volume of litter production, the humus eventually incorporated *in* the soil can vary between three and five per cent., a proportion not dissimilar to that in most temperate forests!

The amount, composition and rate of decomposition of organic matter are reflected in the physical and chemical properties of the underlying forest soils. Figure 56 summarises some of the basic variations in the distribution of organic matter and nutrient cycles in different types of forest ecosystems. Characteristic of *mull* soils (derived from a combination of base-rich parent material and nutrient-demanding species) is a limited A_0 horizon, the mixture of well-decomposed organic matter with the

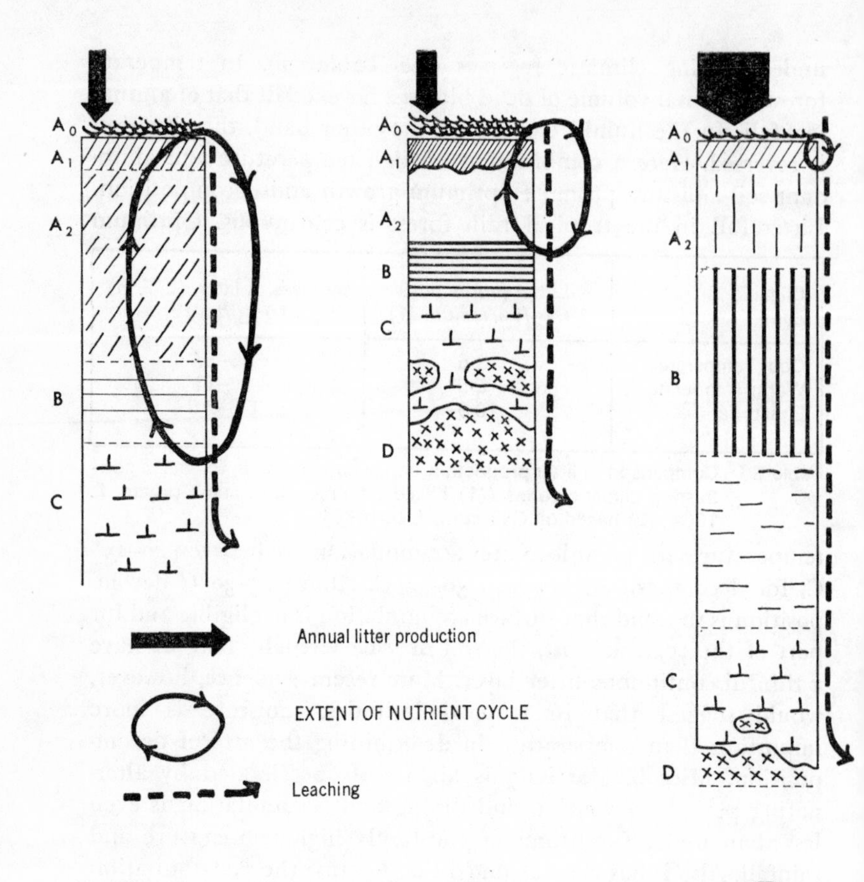

Fig. 56 Diagrammatic presentation of distribution of organic matter and nutrient cycles in three types of forest soil.

mineral material and its gradual dissemination through the A horizon. This is reflected in a gradation from a rich dark-brown near the surface to a lighter colour as the organic matter content decreases from five to six per cent. at the top to less than one per cent. near the base of the horizon. Within this horizon the intimate intermixture of humus and clay promotes a stable crumb to granular structure. An efficient and deep nutrient cycle counteracts the effects of leaching, maintains the nutrient status of the soil, and blurs distinct horizon differentiation. In *mull* soils two factors in particular help to counteract leaching. First, the base-rich organic matter tends to neutralise organic acids produced during the process of humification, and to allow the development of a stable clay/humus colloidal complex. The latter adsorbs and retain the basic cations, released in decomposition, while at the same time maintaining the physical stability of the individual soil particles. Second, the presence of a base-rich parent material and a deeply developed root-system can counterbalance the loss of bases leached from further up the soil profile or released in rock weathering. The latter tends to more than compensate for losses in soil drainage.

In contrast *mor* soils are characterised by the accumulation of partially decomposed nutrient-poor organic matter in a well-developed A_0 horizon and by a limited intermixture of organic and mineral matter in the upper part of the A horizon. Despite the quantity of organic matter the amount of available nutrients is relatively small. Also, the nutrient cycle, restricted in depth by the concentration of the main source of supply at the surface, is paralleled by the distribution of soil organisms and root systems. The effect of leaching is intensified, on the one hand, by the absence of a well-developed clay/humus colloidal complex; not only is the adsorptive capacity of the soil low but mineral and humic matter are more susceptible to mechanical leaching. On the other hand, the production of strong organic acids increases the solubility of inorganic compounds and promotes podsolisation. The latter process results in the chemical dissociation or breakdown of the complex clay-minerals and the removal in solution of compounds of iron and aluminium. The final result is not only a marked differentiation between the organic and non-organic horizons but between a severely leached A_2-horizon whose lighter texture and greyish colour contrast with the heavier texture and iron-enriched B-horizon. In extreme cases the concentration of iron may be such as to form an indurated hardpan which checks root penetration and inhibits free drainage.

Under humid tropical conditions, such as in the tropical rain forest, the production of organic matter may be four to six times that in temperate regions. However, decomposition is very rapid, and so the litter layer is limited and the amount of humus in the soil is small in proportion to that of the organic 'input'. Despite the tremendous volume of plant biomass, rooting systems are characteristically shallow, and while the nutrient cycle is distinguished by an exceptionally large and rapid turnover of nutrients it is limited to the upper few centimetres of the soil profile. The shallow humus-enriched horizon is frequently separated from the parent material by a considerable depth of intensely weathered, highly leached, nutrient-deficient mineral matter. As a result, the most biologically productive forest community in the world is, in effect, living precariously on the products of its own decay.

There is, however, one outstanding feature common to the nutrient cycles of all forest ecosystems. That is, in comparison to other systems, the long-term immobilisation of a large proportion

Crop	Nutrient elements: kg per acre				
	pH	Calcium	Potassium	Phosphorus	Nitrogen
Agriculture (one annual crop) Ebberston	5·8	0·0045	1·25	0·096	0·01
Sitka spruce (70 years old) Thornton Dale	6·4	0·092	5·5	0·7	—
Scots pine (150 years old) Scampston	4·8	0·34	0·40	0·21	0·18
Calluna moor (Silpho)	4·5	0·29	0·31	0·14	0·036
Pine afforested moor (80 years old) Suffield	4·5	2·4	2·4	0·95	0·29
Birch afforested moor (60 years old) Suffield	4·4	·69	3·0	0·42	0·14

Table 37 Ratio of plant nutrient content to available nutrients in top 50 cm of soil in contrasting types of vegetation. (From: Rennie, Peter J. 1956)

of the nutrient capital in the wood. This has far-reaching implications in relation to the use of forest resources. The removal of the forest biomass at a rate greater than it can be renewed has two major consequences. The first is soil-nutrient depletion which because of the cumulative uptake and the high plant: soil-nutrient ratio (see Table 37) of forests can be as detrimental to soil fertility as extractive agricultural practices. The second is the drastic modification of mirco-climatic conditions following the removal of such a massive biomass. Destruction of the forest environment may result in habitat changes inimical to its re-establishment. Nowhere are the effects of forest removal so drastic as in the tropical rain forest. More massive and complex than any other it is nevertheless the most fragile and delicately 'balanced' of ecosystems.

The Forest Resource

The relationship between man and forests has long been ambivalent. On the one hand, forests have, at various times, been regarded as a source of danger harbouring and obscuring wild animals or potential enemies, a barrier impeding ease and rapidity of movement and, not least, a hindrance to the effective production of arable crops. The rise and fall of civilisations have been accompanied by systematic and deliberate forest clearance. As a result, existing forest and woodlands have become increasingly concentrated on land which, for one reason or another, is less suitable for other, and more particularly agricultural uses and which is not necessarily always optimal for tree growth. On the other hand, the forest is one of the most valuable of man's natural resources. The size and complexity of the forest ecosystem is reflected in a variety of products and a diversity of uses unrivalled by any other type of biological community.

The nature and value of forest resources vary, dependent on the type of forest, the relative importance and competition from other resources, and not least the stage of economic and technological development of the areas in which they occur. The original and still the major use of the forest is as a *direct* source of such primary biological products as wood, food and fodder, and a variety of wood extracts among which gums, resin, latex, dyes, tannin, and medicinal substances (e.g. quinine) are the most important. Wood which forms the bulk of the forest biomass, however, still retains top priority; and with the exception of specialised plantations and orchards the vast majority of the world's forests are still exploited and managed for wood production. Until the discovery of coal, wood was the only source of fuel. In many areas which are technically underdeveloped or deficient in other resources it is still the principal source of domestic heat. Wood retained its importance as an industrial fuel well into the nineteenth century before coke replaced charcoal in the smelting of iron ore. In spite of the production of other natural and synthetic substances wood retains its importance as a constructional material. Finally its main constituent, cellulose, is the chemical basis for a wide range of products not least of which are alcohol, paper and rayon. In spite of recent changes in the value of wood as a resource, in 1965 J. D. Ovington estimated that of the world's harvested woodland, forty-two per cent. is used for firewood, thirty-seven for construction, four for pit props, eleven for pulp and six for other purposes.

The main biological problems basic to the exploitation of wood is the length of time required for its formation, and the relatively precarious nature of tree regeneration. The purely extractive process of cutting timber, at a rate greater than natural regeneration and growth, has caused the rapid depletion of forests particularly in the more recently exploited 'virgin' lands of North America and the southern hemishpere. The aim of rational exploitation, or controlled management, of wood production is to achieve a maximum sustained yield of the particular type of wood or wood product required. In this respect the productivity of 'natural' woods can be greatly increased by intensive management. The latter includes the removal of undesirable herbaceous and woody competitors, the harvesting of 'mature' trees whose increment is small in relation to their volume and requirements, and, where necessary, soil fertilisation and drainage. Once attained, yields can only be sustained if the correct balance between cutting and growth is maintained. The 'raising, tending and regeneration of a wood crop' is basic to most silvicultural techniques.

Although in comparison to agriculture, modern silvicultural techniques developed slowly, their origins are nonetheless ancient. Crucial to all systems is the problem of regeneration and one of the oldest methods is that of *coppicing*. This is dependent on the ability of many hardwood species to produce a number of new stems from the stumps or stools of felled trees. The stems can then be harvested on a rotation dependent on the species and size of wood required. This may vary from eight years for species such as hazel, willow, etc. to thirty years for oak. Coppicing is undoubtedly one of the most effective and rapid methods of production and provides a high yield of small wood for a variety of purposes such as firewood, fencing, bark for tannin, hurdles, charcoal, etc. Although some coppices originated from natural woodland the majority were deliberately planted. As well as the pure or simple coppice, *coppice-with-standards* was also established. The standards were scattered trees—often of either oak or ash in Britain—which were allowed to attain full maturity either for timber or a seed-crop. Akin to coppicing is the practice of *pollarding*—whereby trees are 'beheaded' in order to promote a bushy crown of small stems at a height safe from browsing by animals.

Although coppicing and pollarding for productive reasons is much less common than previously, the form of existing woodlands still testifies to its widespread use throughout western Europe. In Britain the system was most prevalent from medieval times to about the mid-nineteenth century. It was regularised at the end of the fifteenth century when a statute was passed authorising the enclosure of woodlands for a term of seven years after cutting. The obvious disadvantages of coppicing are the limited size of wood that can be produced and the inevitable drain on soil nutrients that such a continuous monoculture must effect.

Decrease in demand for small wood and an increasing market for large constructional lumber and pulpwood resulted in a decline in the coppice system. It raised the more complex problems of the management of even longer rotations, and of predominantly soft-wood tree crops. Silvicultural methods which would ensure the reproduction of the desired species from seed had to be developed. These are determined by methods of harvesting. *Selection cutting* involved the continuous felling throughout the forest of trees as they attain 'economic maturity' or 'maximum felling size'. The result is a stand of uneven age in which there is a mixture of trees of all sizes and often of several species. This system maintains the forest in a 'near natural' condition and is claimed to make the greatest use of and afford maximum protection to the site, to be of higher amenity value, and to promote a more stable ecosystem than the single-species plantation. Its management, however, demands a high degree of technical skill and it is questionable whether it is economically suited to the requirements of commercial wood-production.

More commonly employed is some method of *clear cutting or felling* whereby all the trees over a given area are removed at the same time. The result is even aged, single-dominant stands. In the interests of natural regeneration clear felling may be modified by variants of what is called a *shelterwood* system of cutting. In such cases the stand is opened up gradually in narrow strips or blocks in order to give maximum protection to the new seedlings. Successful natural regeneration of the commercially dominant trees requires particularly careful regulation of light and soil conditions, consequent upon cutting, in relation to the particular ecological requirements of the species in question. Artificial regeneration by planting out nursery grown seedlings tends to be more successful and less expensive in the long term. It allows the introduction and establishment of those species for which there is the greatest current demand. It is independent of variations in natural seed production and establishment. When combined with fertilisation and weeding it ensures a more uniform quality crop of even age.

The value of forests as a source of primary food is small in relation to their high level of biological productivity. Relatively little of their total biomass is edible by man and what there is is virtually confined to fruits and seeds. Trees make a relatively meagre contribution to total food supplies at present. The use of forests as a source of fodder for domestic animals, under present systems of management, conflicts with and is handicapped by that for wood production. Forests have long been an important source of animal forage supplied either by the herbaceous undergrowth or by the leaves of woody plants. Rights to common pasturage for pigs and goats in woodland was a characteristic feature of medieval agricultural systems in Britain; and there are still many parts of the world—in Mediterranean and tropical savanna regions for instance—where trees and shrubs provide *most* of the fodder for domesticated livestock. Indeed some authorities would maintain that they probably supply more forage for the world's population of ruminants than does grass. In very few cases, however, has this use been systematically regulated to ensure a sustained forage-production. The uncontrolled grazing and browsing by domestic herbivores has not only curtailed forest growth but, through the destruction of seedlings, prevented regeneration. Forestry, pastoralism and agriculture have long been regarded as incompatible types of land use. However, the necessity to exploit the biological potential of forests more fully has initiated attempts to develop what has been called 'three-dimensional forestry', a form of silviculture designed to combine the production of wood, food and animal forage simultaneously from the same area. The concept originated in the nineteen-thirties in Japan. It was stimulated by the need to rehabilitate rural communities in mountainous terrain devastated by deforestation and consequent severe soil erosion. High-yielding varieties of walnut trees were planted which provided protein-rich fodder for pigs and eventually high quality timber. Sustained production was effected by rotational harvesting and replanting. More recently, in 1957, a similar type of silviculture was introduced to the Limpopo Valley in Africa under the auspices of U.N.E.S.C.O. in an effort to increase agricultural production in a semi-arid area where mining operations had, as a result of deforestation, led to serious soil-water deficiencies. In this case the scheme involved the establishment of plantations of such drought-resistant trees as the mesquite (*Propopis juliflora* and carob (*Ceratonia siliqua*). These species give particularly high yields of protein-rich seeds or beans. Finally the possibility of using forests more effectively as a direct source of foodstuffs—particularly of edible protein—is being given serious consideration. One of the as yet untapped sources is *leaf protein*; leaves from a hectare of forest can contain one to three tonnes. Extraction and processing from fresh leaves has been carried out and the product is as nutritionally valuable as animal protein and is superior to seed protein. Potential sources are available in forests presently being felled for timber and pulpwood while a revival of coppicing could give a continuous production from forests in which wood production is uneconomic. As important as the direct products of the forest ecosystem are the less tangible benefits—the indirect ecological uses—which are assuming ever increasing recognition and importance. Not least is the manipulation of forest and woodland to modify and control local and microclimatic conditions in the interests of other types of land use. Their main use in this case is to provide shelter for agricultural crops, domestic livestock, buildings and, more recently, highways, from the adverse effects of wind, sun and snow. The use of trees as shelterbelts and wind-breaks and the appreciation of their biological benefits long pre-dated modern quantitative analysis of their specific effects. In Scotland many such uses originated during the period of accelerated agricultural improvement in the eighteenth century. Shelterbelts form a characteristic landscape feature in an environment where exposure is one of the main climatic hazards to life.

The principal effect of the shelterbelt is to reduce wind-speed in its lee, with a consequent reduction of evapo-transpiration and an increase in humidity and temperature. And while there is an inevitable decrease in crop yields in the immediate vicinity of the forest-edge as a result of shading and root competition, this is more than compensated by higher yields and often earlier commencement of growth than in unsheltered areas (see Figs. 57 and 58). Data from Germany and Denmark indicate increased crop yields of ten to twenty per cent.—and sometimes as much as fifty per cent.—higher than average within a strip extending a distance equal to ten to twenty times the height of the shelter belt (i.e. 10h–20h) in the lee of shelterbelts. The effect on yield, however, depends on the season and the direction of prevailing or dominant winds. This is strikingly demonstrated by a particular example in Germany. It was found that on the west side of a wind-break oat *grain* gave, on average, a nine per cent. increase in yield up to a distance of 30h, while on the east-side oat and barley *straw* an increase of over ten per cent. for a distance of 15h. The differences reflect seasonal variations in prevailing winds; protection from those from the south-west in late spring account for the increased

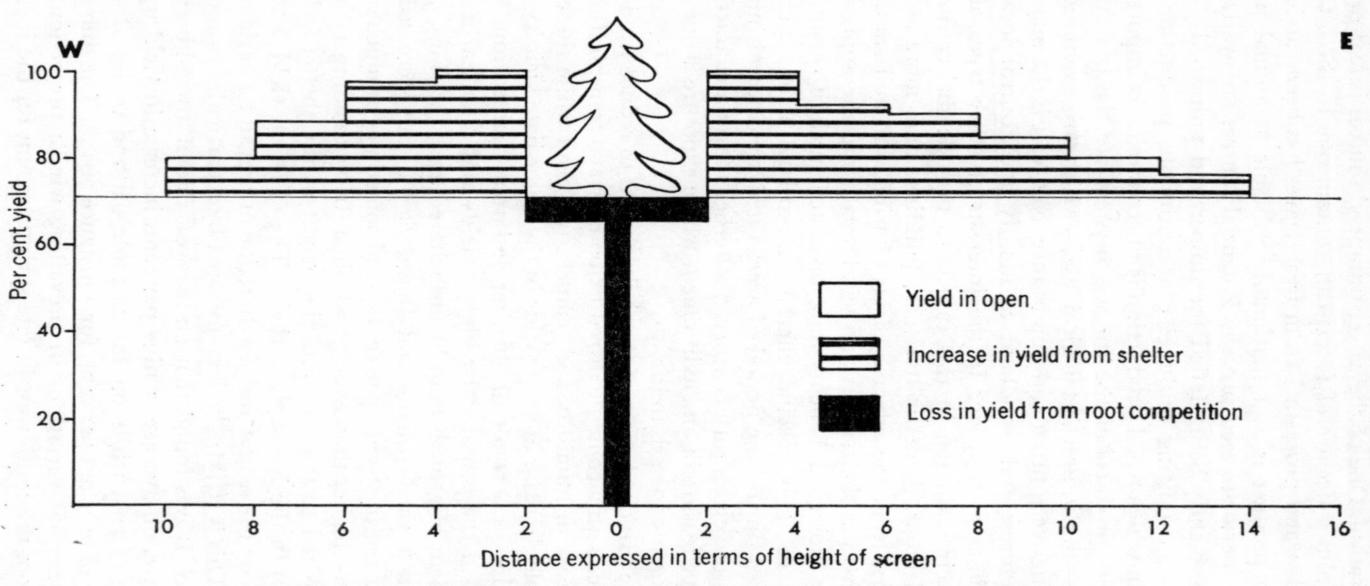

Fig. 57 Effect of shelterbelt on crop yield (From: U.N.O./F.A.O. 1962)

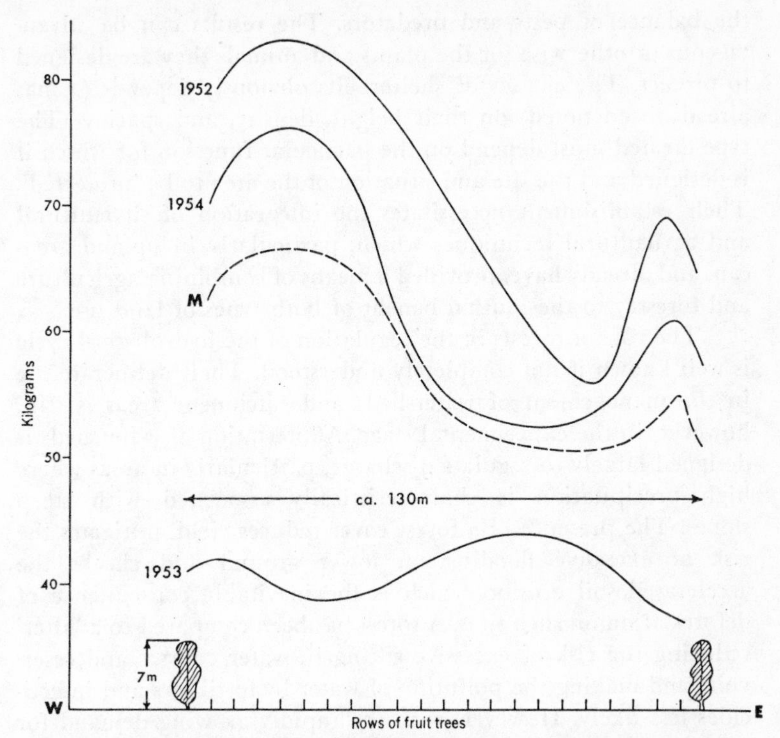

Fig. 58 Effect of wind-break on annual yield in apple orchards in the Dutch coastal province of Leeland. (After: R. J. van der Linde, 1955)

yield in vegetative material, in contrast to those from the east in July when seed maturation is active. In addition, shelterbelts afford protection to plants from physical damage as a result of high wind-force, as well as reducing heat loss from animals. Under certain circumstances it has been suggested that the greater number and efficiency of pollenating insects, when protected from excessive exposure, may be another advantage.

The provision of shelter is also important in silviculture, whether for the protection of forest nurseries or high-value orchards against gale damage and excessive evapo-transpiration. Many foresters, notably the late Professor Mark Anderson, have advocated the use of shelterbelts to make possible the extension of commercial plantations on to some of the higher and more exposed sites in Britain where at present yields are insufficient to give an economic return.

The creation of shelterbelts not only modifies the microclimatic conditions of the existing vegetation, but they also create within themselves a completely different habitat which may alter the balance of pests and predators. The results can be advantageous or otherwise for the plants and animals they are designed to protect. The efficacy of shelterbelts obviously depends (as has already been noted) on their height, density, and spacing. The type created must depend on the particular function for which it is designed and the site and situation of the area to be 'protected'. Their establishment necessitates the integration of silvicultural and agricultural techniques which, particularly in upland areas can, and already have, provided a means of combining agriculture and forestry to the mutual benefit of both types of land use.

The rôle of forests in the regulation of the hydrological cycle is well known if not completely understood. Their deliberate use in the management of watersheds and catchment areas is still, however, in the experimental stage. Afforestation of watersheds is designed largely to regulate discharge, particularly in areas where high precipitation is characteristically associated with steep slopes. The presence of a forest cover reduces yield, mitigates the risk of excessive flooding on lower ground and checks the accelerated soil erosion which is the inevitable consequence of deforestation on such sites. A forest has been compared to a 'filter' reducing the risk of excessive silting in water courses and reservoirs and making the pollution of water by fertilisers and insecticides less likely. However, with the rapidly growing demand for water the comparative benefits of forests in water regulation must be carefully assessed. As J. D. Ovington points out 'afforestation of water catchments is probably of greatest value where the problem is too much rather than too little, so that the basic needs are to reduce water-yield and diminish peak flow'. In arid and semi-arid areas, however, the vital need is to maximise water-yield while controlling as far as possible the less desirable effects of rapid run off and erosion. Nowhere is this problem more pressing than in the highly-populated, affluent, water-deficient areas of the south-west of the U.S.A. In the Fraser Experimental Forest in Colorado, research is being concentrated on watershed management designed to improve and, above all, *increase* water-yield. Records indicate that in this forested mountain catchment area about fifty per cent. of the annual precipitation becomes stream-flow; of this some seventy per cent. is contributed by melting snow in the period April to June. Hence experiments are being directed to the most efficient means of opening up the forest in such a way as to ensure that the maximum amount possible of the snowfall may reach ground level and hence augment stream-flow.

Since the Second World War the recreational and aesthetic

value of forest resources have become increasingly important. Indeed until the nineteenth century the management of forests for sporting and purely 'decorative' purposes was the prerogative of the land-owning aristocracy in Europe. The Royal Forests in Britain were originally established as hunting preserves and in many cases the term 'forest' was retained—as in the deer 'forests' of Scotland—and became synonymous with this form of land use long after the destruction of the original vegetation cover. Until the acquisition of publicly-owned land, the maintenance and management of forests on a significant scale was dependent on the capital investment of large land owners. This was motivated however not only by economic considerations but, particularly from the mid-seventeenth to mid-nineteenth century, by a desire to enhance the appearance of their estates. In Britain the great planting movement of the eighteenth century was influenced by the development of landscape gardening and found a wider and less formal expression in the planting of exotic trees, the creation of wooded parklands, tree-lined avenues, copses and hedges, which still persist as an integral visual characteristic of our rural landscape. Indeed the attractive and often seemingly 'natural' diversity of our present landscape owes much to the genuine interest in and concern for amenity by the 'improving landlords' of the eighteenth century.

Today there is a resurgence of concern about the appearance and despoilation of our landscape. Initially this was expressed in public reaction *against* the new Forestry Commission plantations; rectangular blocks of single species tended to produce monotonous uniformity. Amenity has now become an integral part of the Commission's management policy. The services of a landscape architect have been employed; and increasing attention has been given, with conspicuous success, to the composition and form of forests particularly in National Parks and areas of high scenic beauty. The concept of managing forests for amenity has recently received further statutory recognition in the Countryside Act of 1968. In some cases amenity may be the major justification for woodland and forest establishment in order to rehabilitate derelict mineral workings or 'camouflage' industrial wastelands.

The modern demand for recreational land is a direct result of the rapid increase of a more affluent and leisured population, particularly in highly urbanised and industrialised countries. The value of forests for recreational pursuits is a direct reflection of their diversity and complexity. Usually associated with rough and high terrain their coolness and shade are among their major attractions in countries with markedly continental summers. Their natural shelter provides the most desirable camp and picnic sites; and a wide range of outdoor pursuits can be pursued in an attractive and biologically varied habitat. The management of forests for recreational purposes, however, raises a number of problems which have yet to be solved. In the first place, it is not always easy to reconcile this use with the requirements of commercial forestry. Humans are in danger of assuming the proportions and characteristics of 'pests' in a forest crop in many areas. Uncontrolled access and movement increases the danger of serious damage by fire, and of seedling destruction, as well as increasing the difficulties of forestry operations. In the second place, the diversity of plant and animal species which enhances the recreational value of a forest is, from the point of view of commercial forestry, neither so productive nor economic as the monoculture of one tree species free of competition from weeds and pests. As important as the reconciliation of these conflicts, however, is the problem of how to sustain the maximum recreational use without detracting from the intrinsic value of the resource. This is dependent upon a knowledge of the 'carrying capacity' of particular forests for different types of pursuits. The solution to this problem is urgent.

Characteristic of the forest ecosystem is its rich and diverse resource potential. In view of the increasing demand for and pressure on a static if not dwindling amount of forest land the necessity to reconcile what in the past have become exclusive, incompatible and conflicting uses has been recognised by both ecologists and foresters. It led to the concept of *multiple land-use* as a means of reconciling the variety of uses to which any one stand of forest might be put. Multiple-use management is probably more highly developed in the national forests of the United States and Holland than elsewhere. In the former any one forest may be managed for a variety of purposes from timber-production, grazing, watershed management to wild life conservation and recreation with the emphasis varying according to site conditions and demand. The need for multiple land-use is even greater in Holland where it has also received official recognition. Dutch forests are 'zoned' with recreation concentrated on the poorest, and wood production in the best timber-growing areas. Whatever the means of carrying it out the aim of multiple-purpose use must be to attain maximum sustained use of the whole ecosystem.

References

ANDERSON, M. L. and EDWARDS, M. V. 1955. Possible limits of afforestation in Scotland. *Scott. Geogr. Mag.*, **71** (2): 94–103.

ANDERSON, M. L. 1950. *The selection of tree species.* Oliver & Boyd, Edinburgh.

BARR, J. 1968. If you go into the woods today. *New Society*, 29 August. 302–304.

BRAY, R. J. 1964. Primary consumption in three forest canopies. *Ecology*, **45** (1): 165–167.

BRAY, R. J. and GORHAM E. 1964. Litter production in the forests of the world. *Adv. Ecol. Res.*, **2**: 101–157.

CABORN, J. M. 1957. *Shelterbelts and microclimate.* For. Comm. Bulletin No. 29. H.M.S.O., Edinburgh.

CHAMPION, H. G. 1954. *Forestry.* Oxford University Press.

CROWE, S. 1966. *Forestry in the landscape.* For. Comm. Booklet, No. 18. H.M.S.O. London.

DOUGLAS, J. S. 1967. Men, animals and trees, *New Scient.*, August 24: 382–384.

EDLIN, H. L. 1956. *Trees, woods and man.* Collins, London.

GEIGER, R. 1950. *The climate near the ground.* Harvard Univ. Press.

GLOYNE, R. W. 1955. Some effects of shelter belts and wind breaks. *Met. Mag., Lond.*, **84** (998): 272–281.

GOUROU, P. 1961. *The tropical world*, 3rd. ed. Longmans, London.

GREENLAND, D. J. and KOWAL, J. M. L. 1960. Nutrient content of the moist tropical forest of Ghana. *Pl. Soil*, **12** (2): 154–174.

GRIGGS, R. F. 1934. The edge of the forest in Alaska and the reasons for its position. *Ecology*, **15** (2): 80–96.

HADEN-GUEST, S., WRIGHT, J. K. and TECLAFF, E. M. (Eds.). 1956, *A World geography of forest resources.* The American Geographical Society, Ronald Press, New York.

HANDLEY, W. R. C. 1954. *Mull and mor formation in relation to forest soils.* For. Comm. Bulletin. No. 23. H.M.S.O. London.

HUGHES, J. F. 1949, 1950. The influence of forests on climate and water supply. *For. Abstr.*, (**11**); also Commonwealth Agric. Bureau, London.

JENNY, H. *et al.* 1949. Comparative study of decomposition rates of organic matter in temperate and tropical regions. *Soil Sci.*, **68**: 419–32.

KITTERIDGE, J. 1948. *Forest Influences*, McGraw. Hill, New York.

KRAMER, P. J. and KOZLOWSKI, T. T. 1960. *Physiology of trees.* McGraw-Hill, New York.

LAURIE, M. V. 1958. The present productivity of forests. In *The biological productivity of Britain.* Symposia of the Institute of Biology, No. 7, pp. 73–82. Eds. Japp, W. B. and Watson, D. J.. London.

LAW, F. 1956. The effect of afforestation on yield of water catchment areas. *J. Br. Wt Wks. Ass.*, **38**.

LOVE, L. D. 1960. *The Fraser experimental forest . . . its work and aims.* Station, Paper No. 8, May, 1952. Revised June 1960. Forest Service U.S.

Dept. Agriculture. Rocky Mountain Forest and Range Experimental Station, Fort Collins, Colorado, U.S.A.

LUTZ, H. J. and CHANDLER, R. F. 1946. *Forest soils*, New York.

MACHATTIE, L. B. and MCCORMACK, R. J. 1966 Forest microclimate: a topographic study in Ontario. *J. Ecol.* **49** (2): 301–326.

MADGE, D. 1966. How leaf litter disappears. *New Scient.*, October, pp. 113–115.

MINISTERIE VAN NATIONALE OPVOEDING EN CULTUUR (BRUSSELS). *Ecosystem en Biosfeer*, Tweede Boek. De Ecologie. Moderne synthese-wetenschap. *Documentatie*, 23, n.d.

MUNN, R. E. 1966. *Descriptive micrometeorology*. Academic Press, New York.

MUTCH, W. E. S. 1968. *Public recreation in national forests: a factual survey*. For Comm. Booklet, No. 21, H.M.S.O. London.

NYE, P. H. 1960. Organic matter and nutrient cycles under moist tropical forest. *Pl. Soil*, **13** (4): 333–346.

OVINGTON, J. D. and MADGWICK, H. A. L. 1957. Afforestation and soil reaction. *J. Soil Sci.*, **8** (1): 141–149.

OVINGTON, J. D. 1958. Some biological considerations of forest production. pp. 83–90 In *The biological productivity of Britain*, Symposia of the Institute of Biology, No. 7. Eds Japp, W. B. and Watson, D. J., London.

OVINGTON, J. D. and HERTKAMP, P. 1960. The accumulation of energy in forest plantations in Britain. *J. Ecol.*, **48** (3): 639–646.

OVINGTON, J. D. 1962. Quantitative ecology and the woodland ecosystem concept. *Adv. Ecol. Res.*, 1962 (1): 103–192.

OVINGTON, J. D. *et al.* 1963. Plant biomass and productivity of prairie, savanna, oakwood and maizefield ecosystems in Central Minnesota. *Ecology*, **43**: 52–63.

OVINGTON, J. D. 1964. Prairie, savanna and oakwood ecosystems at Cedar Creek. In *Grazing in terrestrial and marine environments*. British Ecological Society Symposium Number 4, Ed. D. J. Crisp. Blackwell, Oxford.

OVINGTON, J. D. 1965. *Woodlands*. English University Press, London.

PARDÉ, J. 1965. *Arbres et forêts*. Armand Colin, Paris.

PEARS, N. V. 1967. Wind as a factor in mountain ecology: some data from the Cairngorm Mountains. *Scott. geogr. Mag.*, **83** (2): 118–124.

PIRIE, N. 1968. Food from the forests. *New Scient.*, November, **21**: 420–422.

RENNIE, P. J. 1956. The uptake of nutrients by mature forest growth. *Pl. Soil*, **7** (1): 49–95.

RICHARDS, P. 1952. *The tropical rain forest*. Cambridge University Press.

ROBBIE, T. A. 1955. *Forestry*. English Universities Press, London.

RUTTER, A. J. 1958. The effects of afforestation on rainfall and run-off *Instn. publ. Hlth. Engrs. J.*, **57** (3): 119–138.

RYLE, G. 1969. *Forest service: the first forty-five years of the Forestry Commission of Great Britain*. David and Charles, London.

STEVEN, H. M. 1963–4. The beneficent forest. *Adv. Science*, **20** (86): 345–351.

TIKHOMIROV, B. A. 1962. The treelessness of the tundra. *Polar Rec.*, **11** (70): 24–30.

TINKER, J. 1969. Marrying wild life to forestry. *New Scient.*, June 5: 518–520.

TITTENSOR, R. M. 1970. History of the Loch Lomond oakwoods *Scott., For.*, **24** (2): 100–118.

WHITTAKER, R. H. 1961. Estimation of net primary production of forests and shrub communities. *Ecology*, **42** (1): 177–180.

WILDE, S. A. 1946. *Forest soils and forest growth.* Chronica Botanica, New York.

WITKAMP, M. and DRIFT, J. VAN DER. 1961. Breakdown of forest litter in relation to environmental factors. *Pl. Soil*, **15** (4): 295–311.

WRIGHT, T. W. 1954. Effect of tree growth on soil profile development. *Report on Forest Research* for year ending 1953, pp. 103–5. For. Comm., H.M.S.O., London.

UNIVERSITY COLLEGE OF WALES, ABERYSTWYTH. 1959. *Shelter problems in relation to crop and animal husbandry.* (*mimeo*). Memo No. 2.

UNITED NATIONS FOOD AND AGRICULTURAL ORGANISATION. 1962. *Forest Influences.*

U.S. DEPARTMENT OF AGRICULTURE. 1949. *Trees.* Yearbook of Agriculture, Washington, D.C.

VEDAL, H. and HANGE, J. 1960. *Trees and bushes in wood and hedgerow.* Methuen, London.

YEATMAN, C. W. 1954. *Tree root development in upland heaths. For. Comm. Bulletin*, No. 21, H.M.S.O., London.

13 Grasses and the Grassland Ecosystem

The grass family (*Graminaceae*) has attained an ecological and economic significance unsurpassed by any other genetically distinct group of plants. It forms the most important component of herbaceous vegetation. Grass-dominated ecosystems are second only to forests in extent and far exceed them in economic importance. The grass family is the principal direct and indirect source of man's food. The Biblical statement that 'all flesh is grass' only falls a little short of literal interpretation; less than twelve species of domesticated grasses furnish the bulk of man's food supply, while these and other grasses provide more animal fodder than do all other types of plants combined. The grass family has played a major rôle in the evolution and spread of modern civilisation. Nowhere have the extent and implications of man's present dependence on the grass family been more vividly illustrated than in John Cristopher's now-classic science fiction *The Death of Grass*!

Despite the relatively late evolution of grasses (in the Tertiary era) the family is large in number of genera and species. In the former case it is exceeded only by the *Compositeae* and *Orchidaceae*, in the latter by these, the *Leguminoseae* and the *Rubiaceae*. But in total population grasses must far outstrip any other family. Furthermore, the range of the family is wider and more truly cosmopolitan than that of any other group of flowering plants. Members of the grass family occur in all types of habitat and their range of tolerance of environmental conditions is exceeded only by lichens and algae. There are few types of vegetation in which grasses are not represented. Beyond the limits of forest and woodland, grasses have achieved an ecological dominance unrivalled by any other type of herbaceous plant.

The grass family owes its ecological success and economic significance to certain distinctive morphological and physiological

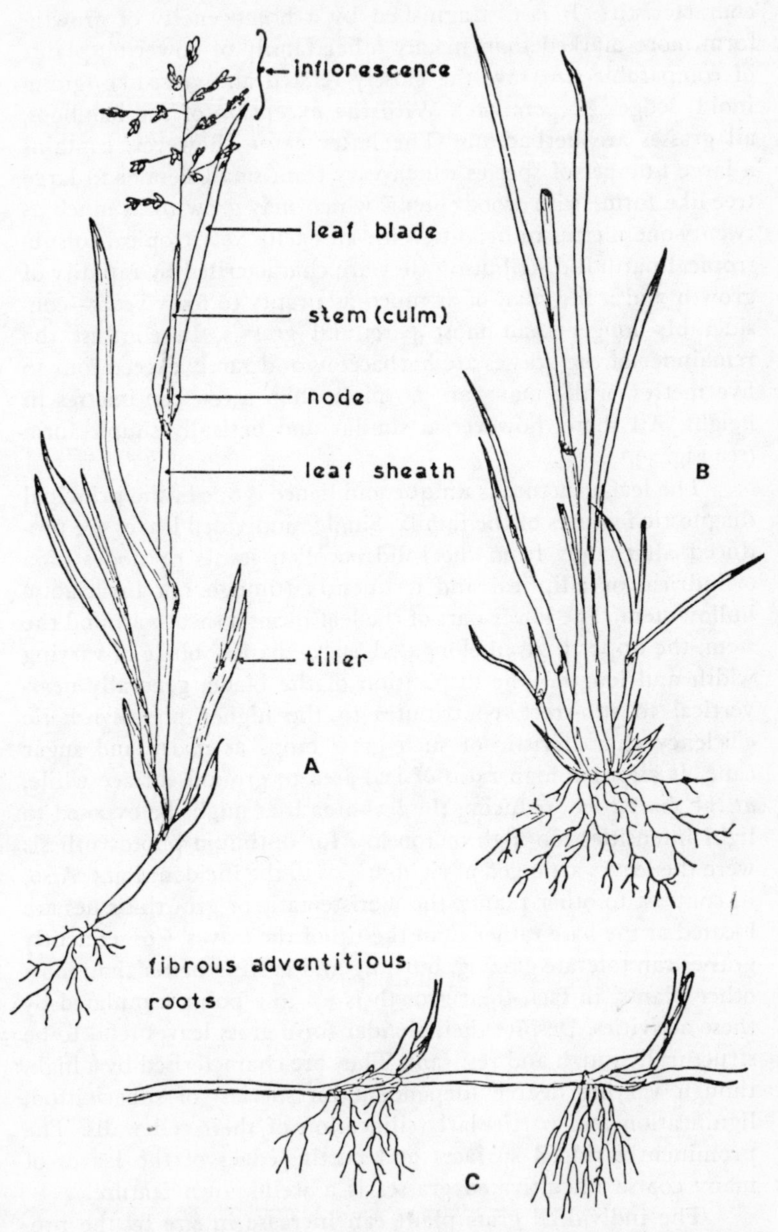

Fig. 59 Basic characteristics of grass form and habit: A. Typical annual grass (e.g. *Poa annua*); B. Tussock habit; C. Rhizomatous habit, creeping underground stem with leafy shoots and adventitious roots developed at nodes.

characteristics. It is distinguished by a homogeneity of growth-form more marked than in any other family of flowering plants of comparable size, save the closely related and grass-like (graminoid) sedges (*Cyperaceae*). With the exception of the bamboos, all grasses are herbaceous. The latter genus, however, contains a large number of species which vary from small slender to large tree-like forms with woody stems which may grow to as much as twenty-one metres in height. With an exclusively tropical to subtropical natural distribution they are characterised by rapidity of growth and a life span of as much as twenty to forty years—considerably longer than most perennial grasses. In contrast the remainder of the grasses are herbaceous and rarely exceed four to five metres, while many are no more than a few centimetres in height. All share, however, a similar and basically simple form (see Fig. 59A).

The leaf structure is unique and hence is one of the principal diagnostic features of the family. Single, undivided leaves are produced alternately from the solid, swollen joints or nodes of a cylindrical or elliptical and frequently (though not invariably) hollow stem. The lower part of the leaf forms a sheath around the stem, the upper part an elongated 'spear-shaped' blade of varying width and length. The disposition of the blade, generally near-vertical to sub-erect, contributes to the higher photosynthetic efficiency characteristic of such grass crops as maize and sugar cane. It allows a high ratio of leaf-area to ground surface while, *at the same time,* reducing the leaf-area that might be exposed to light intensities too high or too low for optimum photosynthesis were the leaves arranged at right-angles to the incident light. Also, in contrast to other plants, the meristematic or growth-tissues are located at the base rather than the tip of the leaves. Consequently grasses can tolerate grazing, burning and cutting better than most other plants; in fact, their growth is up to a point stimulated by these activities. Despite their slender form grass leaves tend to be structurally tough and resistant. They are characterised by a high, though varying degree (dependent on species) of suberisation, lignification and particularly silification of their cell walls. The prominently ridged surfaces and cutting-edges of the leaves of many coarse uncultivated grasses is a well-known feature.

The individual grass plant can increase in size by the production of lateral shoots or 'tillers', from axial buds produced at the base of the stem. The degree of tillering varies with species but can be promoted and intensified by the damage or removal (as in grazing or cutting) of the main shoot. Growth buds of grasses tend, however, to be close to the ground (a 'hemi-cryptophytic' life-form which protects the plant from environmental hazards) and stems, other than the flowering stems, remain short in spite of an increase in the number of leaves and shoots. Grasses have two principal growth-forms (see Figs. 59B and 59C) dependent on their life span and manner of branching. The tufted, tussock or 'bunch' grasses include all annual and many perennial species. In the former, new lateral shoots arise from the basal node of the main stem with the resultant crowding or bunching of tillers near ground level. Annual species do not persist long enough to be able to form the large prominent tussocks characteristic of certain perennial grasses. The latter possess underground stems or rhizomes from whose basal buds new shoots are produced. In those species, the internodes of which are short and crowded together, the shoots become concentrated into a tight densely-packed bunch or tussock. With continued growth of new shoots and the accumulation of dead and decaying ones, such tussocks may attain a considerable height and diameter. In contrast other perennial grasses tend to be mat- or turf-forming; they can propagate themselves vegetatively by means of creeping, elongated underground (rhizomes) or surface (stolons) stems from whose more widely spaced nodes new shoots are produced at regular intervals.

Characteristic of all grasses is an underground root-system which is large and extensive in proportion to the size and volume of the shoot. Lacking tap roots, grasses have, however, the ability to produce rapidly a large volume of adventitious roots from the base of the stem in annuals or from the nodes of rhizomes and stolons in perennials. These are constantly being renewed and, in addition, further root growth can be stimulated by splitting or damage of the stems. The roots are—with few exceptions—characteristically thin, fibrous, freely branching and are all of comparable size. They are covered throughout their length by absorbent root hairs. They form a dense ramifying network which, although individual roots can extend to considerable depths, tends to be concentrated in the surface layers of the soil. The efficiency with which this root system can take up soil-water and nutrients from a large volume of soil gives the grass species a competitive advantage over plants of similar rooting depth. The density and rapidity of growth of their roots and underground stems make the grasses the most effective colonisers and stabilisers of unconsolidated, mobile sediments. At the same time the density of the mat or sod which they form can, once established, check the penetration of the seedlings of other less aggressive species.

The reproductive and dispersal capacity of the grass family is high, a factor which has undoubtedly contributed to the extent of its range and dominance. The grass panicle or inflorescence carries a large number of small, inconspicuous greenish flowers. Most grasses flower every year though some perennials, particularly of salt marshes, may do so more irregularly. Tropical grasses flower less frequently; bamboos, for instance, do so only at intervals ranging from a few to many years. Photo-periodism and the change from the vegetative to the flowering stage in response to day-lengths is a characteristic of the grass family. Wind pollinated or self-fertilising grasses are independent of insect ranges. Seed production is abundant and the characteristically small, light seeds are readily and widely dispersed. They are easily carried by wind, while many possess spines, barbs or sharp points which adapt them to ease of transport by animals and man. Furthermore many grass seeds are particularly resistant to unfavourable soil or climatic conditions and can remain dormant for several years before germination. However, according to J. R. McIllroy, it would appear that tropical grasses do not seed as freely as those of more temperate and cold regions and when they do their seed is often of lower viability. Grasses can be annual, biennial or perennial in habit. In biennial species the seed germinates in late summer or autumn, the plant flowering and setting seed the following year. Perennial grasses can reproduce new shoots each year from surface or subterranean storage-organs. Unlike biennials and annuals, which bear flower heads on all or most of the shoots, perennials produce both flowering and vegetative shoots, the proportion of which depends on the duration of the particular species.

Finally, the form and composition of the grass fruit or 'grain' has made it one of the most valuable sources of human food. The seed, with only a thin, protective outer coating, comprises the bulk of the fruit. In the cereal grains this contains, in varying proportions, all the main food elements necessary for human nutrition. Hence the grass family has provided not only one of the most highly concentrated forms of food but one which because of its low water-content has good keeping qualities and is also economical to store and transport.

The Grassland Ecosystem

The volume and aggressiveness of their root systems combined with their natural adaptation to grazing and burning pressures have given the grasses a competitive advantage over the majority of forbs. When environmental conditions are suitable for their growth and in the absence of competition from woody plants, the grass-dominated community or 'grassland'—either cultivated or uncultivated—is the most common type of herbaceous vegetation. Despite the considerable diversity of species-composition and form of communities, there are certain features common to the grassland ecosystem which contrast markedly with those of the forest. In the first place the former is much less massive and occupies a comparatively restricted physical environment, at most probably

	Oven dry weight plant material (kg × 10³/ha)			
Prairie	*Minimum*		*Maximum*	
Living plant material above ground	0·1	*(Apr)*	0·9	*(Sept)*
Dead plant material	2·0	*(May)*	3·8	*(Nov)*
Roots and subterranean stems	3·5	*(Aug)*	6·7	*(Apr)*
Total plant biomass	6·1	*(May)*	9·7	*(July)*
Savanna				
Living plant material above ground	30·2	*(Apr)*	35·0	*(Aug)*
Dead plant material above ground	10·9	*(Jun)*	16·6	*(Oct)*
Roots and subterranean stems	11·7	*(Nov)*	14·9	*(Sept)*
Total plant biomass	54·4	*(Apr)*	63·2	*(Sept)*
Oakwood				
Living plant material above ground	161·2	*(Apr)*	165·8	*(Sept)*
Dead plant material	56·1	*(Apr)*	72·8	*(Jun)*
Roots and subterranean stems	10·1	*(Nov)*	20·7	*(July)*
Total plant biomass	224·2	*(Aug)*	257·1	*(Jun)*

Table 38 Maximum and minimum values of total plant biomass, and major components, during sampling months April-November inclusive, in three contiguous ecosystems in east-central Minnesota, U.S.A. (After: Ovington, J. D. 1964)

no more than a tenth of the space occupied by the most luxuriant forest. On comparable sites (as is illustrated by J. D. Ovington's measurements in central Minnesota) the total prairie biomass is less than five per cent. that of adjacent oak-woods (see Table 38). The difference in the amount of living plant material above ground is much greater than that between other types of biomass in the two ecosystems. In addition the variation between maximum and minimum values of living surface vegetation, during the period April to November, is proportionately greater in the grassland than in the forest. In the former the minimum biomass is only eleven per cent. of the maximum recorded; in the latter it is over ninety per cent.

The structure of the grassland plant community is much simpler than that of the forest. It is dominated by a 'field' or herbaceous stratum. Even under the most favourable conditions in the humid tropics this rarely exceeds four to five metres at the period of maximum growth; in more rigorous habitats, particularly those subject to heavy grazing pressure it may attain no more than a few centimetres in height. Also the dominance of grasses in terms of cover-abundance is less complete than that of trees in the forest ecosystem. It is dependent more on the abundance and volume of grasses in the total vegetation than on their height and persistence. The brilliantly coloured flowers of forbs, in comparison to the inconspicuous grass inflorescences, often gives a misleading impression of both the relative abundance and dominance of the component species in a grassland. T. L. Steiger's studies of remnants of near 'natural' prairie in east Nebraska suggest that the number of species of plants in grassland communities can be very high; in comparable habitats lack of shade and competition normally allows greater species diversity than in forests. Of the 237 regular elements only thirty-eight were grasses (primarily perennial) and eighteen sedges which together represented some twenty-six per cent. of the total flora; these, however, comprised at least ninety per cent. of the volume of the vegetation. The commonest forbs belonged to the *Compositeae* (forty-six species) and the legumes (twenty species); there were only five other families with more than five species represented. Stratification, however, is less highly developed and much less apparent in the grassland than in the woodland community. In some cases an open, discontinuous stratum of emergent trees or shrubs may be present as in the case of savanna formations. A ground layer of small plants with 'rosette' forms, mosses and lichens is a more consistent feature. However, although the herbaceous stratum is frequently composed, of grasses, forbs and even dwarf-shrubs of varying growth-heights the resultant stratification tends to be obscured by seasonal variation in growth and flowering. Composed of a wide variety of species with differing ecological requirements and life-cycles, variation in species dominance and grassland physiognomy tends to give a greater and more prolonged but more subtle gradation in aspect than in a forest community.

A distinctive feature of the grassland plant community is the high proportion of the total biomass that is contained in the soil. It has been suggested that at the period of maximum development, above-ground vegetation accounts for only a fifth of the total weight of living plant matter, while as is shown in Table 39 the average for the growing period can be less than ten per cent.

Vegetation sample	Ecosystem and Sample Period		
	Prairie (Apr–Nov)	Savanna (Apr–Nov)	Oakwood (Apr–Nov)
Herbaceous layer	449	770	88
Shrub "	10	47	512
Tree "	0	31 223	163 076
Roots and subterranean stems	4 824	11 789	14 997
Total living vegetation	5 283	43 829	178 673
Litter on ground	2 788	9 625	36 735
Total dead plant material	2 788	13 650	58 572
Total plant material	8 071	57 479	237 245

Table 39 Plant biomass (oven dry) in three ecosystems expressed in kilograms per hectare (average of all observation in sample period). (From: Ovington, Lawrence and Heitkamp. 1963)

One of the outstanding features of the grassland ecosystem, in comparison to that of woodlands, is the correspondingly larger contribution made by roots and stems to the dead and decaying organic matter below ground.

The nature and distribution of plant biomass is reflected in that of the associated animal population. The ratio of the weight of soil animals—particularly decomposers—to others is even greater in grassland than forests. The energy value of the stock of decomposers alone can equal that of the surface vegetation (see Fig. 60). This distribution is a function of the distribution of food resources and of environmental conditions. Many grasslands, and more particularly uncultivated 'range' lands of cool and semi-arid climates, are lacking in cover and the protection it affords from predators. They are, as a result, more exposed and subject to more extreme climatic variations than in forests where climatic modification extends through a greater vertical distance. Most vulnerable are the larger mammals which tend to counteract these disadvantages by either a cursorial or a burrowing habit (see Table 40).

The former is a feature of the ungulates native to grassland or more open savanna vegetation. These (such as the bison, antelope or horse) also exhibit 'flocking' and migratory habits which are much less characteristic of forest inhabitants. Smaller mammals,

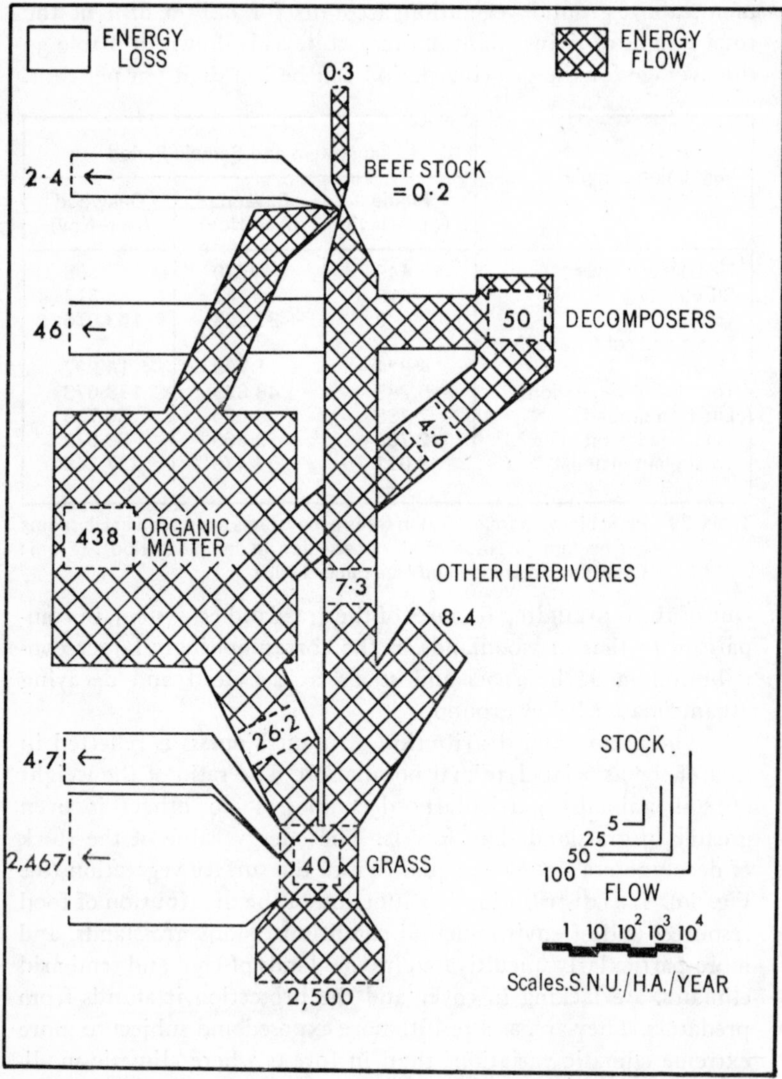

Fig. 60 Energy-flow model of meadow subject to grazing by beef cattle. Mean stock (biomass) indicated by quantities in open squares. Flow channels logarithmically proportional to mean annual energy flow. S.N.U. = Standard Nutritional Unit = 10^3Cals. (From: Macfadyen, A. in Ed. Crisp, D. J. 1962)

Per cent. mammals which use substratum	*Grassland*	*Forest*
— for breeding	70	42
— for habit of life	47	6
Per cent. animals with 'flocking habit'	47	10

Table 40 Comparison of selected habits of grassland and forest mammals. (From: Allee, C. W. et al. 1949)

the most numerous of which are rodents such as mice, voles, rabbits and gophers breed and live below but feed above ground. In addition, the strong well-developed incisors, broad roughened molars and the continuously growing teeth of 'grassland' rodents and some ungulates are thought to be an adaptation to the mastication of tough grass herbage.

Under comparable environmental conditions (as Ovington has further demonstrated) the annual productivity is eight to nine times greater in woodland than in adjacent prairie (see Table 41).

	Prairie	Savanna	Oakwood
Herbaceous stratum	920	1 886	182
Shrubs	10	41	389
Trees			
Current shoots		2 833	4 046
Older ,, (e.g. trunks and branches)		503	3 576

Table 41 Annual net production in kg per hectare of above-ground plant matter in three contiguous types of vegetation in east-central Nebraska. (From: Ovington, J.D. 1964)

That the nutrient content of the oakwood is about twenty-five times that of the prairie is not surprising given the difference in biomass. But the percentage content of nutrients per hundred grams dry weight is also greater in both the oakwood and savanna than in the prairie—particularly in relation to potassium, magnesium, phosphorus and nitrogen. The basic difference in both energy flow (see Fig. 60, p. 302) and nutrient circulation between wood and grassland is that in the latter neither the nutrient 'capital' nor the energy equivalent of the plant biomass shows much annual variation in comparison to the annual accumulation in the wood of the forest ecosystem. The rate of nutrient turn-over in the grassland is relatively high, and with controlled management of domestic livestock, grazing can be regulated to give (because of the continued growth of the grass leaf) higher yields and turnover than would otherwise occur.

Grassland Influences

The extent to which a grassland community modifies its physical habitat is dependent on the specific composition, height and density of vegetation cover. Irrespective of these variations, however, its influence is much less than that of the more massive plant biomass of the forest. Also, in contrast to the latter the grassland exerts a much greater effect on the soil than on the atmosphere of the particular site it occupies. In this respect it differs from the forest in its shallower 'canopy' and also in the absence of a micro-climatically distinct zone (i.e. trunk space) between the canopy and the ground. While a forest can be thought of as a blanket of varying thickness and density suspended *above* the ground, grassland forms a blanket lying *on* the ground.

The herbaceous vegetation of a grassland affects the properties of the air it encloses as well as the physical and chemical condition of the soil. Available data indicate that the albedo for white light from field crops and meadows is of a higher order (fifteen to thirty per cent.) than from forests (five to eighteen per cent.). This difference is presumably related to the generally darker colour and hence greater light absorption of the tree crowns. The presence of the herbaceous layer however protects the surface from the direct insolation to which bare ground would otherwise be subjected, as well as impeding heat loss by outgoing radiation. Rudolph Geiger has demonstrated, in his micro-climatic studies, that in temperate meadowland the greater part of insolation not lost by reflection is absorbed before it reaches the ground (see Fig. 61). The amount and vertical variation in absorption, and its effect on diurnal and seasonal temperature regimes within the vegetation cover are dependent on the height and density of the stand. There is a considerable difference between the cultivated 'grass' community such as a field of corn with vertically disposed leaves and little ground cover, and a mixed meadow or prairie with a varying proportion of forbs with horizontally arranged leaves in which high density is combined with a well-developed stratification and a continuous ground-cover.

In general maximum daytime temperatures occur within the plant cover at a level below the surface of the vegetation dependent on its density. The higher the density the greater the absorption near the top of the cover and the cooler the atmosphere in the lower layers of air near the ground. The absorption of heat gives rise to differences between the temperature in the herbaceous cover and the free atmosphere which can be easily felt

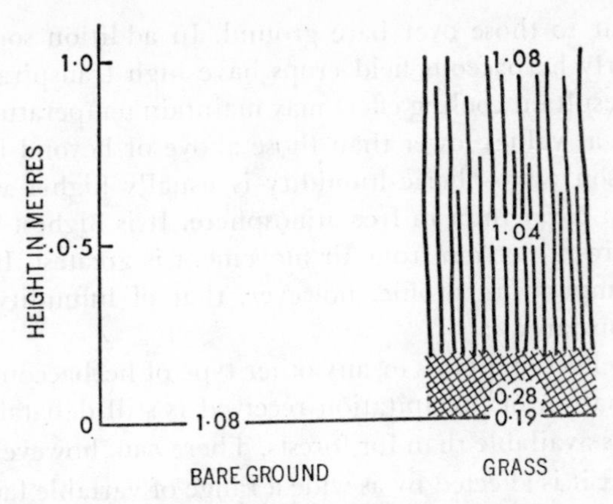

Fig. 61 Absorption of radiation in a meadow one metre high compared with bare ground. Insolation expressed in calories per sq. cm; degree of shading indicative of density of grass. (After Angstrom, from Geiger. 1950)

in a particularly tall, lush meadow, a corn-field or stand of rushes. During the day therefore temperature inversions occur up to the level of maximum temperatures, above which normal lapse rates prevail from the vegetation into the free air. At night, however, temperatures near the ground tend to remain higher than towards the top of the stand, provided the outgoing radiation which affects the upper layer is not accompanied by sinking of the cooled air. At night, therefore, the daytime temperature pattern is reversed with normal lapse conditions occurring up to the level of minimum temperatures, above which inversion may occur (see Fig. 62). However, where the individual plants are so widely spaced that insolation can penetrate easily to the ground, surface temperature profiles

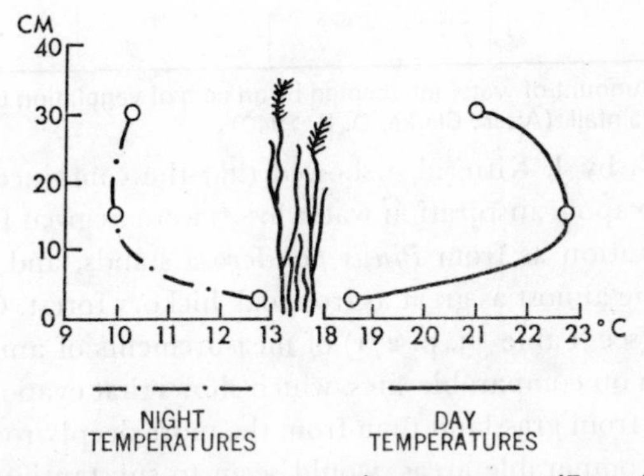

Fig. 62 Night and day-time temperatures in a grass cover. (From: Waterhouse. 1955)

are similar to those over bare ground. In addition some plants, particularly herbaceous field crops have high transpiration rates and the resultant cooling effect may maintain temperatures within the stand at values lower than those above or beyond it. For the same reason atmospheric humidity is usually higher within the vegetation cover than in free atmosphere. It is highest at ground level where protection from air movement is greatest. In contrast to the temperature profile, however, that of humidity tends to remain constant.

The effect of a grass or any other type of herbaceous cover on the destination of precipitation received is still debatable. Much less data is available than for forests. There can, however, be little doubt that it is affected by as wide a range of variable factors as in the latter system. All other factors being comparable, there are indications that at the period of maximum growth interception by a grass cover may, for a short time, be as effective as a tree canopy. O. R. Clarke, for instance, notes that because of the greater surface area and strata development, it is greater in taller and denser than in lower and more open 'grass' stands (see Table 42).

Date	Vegetation	Rainfall in inches	Per cent. Interception
June 22	Bluestem grass	1·78	68
July 2	Bluestem grass	1·61	68
July 19	Bluestem grass	0·58	73
Aug 15	Bluestem grass	1·20	63
June 15	Wheat	0·08	51
June 22	Wheat	1·48	33
July 1	Alfalfa	0·84	36
July 2	Alfalfa	1·80	26
July 2	Oats	0·74	45
July 14	Slough grass	0·45	73

Table 42 Amount of water intercepted by an acre of vegetation under natural rainfall. (After: Clarke, O. R. 1940)

Early work by J. Kitteridge showed that the combined interception and evapo-transpiration water losses were as great from short-grass vegetation as from *Pinus ponderosa* stands, and from tall-grass prairie almost as great as from oak-hickory forest. Ovington's summary (see Table 33, p. 271) of measurements of annual water circulation on comparable sites, which shows that evapo-transpiration is less from grassland than from the more deeply rooted forest stands in comparable areas, would seem to substantiate the relative importance of interception losses in grassland. R. H. Burgy and C. E. Pomeroy observe that small watersheds converted from brush to grassland have yielded significant increases in total run-off; however the extent to which reduction of interception, decrease in evaporation or increase in percolation contributes to this change is extremely debatable. The efficiency of percolation in grasslands as compared with other types of cover still awaits more detailed investigation. The grass form is thought to facilitate stem-flow and funnelling of water. On the other hand *surface* run-off is thought to be greater from grass-covered than from forested slopes. In addition it has been suggested that percolation is greater in summer and less in winter than under a forest cover. Particularly in areas with rigorous winters several authors have noted that easier and deeper frost-penetration under grassland inhibits percolation in winter and early spring.

Studies of the effect of grassland on micro-climatic conditions have been largely confined to temperate climatic regimes. There is a dearth of comparable data about tropical grasslands. This is no less true in respect of grassland soils. In temperate regions, with up to ninety per cent. of its biomass below the surface, the grassland ecosystem has its greatest effect on soil formation. While its subterranean organs do not tap as deep a zone as in temperate forests, the network of fine, fibrous roots ramify and exploit a greater surface area in the uppermost soil horizon. Roots are short-lived, dying and being continually renewed practically every year. Together with underground storage organs and basal stems, they result in the direct incorporation in the soil of a large quantity of organic matter. On well-drained, nutrient-rich parent material humification is rapid and there is an intimate intermixture of the humus and clay fractions to form a well-developed colloidal complex. This is facilitated on the one hand by the density of the root system and, at a later stage, by the activities of worms and other burrowing animals. The presence of this well-developed clay/humus complex (whose high adsorptive capacity helps to check excessive leaching and base desaturation) contributes to the flocculation and binding of soil particles into a 'crumb' structure. The latter provides a particularly favourable physical and chemical environment for soil life. Good aeration is combined with high water availability. Also associated with the clay/humus complex is a slow mineralisation of humus which allows for efficient nutrient uptake by plants and contributes to the high proportion of humus associated with nutrient-rich grasslands.

Origin and Ecological Status of Grasslands

Various criteria have been used to differentiate types of grassland formation. The most commonly employed are the absence or presence of trees and/or shrubs; the luxuriance and growth-form of the dominant grass species; and the nature of the physical habitat. In the first case a basic distinction is made between *prairie* and *steppe,* in which woody growth is absent or negligible and *savanna,* wooded grassland or parkland characterised by a discontinuous tree/shrub stratum but in which the herbaceous field-layer is ecologically dominant. These terms are used by plant ecologists in a strictly morphological sense. Their origins, however, are geographical and they have for long been descriptive of temperate and tropical grasslands respectively. The French bequeated the term *prairie* (or meadow) to the formerly more extensive grassy plains of central North America; the Eurasian *steppe* derives from the name of one of the commonest genera (*Stipa*) in the grasslands which extend from the Black Sea to eastern Mongolia across Central Asia. Unfortunately the two terms are not completely synonymous. In North America the so-called 'short-grass' formation of the High Plains is not, strictly speaking, covered by the term prairie; while in Russia, steppe denotes all non-forest vegetation including semi-desert grasslands. Within the temperate grassland a further distinction is frequently made on the basis of height and luxuriance of the vegetation between the *true prairie* with varying proportions of tall and medium-sized grasses and forbs; the *mixed prairie* in which the 'medium' grasses are dominant though accompanied by both tall and short species in proportions dependent on site conditions; and the 'short-grass' plains. In tropical regions the term *savanna* includes a wide range of vegetation types from the wooded to the treeless grassland. There are, however, some ten thousand species of grass each with varying ecological tolerances and adapted to a diversity of physical conditions. The corresponding variation in specific composition is implicit in the distinction made, on the basis of habitat between tropical and temperate, lowland and alpine, acid and basic, wet and dry, cultivated and uncultivated grasslands.

Of the existing grasslands some are of relatively recent origin, derived from or replacing a formerly different type of vegetation. Characteristic particularly of humid temperate regions, their establishment is due to deforestation and their maintenance to burning and grazing, or to cultivation. Others are the remnants of the ancient and once more extensive 'natural' grasslands of both

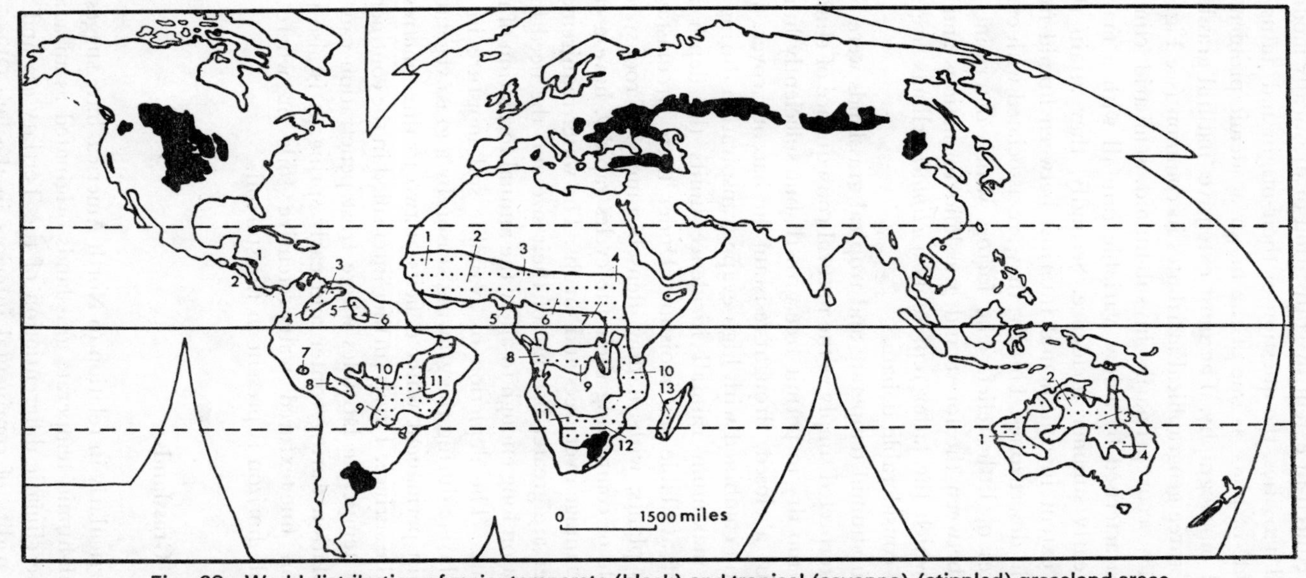

Fig. 63 World distribution of major temperate (black) and tropical (savanna) (stippled) grassland areas Numbers refer to particular localities named by Hills. (Former based on *Goode's World Atlas* 11th Edition. 1960, the latter from Hills. 1965)

the temperate and tropical regions of the world. Though considerably reduced in extent by cultivation or drastically modified by pastoral activities, the question of their origin and status still retains pride of place as one of the major academic problems of traditional biogeography. The most extensive 'natural grassland' areas have three geographical attributes in common (see Fig. 63). Firstly, they occur primarily in sub-humid, semi-arid climatic regions characterised by a low, variable rainfall with a marked spring to early summer incidence. Secondly, they attain their *maximum* extent in continental interiors between humid forest and the arid desert zones. Thirdly, they are associated with extensive surfaces of little relief. The factors which determine the boundary between the forest and grassland formations and the extent to which the latter represents a *climatic* climax are still subjects of considerable debate.

The sub-humid temperate and tropical grasslands were both initially interpreted simply as the natural consequence of climatic conditions too dry to permit tree growth but sufficiently humid to maintain a closed, drought-resistant herbaceous cover. Low precipitation combined with high evapo-transpiration during the period of maximum rainfall incidence limits the depth and duration of available soil moisture. Under these circumstances herbaceous plants, with shallow, densely ramifying root-systems and capable of completing their life cycles rapidly, have a competitive advantage over trees and shrubs. The water requirements of the latter are greater because of their size and the need for a growing season long enough to allow the annual formation of new woody tissues. The depletion of available soil moisture in the upper soil layers during the growing season by a grass vegetation can result in permanent water deficits at two to three metres or less below the surface. This is further intensified in the continental interiors of temperate latitudes where frost penetration can impede percolation during winter and early spring. It is also pronounced on fine-textured water-retentive soils or where an impermeable horizon is present in the sub-soil.

Temperate Grasslands

Studies, particularly in relation to North America, have suggested that the sub-humid temperate grasslands originated as an adaptation to the climatic differentiation of the Tertiary era and the increasing aridity of continental interiors in the late Oligocene and Miocene periods. The evidence, as Dix notes, is mainly 'circumstantial' and palaeontological. The evolution of large herbivorous mammals (characteristic of existing grassland areas) during the Tertiary era was accompanied by anatomical changes thought to be indicative of increasing adaptation to specialised grazing and rapid movement. The fossil series of horses and other cursorial ungulates has been linked with the development of open grassland habitats. It is certainly tempting to apply such a 'symbiotic' evolution to all the major sub-humid grasslands of the world. However both C. O. Sauer and P. V. Wells seriously doubt the existence of a widespread grassland formation in Tertiary North America. In the first place it was a period of considerable geological instability, during which detrital deposits derived from the erosion of the Rockies were accumulating across the area of the Great Plains. In the second, such palaeobotanical evidence as exists is indicative of open woodland rather than treeless grassy plains. It is, in fact, more than probable that the sub-humid temperate grasslands are of relatively recent origin, certainly post-dating the Pleistocene glaciation of North America and probably also of Eurasia. Advancing ice-sheets would have obliterated the pre-existing vegetation, while beyond their limits there is ample evidence of formerly humid conditions in areas now arid or semi-arid. Furthermore present grassland soils have developed on transported parent material much of which was deposited in late glacial or more recent times. Pollen analyses of sediments from Illinois and Iowa to as far south as the Llana Estacado confirm that the emergence of a grass-dominated vegetation was a post-glacial phenomenon. For instance, in the Sand Hills area of Nebraska investigation of lake sediments has revealed that as recently as five thousand years ago the arboreal (chiefly pine) contribution to the vegetation was as important as that of the grasses.

The *extent* to which climatic factors account for the establishment and maintenance of a climax grassland is still open to question. Consideration of the problem is complicated by the fact that much temperate grassland has been removed by cultivation while the last dwindling remnants have been profoundly modified by the intensification of livestock farming within the last century. On the one hand there are grassland areas with climatic conditions quite sufficiently humid for tree growth. On the other, trees are capable of surviving not only in the existing grasslands but under even more arid semi-desert conditions. The former include the prairies of the Mid West of the U.S.A. from eastern Iowa to western Michigan. Here it is thought that the original grassland and its associated soils may well have developed in response to the drier climatic conditions of the post-glacial 'xerothermic' period (about 3–4 000 B.C.). With a later increase in rainfall, re-establishment of a forest cover was impeded partly by soil conditions, partly by man. It has been suggested of the former that given the fine-textured water-retentive and, particularly, loessal soils of the area once a grass cover had developed it would be difficult for tree seedlings to penetrate the sod and compete for water successfully. More important, however, in the maintenance of a grass cover was the early regular use of fire by the North American Indian both for game drives and forage renewal. The presence of man in North America during the Pleistocene has now been established and his emergence as a powerful ecological factor in the Mid West at a time when the climate was somewhat drier than at present would undoubtedly have facilitated the replacement of forest by grassland. Cessation of prairie fires with the spread of white settlement has been accompanied by an extension of woodland (formerly confined to fire-protected valley-side slopes) up on to the interfluvial surfaces. A comparable recent extension of poplar into the black-soil areas of the Canadian prairie region is also considered indicative of a readjustment by vegetation to existing climatic conditions. The degraded chernozem soils which occur along the forest-steppe boundary in south-eastern U.S.S.R. are thought to have resulted from the transgression of forest into a grassland which extended further north during the post-glacial xerothermic era. That the vegetation along the northern margin of the steppe was not in equilibrium with existing climatic conditions probably explains the success of re-afforestation policies, initiated by Peter the Great and resumed at intervals in more recent centuries. The less extensive grassland or former grassland areas of the southern hemisphere, particularly the pampas of South America and the Canterbury Plains of New Zealand defy a climatic explanation. The relative effects of climatic change, soil moisture regimes and man must likewise be taken into consideration.

Indeed there are those who, like Carl Sauer, would doubt whether any extensive area of grassland in the world can be maintained without the aid of man. The early symbiosis of man-fire-grazing in existing grassland areas forced him to the conclusion that 'grasslands are found in plains, subject to periods of dry weather. . . . Their occurrence all round the world points to the one known factor that operates effectively across such surfaces—fire'. More recently P. V. Wells has tended to give further weight

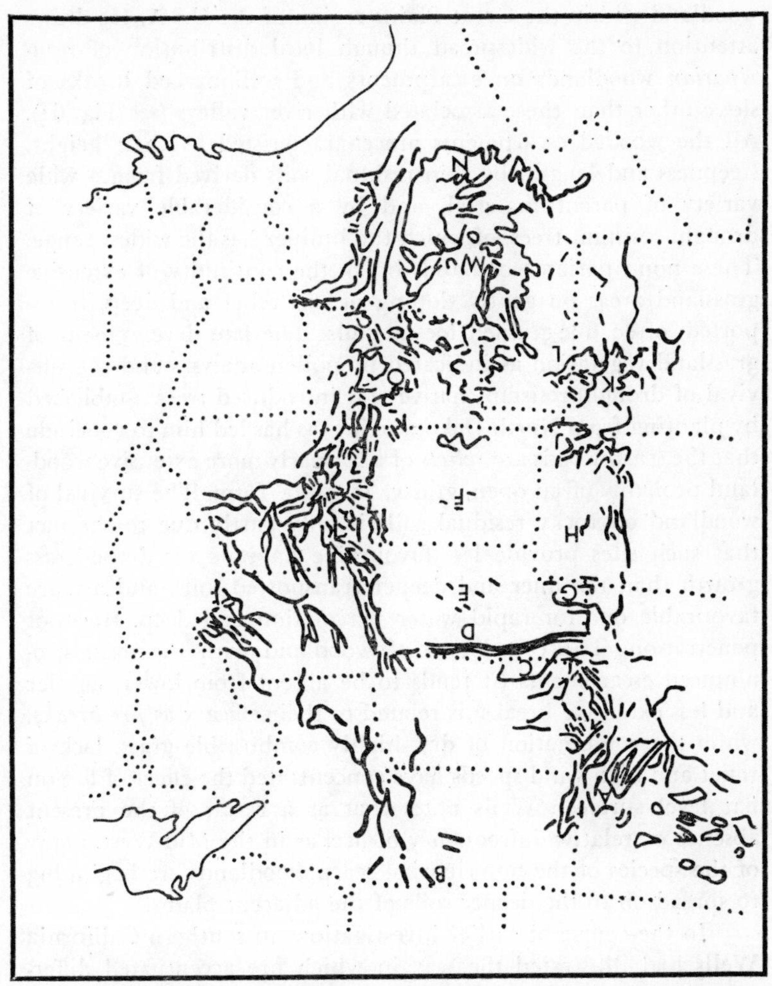

Fig. 64 Non-riparian woodlands in the grassland province of the central plains region of the United States. Black lines indicate location of scarps and other rough, broken or steep topography with indigenous woodland vegetation. Dotted lines are state boundaries. Dominant Species in A, C, G : *Pinus ponderosa, Juniperus scopulorium;* D. *P. ponderosa, J. scopulorium, J. virginiana, Quercus macrocarpa;* E : Sandhills, Nebraska, *P. ponderosa, J. virginiana, Celtis occidentalis;* F : *J. virginiana* H : *P. ponderosa, J. scopulorium;* J, K : *P. ponderosa, J. monosperma;* L : *Llana Estacado*—north-west, *P. edulis, J. monosperma, Quercus undulata;* east-break of plains, *J. pichottii, Q. mohriana;* M, N : *Q. virginiana, Q. shumardii, Q. mohriana, J. ashei, J pinchattii;* O : *Q. stellaria, Q. marilandica, Q. muehlenbergii, Q. shumardii, J. virginiana.* (From : Wells, Philip, V. 1965)

to this hypothesis in his analysis of the nature and distribution of woodland within the Great Plains region of the U.S.A. He draws attention to the widespread though local distribution of *non-riparian* woodlands on escarpments and well-marked breaks of slope other than those associated with river valleys (see Fig. 64). All the wooded escarpments are characteristed by their height, steepness and length, by thin residual soils derived from a wide variety of parent material, and by a considerable variety of drought-resistant trees, of which the juniper has the widest range. These non-riparian woodlands break the continuity of extensive grassland areas on gently sloping or flat relief and deep, transported, often fine-grained loessal soils. The late development of grassland vegetation as indicated by pollen analyses and the survival of drought-resistant native and introduced trees established by planting from North Dakota to Texas has led him to conclude that the scarp woods are *relicts* of a formerly more extensive woodland probably of an open, grassy, 'savanna' form. The survival of woodland on rocky residual soils is only partly due to the fact that such sites provide less favourable habitats for dense grass growth than the finer and deeper transported soils, and a more favourable one for rapid water percolation and deep, tree-root penetration. The correlation of woodland with the most pronounced escarpments (it tends to be absent from lower, gentler and less extensive breaks) is related to their efficacy as *fire-breaks*, while the combination of dry, highly combustible grass, lack of relief and high wind speeds have concentrated the effect of fire on flat level surfaces. Wells notes that as a result of the present absence or relative infrequency of fire, as in the Mid-West, many of the species of the non-riparian scarp woodlands are beginning to spread on to the deeper soils of the adjacent plains.

In the course of earlier investigations in southern California Wells had illustrated the way in which fire accentuated differences in the substratum by destroying a forest canopy and opening up the way for other and more diverse forms of vegetation. In the first instance the increase in herbaceous and shrubby species in a 'droughty' climate provides a fire-conducting matrix whose regeneration is usually sufficiently rapid to suppress tree regeneration. Once trees are eliminated from an extensive area the more difficult re-establishment becomes. This process favours grassland particularly on deeper soils of a fine texture, where, because of surface water-retention, grass/herb regeneration is rapid and recurrent fires frequent. In contrast on coarse, permeable parent-material, where water percolation is rapid, herbaceous growth is less well-developed and after burning does not provide sufficient 'fuel' to cause fires of such intensity or duration as to prevent the regeneration of woody growth.

His conclusion in the Great Plains that 'a combination of a fire-conducting ground-cover of seasonally dry grasses with extreme flatness and continuity of topography has been a hazardous environment for woody plants in a region of droughty climate and strong winds where the incidence of fire has undoubtedly been increased for at least the last 11 000 years by the presence of man' re-echoes that of Sauer. Finally, the influence of the extensive herds of buffalo (the mainstay of the Plains Indians' economy) whose range coincided with that of the grasslands must not be overlooked as an important ecological factor that may well have been 'co-dominant' with fire. The short grass plains, formerly attributed solely to increasing aridity westwards in the U.S.A. are thought, by Floyd Larson, to have been developed and maintained by grazing pressures effected in the first instance by the large herds of indigenous 'wild' herbivores and later by the introduction of cattle and sheep. The 'short' grasses are better able to withstand grazing pressure; and there is evidence that when this is eliminated or during cycles of wetter years taller, particularly medium-grasses, reappear and become dominant.

Tropical Savannas

The problem of the origin and status of tropical savannas is even more complex than that of temperate grasslands. Also, it is of more than academic interest. A large proportion of tropical regions is still covered with some type of 'semi-natural' savanna vegetation. It constitutes a major resource in those areas where the rate of population growth is fast outstripping that of food production. A greater understanding of the savanna ecosystem is essential in order to ensure its efficient use and the conservation of the highest levels of productivity possible.

Savanna is characterised by a much greater diversity of composition, form and habitat than is found in the sub-humid temperate grasslands. It is composed of an ecologically dominant herbaceous stratum in which more or less xerophytic perennial grasses and sedges with a pronounced tussock habit are the principal and often the only components. In addition savanna may include varying proportions of drought-resistant woody plants varying from low shrubs to quite tall trees. While the cover of the latter may be as high as fifty per cent. it is essentially open and discontinuous, and in some cases may be completely lacking. Savanna is generally associated with areas characterised by a marked seasonal drought; however, there is little correlation between its limits and either the amount of precipitation or the duration of the rainy season. It occurs under a variety of tropical climates, from those with 500 to 3 750 mm of rain per annum and from eight months consecutive drought to none. Soils associated with savanna vegetation are as varied as the climatic conditions. Further savanna exists in close, though usually sharply delimited, juxtaposition with every type of tropical forest formation from the ever-green rain forest to the lighter deciduous woodland.

With increasing appreciation of the ecological diversity of savanna vegetation has come greater awareness of the probable diversity of its origins. It is hardly surprising that the theories proposed have been influenced by the academic training of the researchers and the particular aspect of the savanna habitat or region that has been studied. The earlier 'zonal' plant geographers such as A. Engler and A. W. F. Schimper interpreted savanna 'grassland' as the climax vegetation of tropical regions too dry to allow the successful establishment of a closed tree cover. Abortive attempts were made to define and characterise the tropical grassland climate. The lack of a clearly defined and consistent zonation of vegetation, the persistence of woody shrubs well into desert environments, and the existence of treeless savannna in areas sufficiently humid to support a forest vegetation have made the concept of a climatic climax untenable for most, if not all, savanna vegetation. On the other hand, the distinctiveness and stability of the flora of the wooded Sudanese savanna in the drier areas of West Africa are still considered by some botanists to be indicative of a climatically determined or, at least, *climax* formation.

However the widespread occurrence of savanna on plains and plateaus of little relief has directed attention to the possible rôle of soil—particularly soil fertility and drainage—as the dominant ecological factor determining its distribution. It has been noted that savanna is frequently localised on substrata inimical to the optimum development of forest. This is particularly the case where impermeable and/or indurated horizons or lateritic 'crusts' occur at or just below the surface of the ground. Under a markedly seasonal rainfall regime there is an alternation of soil saturation (or even temporary inundation) and desiccation, with which only the xerophytic savanna grasses and shrubs can cope. In other instances, however, there is no evidence of impeded drainage, and deeply rooting shrubs and low trees can maintain themselves, as in the wooded savannas of South America (e.g. cerradão and cerrado). The lack of productivity, the stunted growth and the xeromorphic characteristics of the woody vegetation has been attributed, by South American ecologists, to a deficiency of soil nutrients, particularly of phosphates and nitrates.

The distribution of savanna areas coincident with discontinuous, high, plateau surfaces, and the frequently abrupt vegetation change between them and surrounding dense humid forest on steep slopes have been noted by several authors, particularly in tropical South America, and latterly in Africa south of the equator. Furthermore it has been observed that they share a floristic composition quite distinct from that of the surrounding forests, but with marked relationships to that of savanna vegetation in the drier areas beyond the main forest zone. Botanists have established that this savanna *flora* is an ancient one which can be traced back to at least the beginning of the Tertiary era and was probably once more widespread. Some authors have interpreted these savanna 'enclaves' as relicts of a drier climatic period (of either Tertiary and/or Quaternary date) which have persisted as a result of edaphic conditions unfavourable to the subsequent establishment of a humid forest. In some cases protection from anthropogenic activities seems to confirm their floristic and structural stability; in others, there is evidence of a slow invasion by forest vegetation.

More recently M. M. Cole has proposed, on the basis of these data, an explanation of the floristic, vegetational and pedological characteristics of such savanna areas in South America and Rhodesia in terms of their geomorphological evolution. In her view the plateau surfaces present uplifted and undissected remnants of erosion surfaces. In many cases their preservation has undoubtedly been aided by the presence of resistant lateritic crusts. A long period of evolution has been accompanied by progressive soil degradation and hence vegetation impoverishment. The wooded savannas can therefore be regarded as edaphic climaxes, reduced in many instances by burning and grazing to treeless grassland (e.g. *campo limpo* of Brazil). Subsequent dissection of the formerly continuous plateau surfaces and the presence of well-drained, relatively youthful, and hence more fertile soils, is reflected in the presence of the dense semi-deciduous forest characteristic of the steeper slopes and valleys which bound the plateau surfaces (see Fig. 65).

There is, however, also evidence that other savannas are of *secondary* origin, *derived* from a pre-existing forest cover either as a result of deforestation and/or burning (see Fig. 66). In the humid tropics shifting cultivation is (given the existing techniques and knowledge of the indigenous populations) an adaptation to soils of low inherent fertility. In humid and formerly forested

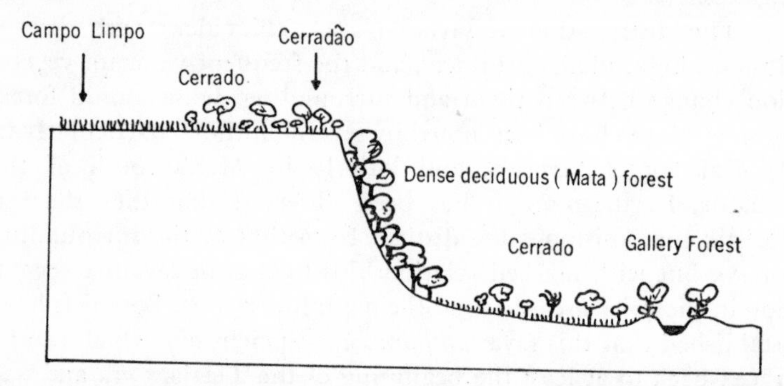

Fig. 65 Diagrammatic cross-section of a catena in the interior of Brazil showing relationship of savanna and forest vegetation to landform. *Campo limpo*—treeless savanna; Cerrado—wooded savanna, total cover xeromorphic trees and shrubs less than 50%; *Cerradão*—wooded-savanna, arboreal cover over 50%. (After: Birot, P. 1965)

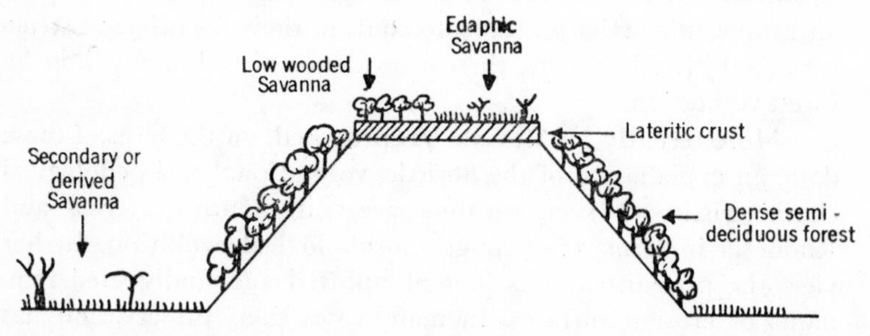

Fig. 66 Cross-section of Mt. Orombo-Boka (Ivory Coast) showing relationships of edaphic and derived savanna to slope. (From: Lemeé, G. 1967)

areas progressive shortening of the fallow period not only prevents the complete re-establishment of a vegetation cover but may expose the soil to intense desiccation. As a result the re-establishment of forest is made difficult and a herbaceous savanna grassland, of which one of the main components in Africa is the intractable weed, spear-grass (*Imperata cylindrica*), becomes established. Regular burning and grazing may then maintain permanently such post-cultural savannas. In drier areas or on soils which would normally carry a light xerophytic woodland burning alone can cause degradation of vegetation to savanna. Most savanna areas are burned frequently—in some areas every year or eighteen months—to stimulate fresh forage growth. The high proportion of pyrophytic plants is indicative of a long period of adaptation to fire. There are many botanists and geographers who consider most, if not all, savannas to be anthropogenic in origin. Whatever its origin, however, the present ecological importance of fire in the *maintenance* of savannas probably more than justified their designation by Richards as a *fire-climax*.

It is clear now that, in both temperate and tropical regions, the nature and distribution of grassland ecosystems are the result of a variety of interacting factors whose relative importance has varied in time and place. But, as Hills remarks in his review of savanna research, confusion has arisen in the interpretation of the origin and status of tropical grasslands 'because of the failure to distinguish clearly between pre-disposing, causal, resulting and maintaining factors'—a statement no less true of the studies of temperate grasslands. While it is not possible to explain the distribution of 'natural' grasslands in terms of climatic parameters alone, their extensive development in areas characterised by marked seasonal drought and subjected over a long period of time to alternating periods of greater or lesser precipitation cannot be dismissed as fortuitous. Climatic and geomorphological evolution have resulted in a soil-water regime and/or nutrient status inimical to the optimum development of a closed forest ecosystem. Within what must have been a constantly expanding and contracting ecotone between humid and arid habitats, the open woodland or wooded grassland formation and its characteristic fauna of large mobile grazing herbivores would have evolved contemporaneously. The relative ecological dominance of the herbaceous and tree strata would have been, and it still must be, dependent on a number of ecologically differentiating factors susceptible to variation through time in a constantly fluctuating physical environment. Climate, relief and the original character of the vegetation provide ideal conditions for the propagation of widespread fires. Propagated early by man it would appear that in the more arid areas, or in periods of more arid climatic conditions, fire 'tipped the balance' in favour of the pyrophytic plants of which grass is the most prolific; and together with the increase in wild and domesticated herbivores has served to maintain grassland ecosystems. Once established however, grasslands have revealed a high degree of stability even in face of changing environmental conditions. Whether it be increase of rainfall or cessation of pastoral activities, in both temperate and tropical grasslands, the invasion of trees is often retarded, if not inhibited, by soil conditions unfavourable to the establishment of tree seedlings.

Improved or Cultivated Grassland

Grass when grazed is one of the cheapest and most easily exploited sources of fodder for ruminant herbivorous livestock. Under optimum environmental conditions the normal range of plant foods can be used more efficiently and a more rapid cycling of nutrients can be effected than by any other type of agricultural crop. Grass production is immediately available since it can be continuously cropped while growing. The *potential* productivity and nutritive value of grass is exceptionally high. When this is fully developed it has been noted by R. J. McIlroy that yields of *dry* matter of over 10 000 kg/ha/annum in temperate and twice this amount in tropical regions, may be obtained, given adequate water and heavy application of fertilisers. In addition, the starch and protein equivalent per hectare is higher than for any other agricultural crop. Young growing grasses in temperate regions can contain a percentage of digestible protein comparable to that in linseed. In Britain good grassland, if properly grazed in order to maintain high levels of productivity, may have a starch equivalent[1] of sixty-six per cent. and of protein of fifteen per cent.

The grazing animal however is a relatively inefficient food-converter. The ratio of total digestible nutrients to total yield of dry weight of fodder is relatively small (often less than a fifth). Despite the inevitable cumulative drain of nutrients in meat, wool, hides etc., a higher proportion of these is returned to the soil than under other cropping systems, so that the maintenance of fertility is made easier. As a result of the high contribution of organic matter *within* the soil, and the consequent effects of this on the development of a crumb structure grassland have long played an important rôle in the restoration of soil fertility and maintenance of its stability.

As has already been noted, in comparison to the cereals the

[1]Starch equivalent = number lbs starch equivalent in production energy to 100 lbs of feeding stuff.

cultivation and domestication of fodder grasses is a product only of the last two hundred years. Of the ten thousand known species only some forty are used to any appreciable extent in sown pastures and those of a high agricultural value—dependent upon productivity, palatability, nutritive value, as well as adaptation to physical conditions and type of farming—are even fewer. Britain has undoubtedly played a leading world rôle in the cultivation,

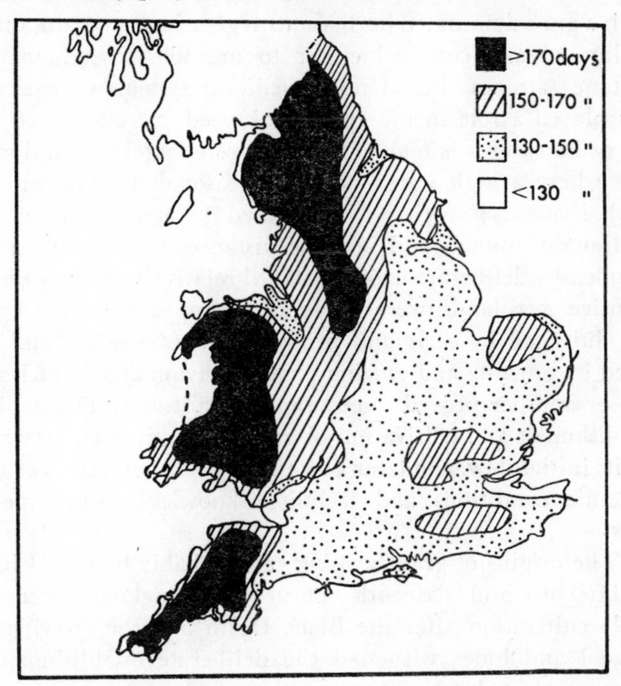

Fig. 67 Grass-growing days (i.e. when calculated soil moisture *deficit* was 2 inches or less) in England and Wales April to September, 1962. (From: Hurst, G. W. and Smith, L. P., in Taylor, J. A. 1967)

use and breeding of the grass 'crop'. To this end it has proved a peculiarly suitable area (see Fig. 67). In the absence of competition, humid climatic conditions, combined with a thermal growing-season of as much as nine months in lowland areas, are particularly favourable to luxuriant grass growth, as every gardener knows to his cost. During mild winters, perennial grasses can resume growth when daytime temperatures rise above 6°C. The relative 'greenness' of our lowland pastures throughout the year is in contrast to that of more continental regimes with severe winters and/or summer drought; it is a feature of the landscape which makes a vivid impression on the visitor new to these islands. Today grass is the most extensively managed crop and makes the greatest single contribution to farming income. Nine-tenths of British farmland is under a grass cover of one type or another and it provides two-thirds of the starch and even more of the protein needs of all cattle, sheep and horses.

In humid temperate regions, such as Britain, grass is essentially an unstable seral stage in the vegetation cover. Habitats in which a grass-dominated herbaceous vegetation can maintain itself are limited and confined either to unstable coastal, marsh or montane (alpine) sites. Under medieval systems of agriculture grass played a rôle in the ephemeral weedy vegetation of fallow land or in *water meadows*. The latter were largely natural marshy areas where a high water-table and seasonal flooding promoted lush herbaceous growth. They provided the main source of fodder for draught animals and in some instances they were artificially extended by deliberate flooding. Until relatively recently the most extensive grasslands were those of unimproved *rough grazing* land, initiated by early and progressive deforestation and maintained by burning and grazing particularly on areas which at one time or another proved uneconomic to cultivate. The deliberate establishment and cultivation of grassland is, in fact, very recent, and its initial slow, and later very rapid, evolution reflect changing economic conditions and technical knowledge since medieval times.

The origins of grass farming can probably be traced back to the fifteenth and sixteenth centuries in England. Decrease in arable cultivation after the Black Death and the growing trade in wool and hides witnessed the deliberate establishment and enclosure of lowland pastures. Although the subsequent increase in wheat yields when old enclosed pastures were ploughed up, in comparison to those from the continuously cropped open fields was recognised, it was some time before grass as a crop became an integral part of the arable farming system. The actual *cultivation* of grasses, as distinct from allowing an area to 'tumble down' to pasture, is much more recent. It was, in fact, preceded by the use of legumes, now integral constituents of cultivated grassland. The earliest, lucerne (alfalfa) and red clover (originally domesticated in Mediterranean Europe in Roman times) were introduced into Britain from Holland in the seventeenth century. The first records of the sowing of indigenous grasses date from later in that century. But the use of these fodder plants was haphazard and it was not until nearly a century later that the modern concept of 'ley' farming really developed. Among the important innovations of the eighteenth-century technical 'revolution' in agriculture was the replacement of the old medieval system, based on the alternation of cereal cultivation and fallow, by the modern methods of rotation farming which permitted the integration of arable and livestock farming.

The 'ley' was initially a one-year crop of red clover which together with turnips and grain formed the basis of the early Norfolk four-course rotation. Later this was extended to a two- to three-year grass/legume mixture which fulfilled the dual rôle of providing fodder and building-up soil fertility. It initiated the importation of grass seed (Cocksfoot—*Dactylis glomerata* from Virginia and Italian rye grass, *Lolium italicum* from the continent) as well as encouraging the collection of 'good seed'. The development of ley farming stimulated a continuing interest in grassland improvement during the nineteenth century. To this end three factors made an important contribution. One was the increase in, and realisation of the value of, the indigenous wild white clover. Another was the use of a wider range of grasses and mixtures of these with both red and white clover for the purpose of establishing permanent pasture. Finally the use of basic slag, as a major source of phosphatic fertiliser, promoted the vigorous growth of wild white clover and increased the stock-carrying capacity of pastures. The result was the emergence in 1910 of the long-duration ley (four to fifteen years) particularly suited to areas with cooler and more humid climatic conditions. Developments in grassland farming during the last fifty years have been concerned primarily with the refinement and simplification of grass-seed mixtures and the breeding of strains (or ecotypes), of a smaller number of species designed for particular uses and habitats. Higher yielding and more nutritive grasses have been combined with the increasing use of inorganic fertilisers and more efficient methods of grazing and storage. Methods of intensive grassland farming designed to achieve maximum production in yield and nutritive qualities are, particularly in dairy farming, already beginning to replace the more traditional rotations.

Although the distinction is much less clear-cut than it was before the Second World War, two categories of cultivated (or 'improved' grassland) can still be distinguished.

(a) *Temporary or rotation grassland* (or 'ley') is a selected mixture of grasses and legumes sown as one of the pre-designed rotation of crops and destined to be ploughed-up and resown after a given number of years. It is composed of varying proportions of varieties of the highest yielding, most nutritious agricultural grasses—which include Italian and Perennial rye grass (*Lolium italicum* and *perenne*), Cocksfoot (*Dactylis glomerata*), Timothy (*Phleum pratense*) and clover (*Trifolium spp.*). The particular mixture is dependent on the use and duration for which the crop is required together with general and local environmental conditions. A characteristic feature of the youthful ley is the complete or almost complete absence of weeds, ensured and maintained today by the use of herbicides. In the words of W. Davies such a grassland remains a ley 'so long as the species of grasses which it contains are directly attributable to the seeds that were initially sown'. After a time however, the ley tends to lose its original characteristics and acquire those of

(b) *Permanent grassland or pasture*. This usually originates from sowing or abandonment to grass colonisation of formerly cultivated land. Such pastures vary in age and some are of considerable antiquity. Many, in England, were established deliberately as permanent pasture, and clauses in farm leases ensuring their perpetuation are not unusual in some parts of the country. They are composed of a greater variety of species —a higher proportion of the poorer-grade grasses and herbaceous weeds than in leys. The best may resemble good temporary grass, the worst are not unlike some types of rough grazing. According to a classification, on the basis of the percentage of perennial rye grass and clover to *Agrostis spp.*, first-grade permanent pasture contains thirty-five to twenty-five per cent. respectively of the former and no more than 'a trace' of the latter species. Such are the 'traditional' fattening pastures of midland England and their distribution[1] reflects a combination of soils of high fertility and good management. Fourth-grade pastures, in contrast, are composed of seventy per cent. *Agrostis spp.* with negligible amounts of rye grass and clover and are a result, usually, of poorer management. In the nineteen-thirties over half the agricultural land in England and Wales was under permanent pasture, only three per cent. of which was of first-class quality. By 1960, however, the increase in arable cropping (including intensive grass crops) had reduced the hectareage to a third while the percentage of first-class permanent pasture had more than doubled.

[1]See O.S. National Atlas of Great Britain, Sheet 1. Grasslands of England and Wales 1 : 633600

In north-west Britain, and particularly in Scotland, high quality permanent pasture comparable to that of lowland Britain does not exist. Under cooler, humid climatic conditions and intensive leaching, maintenance of soil fertility is more difficult. Long-duration leys are more important in livestock farming and the visual and functional difference between improved grassland and permanent rough grassland is sharper than in England and Wales. In Scotland in particular there is no counterpart to the high-quality permanent pasture of the midlands of England.

In contrast to the increasing importance of the cultivated grassland in humid temperate regions the problem of increasing forage production and, particularly, quality in the tropics has yet to be satisfactorily solved. While production in terms of dry weight can be very high during a hot, humid growing-season, in the important agricultural areas this is limited by a period, varying in duration, of serious water-deficit. Tropical grasses, in addition, are low in crude protein and high in fibre compared with temperate grasses at similar stages of growth. Crude protein values can vary from two-thirds to less than half that in the latter; while many of the common tropical grasses have less than half the starch equivalent of temperate species. These differences are thought to reflect lower soil nutrient levels, particularly of nitrogen. Also there are few herbaceous legumes indigenous to tropical regions and soil temperature conditions are frequently above the maximum for nitrifying and nitrogen-fixing bacteria. The introduction of high yielding temperate grasses and legumes—apart from the physical and associated economic problems of heavy fertiliser needs—is inhibited by temperature and day-length tolerances incompatible with those of tropical latitudes. It may well be that species other than grass may have to be found to fill the equivalent agricultural niche in these regions.

References

ALLEE, C. W. *et al.* 1949. *Principles of animal ecology.* W. B. Saunders & Co. London.

ARMSTRONG, S. F. 1950. *British grasses and their employment in agriculture.* Cambridge University Press.

BARNARD, C. (Ed.) 1964. *Grasses and grassland.* MacMillan, London.

BEARD, J. S. 1953. The savanna vegetation of Northern Tropical America. *Ecol. Monogr.*, **23** (2): 149–215.

BIROT, P. 1965. *Les formations végétales du globe.* S.E.D.E.S., Paris.

BURGY, R. H. and POMEROY, C. E. 1959. The interception losses in grassy vegetation. *Trans. Am. geophys, Un.*, **39**: 1095–1100.

CARPENTER, J. R. 1940. The grassland biome. *Ecol. Monogr.*, **10** (4): 617–684.

CLARKE, O. R. 1940. Interception of rainfall by prairie grasses, weeds and certain crop plants. *Ecol. Monogr.*, **10** (2): 244–277.

COLE, M. M. 1960. Cerrado, Caatinga and Pantanal: the distribution and origin of the savanna vegetation of Brazil. *Geogr. J.*, **126** (2): 168–179.

COLE, M. M. 1963. Vegetation and geomorphology in Northern Rhodesia: an aspect of the distribution of the savanna in Central Africa. *Geogr. J.*, **129** (3): 290–310.

CRISP, D. J. (Ed.) 1964. *Grazing in terrestrial and marine environments.* British Ecological Society Symposium, No. 4, 1962. Blackwell, Oxford.

DAUBENMIRE, R. 1968. Ecology of fire in grasslands. *Adv. Ecol. Res.*, **5**: 209–266.

DAVIES, W. 1960. *The grass crop: its development, use and maintenance*, 2nd ed., (revised). Spon, London.

DAVIES, W. and SKIDMORE, C. L. (Eds.) 1966. *Tropical pastures.* Faber and Faber, London.

DE VOS, A. 1969. Ecological conditions affecting the production of wild herbivorous mammals on grasslands. *Adv. Ecol. Res.*, **6**: 137–183.

DIX, R. L. 1964. A history of biotic and climatic changes within the North American grassland. In CRISP, D. J. *Grazing in terrestrial and marine environments*, pp. 71–90.

FRANKLIN, T. B. 1953. *British grasslands from earliest times to the present day.* Faber and Faber, London.

GEIGER, R. 1965. *The climate near the ground.* Harvard Univ. Press.

HANSON, H. C. 1950. Ecology of the grassland, II. *Bot. Rev.*, **16** (6): 283–360.

HILLS, T. L. 1965. Savannas: a review of a major research problem in tropical geography. *Can. Geogr.*, **4**: 216–78.

HOPKINS, B. 1965. *Forest and savanna.* Heinemann, Ibadan and London.

HUBBARD, C. E. 1968. *Grasses*, 2nd ed., Penguin Books.

HURST, G. W. and SMITH, L. P. 1967, 'Grass growing days' in *Weather and agriculture.* Taylor, J. A. (Ed.), Pergamon Press, London.

KITTERIDGE, J. 1948. *Forest influences.* McGraw-Hill, New York.

LARSON, F. 1940. The role of the bison in maintaining the short grass plains. *Ecology*, **21** (2): 113–121.

LEMÉE, G. 1967. *Précis de biogéographie.* Masson et Cie, Paris.

MCILROY, R. J. 1964. *An introduction to tropical grassland husbandry.* Oxford University Press. London.

MALIN, J. C. 1956. *The grassland of North America: prolegomena to its history with addenda.* Lawrence, Mass.

MALIN, J. C. 1952. Man, the state of nature and climax as illustrated by some problems of the North American grassland. *Scient. Mon., N.Y.*, January: 29–37.

MICHELMORE, A. P. G. 1939. Observations on tropical African grasslands. *J. Ecol.*, **27** (2): 282–312.

MOORE, I. 1966. *Grass and grasslands.* Collins, The New Naturalist, London.

Ovington, J. D., Heitkamp D. and Lawrence, D. B. 1963. Plant biomass and productivity of prairie, savanna, oakwood and maizefield ecosystems in Central Minnesota. *Ecology*, **44** (1): 52–63.

Ovington, J. D. 1964. Prairie, savanna and oakwood ecosystems at Cedar Creek. In Crisp, D. J., *Grazing in terrestrial and marine environments*, pp. 43–54.

Rattray, J. M. 1960. *The grass cover of Africa*. F.A.O. Agricultural Studies No. 49, Rome, F.A.O./U.N.O.

Richards, P. W. 1952. *The tropical rain forest: an ecological study*. Cambridge Univ. Press.

Robinson, D. H. 1947. *Good grassland*. English Universities Press, London.

Roseveare, G. M. 1948. *The grasslands of Latin America*. I.A.B. Bulletin No. 36, Aberystwyth.

Sauer, C. O. 1944. Early man in America. *Geogr. Rev.*, **34** (4): 529–573.

Sauer, C. O. 1950. Grassland climax fire and man. *J. Range Mgmt.*, **3**.

Sears, P. B. 1961. A pollen profile from the grassland province. *Science, N.Y.* **134** (3494): 2038–39.

Sprague, H. B. (Ed.) 1959. *Grasslands*. American Ass. Advancement of Science, Publ. No. 53, Washington, D.C.

Steiger, T. L. 1930. Structure of prairie vegetation. *Ecology*, **11** (1): 170–217.

Tiver, N. S. and Crocker, R. L. 1951. The grasslands of south-east Australia in relation to climate, soils and development history. *J. Br. Grass. Soc.*, **6** (1): 29–80.

U. S. Dept. of Agriculture. 1948. *Grass. The Year Book of Agriculture*, 1948. Washington, D.C.

Veasy-Fitzgerald, D. F. 1963. Central African grasslands. *J. Ecol.*, **51** (2): 243–274.

Waterhouse, F. L. 1955. Micro-meteorological profiles in grass cover in relation to biological problems. *Q. Jl. R. met. Soc.*, **81** (347): 63–71.

Weaver, J. E. and Fitzpatrick, T. J. 1934. The prairie. *Ecol. Monogr.*, **4** (2): 108–295.

Wells, P. V. 1962. Vegetation in relation to geological substratum and fire in the San Luis Obispo Quadrangle, California. *Ecol. Monogr.*, **32** (1): 79–103.

Wells, P. V. 1965. Scarp woodland, transported grassland soils and the concept of grassland climate in the Great Plains Region. *Science*, April, **148** (3367): 246–249.

14
Biological Deserts

Absolute deserts—barren, uninhabited, completely devoid of *any* form of life—are in fact relatively restricted in extent. More widespread are areas where plant and animal biomass is meagre and organic productivity low in comparison to those which are more fertile. The relative poverty of biological deserts may be the result of a lack, or deficiency, of one or more of the requirements essential for life: water in arid climatic regions or excessively permeable substrata. On the other hand, life may be inhibited by conditions, such as exceptionally high or low temperatures, extreme wind force, the presence of toxic substances in the soil or air, or the direct or indirect activities of man, which are beyond the range of tolerance of many organisms. 'Desert' conditions may, under some circumstances be a temporary, ephemeral stage in the initial colonisation or early re-colonisation of a bare, inorganic habitat; under others, they may represent the maximum biological productivity possible in the existing physical environment. The most extensive deserts are those where the principal biological limitations are climatically conditioned; either by a scarcity of water as in the 'arid zones' or an insufficiency of heat as in the 'cold deserts' of high latitude (Arctic) or altitude (Alpine) areas of the world.

Despite the marked and obvious differences between the ecosystems of these contrasting areas, the arid and cold deserts have certain basic ecological characteristics in common. In both the physical environment is harsh and 'living conditions' consequently extremely difficult. Climates are extreme; protracted periods of intense drought, heat or cold result in a potential growing season of short and often variable duration. Survival requires a high degree of tolerance or adaptation to these unfavourable environmental conditions. As a result the flora and fauna of deserts tend to be poorer in species but morphologically and physiologically more specialised than those of less rigorous areas. Nevertheless, endemism is not marked. Many organisms which occupy deserts are the more resistant, hardier representatives of families or genera in adjacent non-desert regions. The fact that most endemic taxa are of specific or lower status has been attributed to the relatively late emergence of the present major deserts of the world. The arid zones are thought to have originated as a result of climatic differentiation initiated by the latest Tertiary orogenesis. It is only within the last ten thousand years that the cold deserts of the Tundra have finally emerged from beneath the last of the Pleistocene ice-sheets.

The absence or extreme sparsity of trees is a generally universal feature of cold and arid deserts. In addition both are characterised by plants (with the exception of the 'giant' cacti) of low stature with a much greater volume of subterranean organs than of surface shoots. Physical limitations to photosynthesis are reflected in low annual primary productivity. The vegetation is distinguished by plant communities of a relatively simple structure with a lack of complex stratification and with, frequently, an 'open' or discontinuous ground-cover. The lack of primary food restricts the amount of animal biomass that can be supported. The number of trophic levels is smaller than in richer ecosystems, and animal populations tend to be subject to more violent seasonal and cyclic variations in number.

In addition, there is a close inter-relationship between an often incomplete vegetation cover and immature unstable environments. The latter, in some instances, may be the cause of the former, in others the result! A deficiency of water and/or heat, combined with poverty of soil organic-matter is reflected in the dominance of mechanical over other forms of rock weathering. On the one hand, unprotected or only partially protected unconsolidated mineral debris is very susceptible to movement whether under the force of gravitation, heat-expansion, frost-heaving or wind. As a result highly mobile parent material (e.g. sand-dunes, mud-flats) alternates with bare rock or boulder-strewn surfaces. On the other hand, such immature, skeletal and unstable soils contribute to the continuous cyclic change of vegetation establishment, disruption and re-establishment. The alternation of stable and unstable conditions over a period of time in any one habitat is a well-known feature of all desert ecosystems. The severe limitations imposed on soil and vegetation development maintain the potential climax in a condition of 'protracted immaturity' comparable to the pioneer stages in more favourable habitats.

Because of their inherent instability, desert systems are particularly susceptible to disruption by the activities of man.

An important feature of desert ecosystems, then, is the negligible extent, in comparison to 'closed' and highly stratified communities, to which the vegetation cover can modify the atmosphere and soil it occupies. The nature of the physical habitat exerts a proportionately greater ecological influence in the determination of the composition and form of the climax vegetation than in non-desert environments. As a result both hot and cold deserts are distinguished by a diversity of vegetation—a mosaic of communities, which reflects a corresponding diversity of physical habitats.

Tundra ('cold and polar desert') occupies a zone between the latitudinal limits of tree growth and the polar ice-caps. However the absence of extensive land masses at high latitudes in the southern hemisphere makes the tundra almost synonymous with Arctic habitats. In these areas biological poverty is the result of a deficiency of warmth combined with intensity of cold. For *at least* seven months of the year air temperatures are below zero, while the mean of the coldest month may vary from −5° to −10°C on the southern to −30° to −35°C on the northern margins of the tundra. Absolute minima below −70°C have been recorded. Obviously only those plants capable of enduring protracted periods of intense frost without injury can survive. Tolerance varies with species and stage of vegetative development. Adaptation to sub-zero temperatures would appear to be primarily physiological rather than morphological. It is effected by the seasonal process of 'hardening'; this entails changes in the cell sap and protoplasm which protect the plant from frost injury. Many plants of the tundra can survive periods of varying length when their tissues are frozen solid. In some species cellular ice does not form until temperatures drop below −30°C. Others never freeze and are therefore immune to damage. Among the latter the most notable and characteristic of the tundra are the lichens. They possess a greater cold-resistance than any other type of plant and can endure the harshest of Arctic winters. Also, they can withstand and adjust to rapid and extreme temperature changes. In addition, they are capable of photosynthesis at sub-zero temperatures; Birot notes that at −5°C, in *Cladonia spp.*, the rate is still half the normal, while slight activity has been detected even at −24°C.

As biologically exacting as the intensity and duration of the tundra winter is the short and cool growing-season. It is usually of less than three months' duration and the mean temperature of the warmest months rarely exceeds 10°C. The brevity of the growing-season is to a certain extent mitigated by long day-length. Stomata may remain open for twenty to twenty-four hours and, despite high respiration rates mean daily plant-yields which are twice as high as those in temperate regions can be attained for brief periods. In contrast, however, mean annual productivity is low—no more than a tenth of a temperate forest—and the total plant biomass is correspondingly meagre.

The short duration of the growing-season poses major problems for plant reproduction. Species must be capable of completing their development cycle—setting perennating buds and/or seeds, and completing the process of hardening before killing frosts commence. Plants become active when the snow starts to melt and soil temperatures attain at least 0°C. The rapid commencement of photosynthesis is a well-known feature of tundra species, particularly of evergreen shrubs. There are even some deciduous plants whose perennating buds have already opened and whose leaves have been partially formed in the autumn, which can survive the winter in this condition and hence allow the plant to take immediate advantage of the onset of favourable temperatures in the following spring. Although vegetative reproduction is common, the production of seeds is nonetheless important. Flowers are often brilliant in colour and very large in proportion to the size of the vegetative organs. Seed development, however, tends to be protracted and in some species three years may elapse between the formation of flower buds and seed maturation. Vivipary, whereby germination commences before the seed is dispersed, is another mechanism which, in some species, aids the successful establishment and survival of the young plant.

To the problems created by general climatic conditions are added those associated with the soil. Perennially frozen ground (*permafrost*) and intensive frost action are the two principal edaphic difficulties with which plants have to cope. The depth at which permafrost occurs varies from twenty centimetres to a metre in northern Alaska where, in addition, it is continuous, to 1·5 to over 3 metres in the south where its distribution tends to become discontinuous (see Fig. 68). Since it cannot be occupied or penetrated by living roots permafrost forms a tough, impervious substratum—comparable to a 'hard pan' in other soils. It, therefore, limits the 'effective' soil and root development to the upper 'active layer' which thaws out, to a variable depth and for a variable period, above it during the short Arctic summer. Indeed, resultant restriction of root penetration is considered one of the

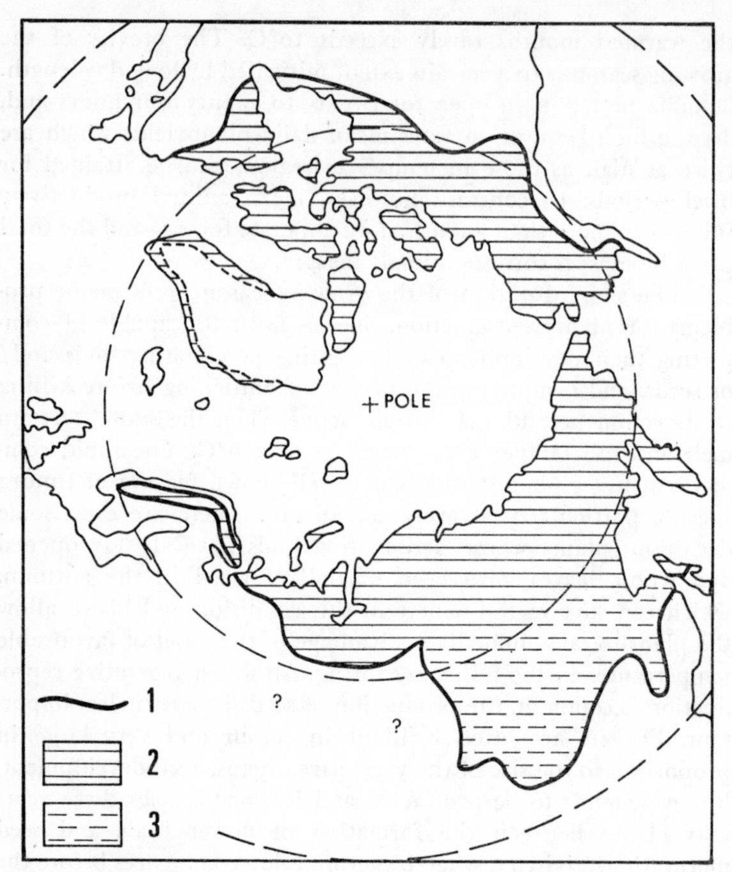

Fig. 68 Distribution of permafrost in the northern hemisphere: 1 = southern limits; 2 = regions of actual permafrost; 3 = regions of fossil permafrost. (After: Tricart, J. 1952)

factors inhibiting successful tree growth. In addition, the presence of permafrost may create problems of either soil saturation or drought during the growing season. The former condition, which often gives rise to peat development, is most frequently associated with abundant soil moisture supplied by snow and ice melt, combined with a fine-textured, water-retentive soil and lack of surface slope; the latter with coarser parent-material and steeper slopes. Also, where the active layer is exceptionally shallow and precipitation very low, the available soil-water released during the thaw may be so rapidly depleted by evapo-transpiration as to create conditions of physical drought during the growing-season.

Plants of the tundra have to be adapted to severe water stresses, created by and intensified by a characteristically high wind-force. Permafrost prevents water absorption during winter; and while many plants possess morphological attributes which may check transpiration, the process of 'hardening' is as important in imparting drought—as well as cold-resistance. Tundra plants are most susceptible to water stress during early summer. Then air temperatures may be above 0°C and evapo-transpiration is accelerated by drying winds at a time when roots may either still be encased in the frozen soil or be drawing water from the vicinity of the permafrost where temperatres are at, or just above, freezing point. An inability to absorb water because of low soil-temperatures, at a rate sufficient to make good losses by transpiration could, if sustained, be fatal. Experiments, however, have indicated that tundra plants are capable of absorbing water efficiently at very low temperatures and hence combatting what has often been referred to as 'physiological drought'. It is now generally conceded that neither physical nor physiological drought is a major obstacle to plant growth. While the low stature and more particularly the 'rosette' and 'cushion' vegetative-form of many tundra species may help the plant avoid the undesirable effects of exposure it is probably more important in allowing an early advantage to be taken of surface and upper soil temperatures which, in the critical period at the beginning of the summer, are somewhat higher than those of the air above. Probably more important than the effect of low soil-temperature on water absorption is that on nutrient uptake and cycling. The rate of mineral absorption is thought to be reduced as is the production of litter. Decomposition, consequent upon retarded bacterial activity, is slow; it has been estimated that in the tundra it may take ten years before leaf litter is completely humified. The release and availability of nutrients is reduced. The deficiency, particularly in nitrogen, of most arctic soils is made the more obvious by the much greater productivity of those heavily manured sites where animals congregate.

Not only are soils skeletal in character; they are also, as a result of the alternate freezing and thawing of the active layer, highly mobile. Seasonal expansion and contraction of soil water causes differential vertical and horizontal movements (and hence sorting) of the mechanically comminuted rock debris. This process of *cryoturbation* is most marked where there is an appreciable proportion of fine, water-retentive silt and clay particles. On sloping ground the effect of gravity is reinforced by the cryoturbation of water-saturated debris overlying permafrost and causes *solifluction* (soil-flow). In summer well-drained, dry, finely comminuted debris is often subject to deflation. Freezing of the active layer, which proceeds downwards from the surface, exerts considerable pressure on the unfrozen material above the permafrost; the former becomes contorted and convoluted and may eventually cause disruption of the surface. Finally, during the winter, when the substratum is completely frozen, soil temperatures may drop below the critical threshold of −4°C at which the contraction in volume of ice causes the ground to crack and fissures appear. The physical stresses to which roots are subjected are obviously very great. Best adapted are those species with shallow, laterally spreading systems, with roots and/or subterranean stems whose growth can keep pace with soil movements, which can withstand constant disruption and may be stimulated by constant tearing and splitting.

The vegetation of the tundra is characterised by an absence of trees and a relative paucity of annual plants. Most abundant are low-growing, woody and herbaceous perennials with well-developed underground storage organs and, particularly, mosses and lichens. Cover is, to a lesser or greater degree, discontinuous. The principal, regional variations in vegetation are related to the reduction in the duration and warmth of the growing-season, combined with an increase in the length and severity of the winter with latitude. On the southern, and particularly more oceanic, limits of the tundra the mean temperature is above freezing point for five months; the mean of the coldest month is seldom below −10°C and snowfall is heavy. Under such climatic conditions dense stands of deciduous dwarf-shrubs (particularly willow, alder and birch) up to two metres in height, attain their maximum development. With increasing latitude the frost-free period becomes even shorter—at the extreme limits less than a month in duration, the mean temperature of the coldest month drops to −30°C that of the warmest is between 1° and 5°C, and precipitation decreases, particularly in the more continental areas. Dwarf shrubs give place to heaths—low woody plants, rarely exceeding thirty to fifty centimetres in height, with small thick and frequently evergreen leaves as in *Empetrum nigrum* or *Vaccinium vitis ideae*, or to 'meadows' dominated by rushes and sedges, in ill-drained depressions, and hardy species of such grasses as *Poa* and *Deschampsia*. Finally, under the most extreme conditions virtually all plants save mosses and lichens are eliminated and the proportion of unvegetated ground is high.

However, within the broad framework of this climatically determined regional zonation, tundra vegetation is characterised by a diversity of communities dependent on local variations of physical habitat. The two most important 'inter-zonal' ecological variables are the depth and duration of snow-cover. Both are a function not only of climate but of local variations of relief, exposure and aspect. They affect the degree of frost penetration, the duration of the vegetative period and the availability of soil moisture. Within a particular climatic regime the distribution of associations is often closely related to the duration of snow-cover. For each species there is an optimum thickness above which the summer thaw is not sufficient to expose the ground for a period long enough for the completion of its growth cycle. Local variations in snow-depth related to variations in altitude and surface micro-relief result in a mosaic of vegetation 'catenas' or snow-patch communities related to this particular ecological gradient (see Fig. 69).

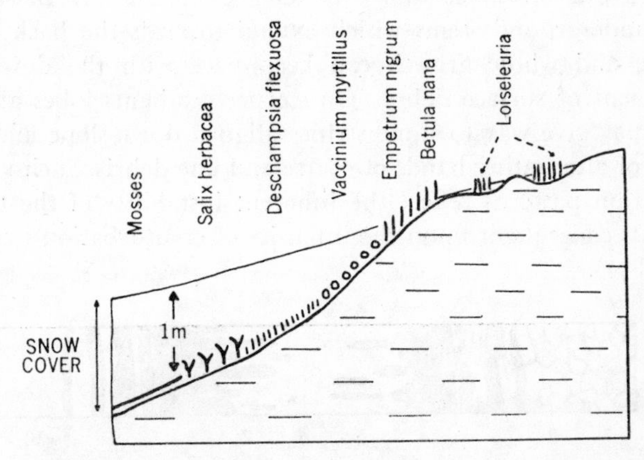

Fig. 69 Snow-patch vegetation 'catena' in the region of Abisko (Swedish Lapland) : an example of a mosaic in the southern part of the tundra near the forest limit with a relatively oceanic climate. (After: Birot, P. 1965)

The hydrological condition of the soil is dependent primarily on the nature of the parent material, texture and slope. Resistant, little-weathered rock surfaces, laid bare by relatively recent glacial erosion, can support little except crustose lichens. On poorly-drained surfaces, in depressions where the active soil layer is shallow or the substratum is impermeable, anaerobic water-logged conditions promote peat development. In contrast, coarse-textured parent material allows the thaw to penetrate deeper, and the permeable well-drained soil provides a drier habitat. Cryoturbation, causing differential movement and sorting of rock debris, is reflected in distinctive soil and associated vegetation patterns. On surfaces of little slope these take the form of polygons varying in diameter from one to four metres. Their borders are formed of coarse stony debris, their centres of finer-textured material. As a result of high water-capacity the latter may be subjected to such intense cryoturbation as to completely inhibit the establishment of vegetation. Plant growth therefore tends to be confined to the coarser and more stable borders which provide a more favourable habitat for root penetration and development. On slopes exceeding 5° solifluction becomes active, and these polygonal patterns are replaced by crescentic lobes or banks of coarse debris which form terraces of varying breadth and continuity roughly parallel to the slope. Vegetation again tends to be concentrated on the coarse material forming the steeper down-slope edge of such features and is frequently dominated by species adapted to well-drained but unstable sites. Characteristic plants are those with long underground stems which extend towards the back of the terrace and whose growth can keep pace with the downslope movement of surface debris. On steeper gradients lobes and terraces may give way to stone stripes aligned down-slope and composed of alternating bands of coarse and fine debris. Such varying vegetation patterns reflect the inherent instability of the tundra habitats consequent upon the intensity of cryoturbation (see Figs. 70 and 71).

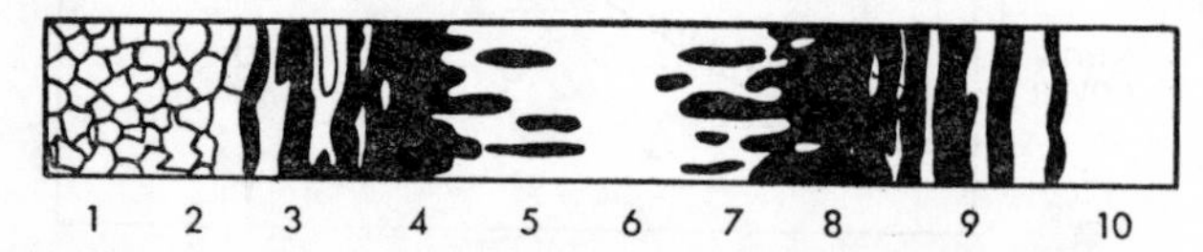

Fig. 70 Diagram showing ideal sequence of soil-vegetation patterns down a hillside.
1. 0–5° = polygons; 2. 5–15° = garlands; 3. 15–25° = terraces; 4. 25–35° = continuous vegetation; 5. 35–40° = stripes; 6. < 40° bare scree; 7. stripes; 8. continuous vegetation; 9. terraces; 10. bare scree. Black areas = vegetation cover. (After: Wilson, J. W. 1952)

Life in the tundra is no less rigorous for animals than for plants. Shortage of food coupled with the intense winter cold limit the numbers and types that such habitats can support. There is a lack of surface shelter, while permafrost and poor drainage inhibit subterranean refuge. Warm-blooded animals must either be protected against excessive heat-loss or avoid the winter cold. In the former instance long woolly coats, thick sub-cutaneous adipose tissues and a low ratio of surface to body volume are particularly useful adaptations to be found in such species as the

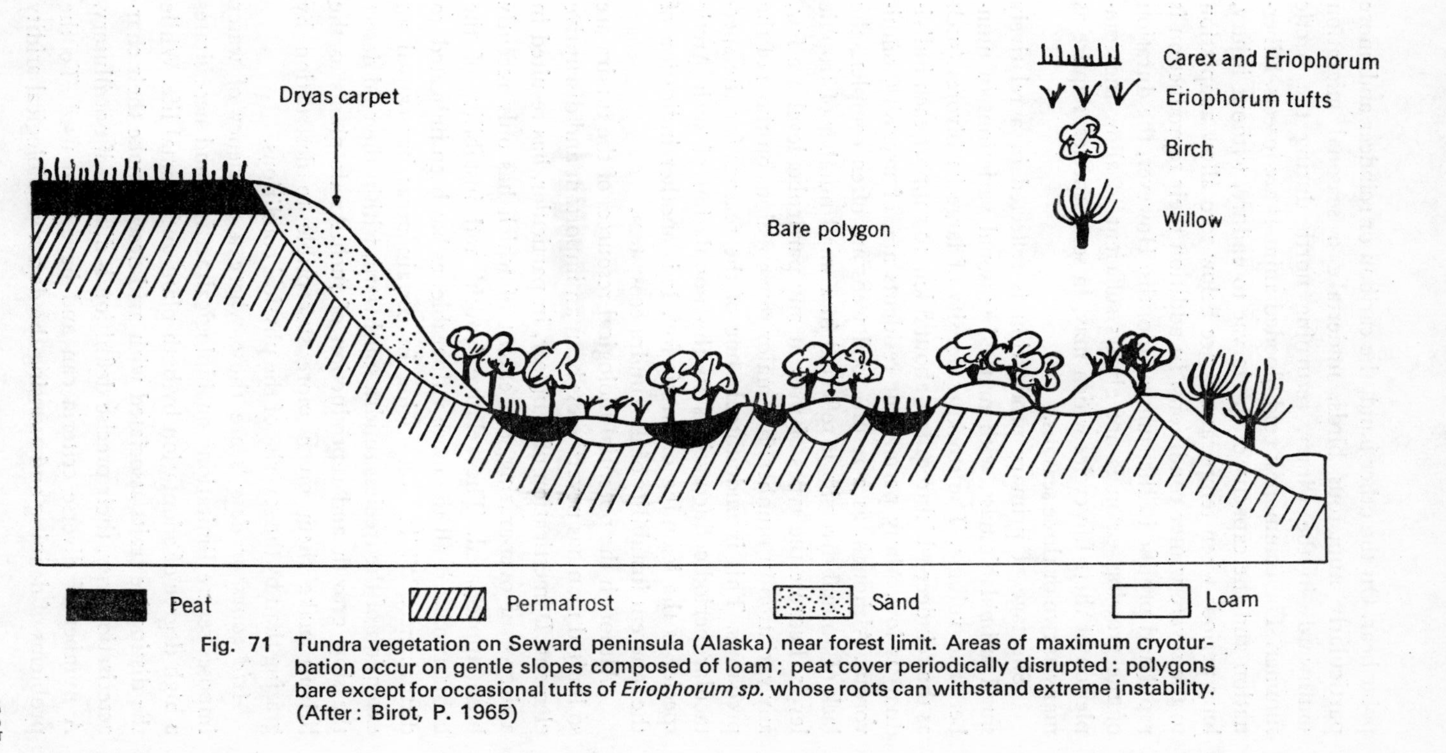

Fig. 71 Tundra vegetation on Seward peninsula (Alaska) near forest limit. Areas of maximum cryoturbation occur on gentle slopes composed of loam; peat cover periodically disrupted: polygons bare except for occasional tufts of *Eriophorum sp.* whose roots can withstand extreme instability. (After: Birot, P. 1965)

polar bear. On the other hand, the caribou or reindeer and, more particularly numerous birds, undertake a seasonal migration southwards in the winter, returning north during the Arctic summer. The commonest cold-blooded animals are insects. Hibernation and the capacity of the larvae to endure, without injury, long periods when temperatures are below zero allow adaptation to the severe winter conditions. In addition their reproduction is rapid and prolific in the summer months. However, the duration of the growing-season is not always sufficient to allow the completion of their life cycles which may, in some species, require as many as two or three seasons.

Shortage of primary production is reflected in a relatively small animal biomass, with marked seasonal variations in numbers and volume. The stocking density of large herbivores (such as the reindeer which requires about 8 km^2 to support each individual) is low. Many permanent residents are, of necessity, omnivorous. Although, as a result, food webs are often complex, the lack of animal diversity is reflected in a small number of trophic levels. Because the types of prey at any particular level are few, any variation in numbers has major repercussions on those of the predators. This is undoubtedly one of the reasons for the spectacular periodic fluctuations in the population of such Arctic species as the lemming or Arctic fox. It is another indication of the inherent instability of the tundra ecosystem.

Although the potential biological resources of the tundra are so limited man has nevertheless been an important and disruptive element. Domestication of reindeer, in particular, has resulted in a degree of over-grazing, the extent of which has only recently been appreciated. The relative 'youth' and instability of the habitat make it all the more vulnerable, as has been indicated, to disturbance. While overgrazing in certain areas has caused an extension of the treeless tundra southward, within the cold desert the slow growth and reproduction of such Arctic plants as the lichens make them much more susceptible to destruction by grazing than by the severity of the physical conditions.

Arid zones or deserts are those where a deficiency of water imposes severe limitations on biological activity and necessitates a high degree of adaptation by both plant and animal life. While the distinctive traits associated with arid deserts make their characterisation easy, their precise definition is difficult if not illusory. A number of diverse criteria can and have been used. To the problems of defining and assessing the degree of biological aridity

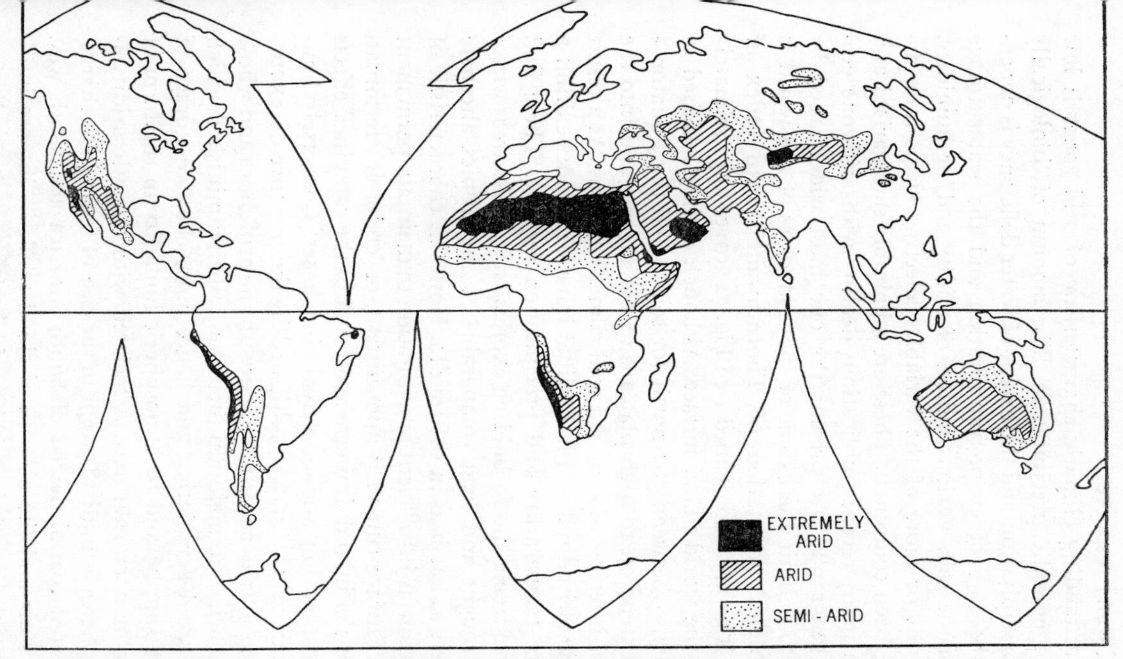

Fig. 72 Delimitation of arid zones using an adaptation of Thornthwaite's index of humidity

$$Ih = \frac{100e - 60d}{ET_P}$$

ET = potential evaporation; e = water surplus given soil retentive capacity 100 cm; d = water deficit. Indices: semi-arid, −20 to −40; arid, less than −40; extremely arid, no seasonal rhythm of precipitation, 12 consecutive months without rain have been recorded, index less than 57. (After: Meigs, Percival. 1953) is added that of continuous variation from humid through semi-arid to arid climatic conditions. Aridity is relative and there are few places where it is absolute. From the biological point of view it is a function of lack of availability, as much as a complete absence, of water resources.

An outstanding climatic characteristic of arid zones is low seasonal or episodic precipitation. The amount of 'biologically effective' precipitation and the extent of water deficiency is dependent on the rate of evapo-transpiration and the water-storage capacity of the soil. Various indices of aridity and/or humidity, based on one or more of the variables affecting evapo-transpiration have been computed. These are admirably summarised and assessed in many recent publications dealing with arid-zone ecology and geography. At present the one most widely used by biologists to delimit the extent of the arid zones of the world is Percival Meigs' adaptation of Thornthwaite's 1948 index of humidity. The areas so defined (see Fig. 72) cover approximately a third of the world's land surface. Although mainly localised in warm regions—between 30° north (or south) and the Equator—they also extend on to high plateaus in the Tropics and into the continental interiors of cool and cold climatic regions. As a result of this wide altitudinal and latitudinal range arid zones exhibit a diversity of temperature and precipitation regimes. These find their most extreme and forbidding expression in those areas where summer drought and heat comparable to the tropics alternate with winters as harsh as in the tundra. However, characteristic of all arid regions is insufficient precipitation to maintain permanent river flow. Surface run-off is intermittent, or 'endoreic', terminating in basins of inland drainage as a result of the combined effects of high evaporation and/or percolation losses. Conversely however, climatic aridity may be mitigated by the presence of 'exotic' rivers whose sources are situated outwith the arid zones and whose supply is sufficient to maintain their flow, albeit diminished, which facilitates rapid percolation and hence more effective storage of the meagre supply; and by atmospheric humidity in the form of coastal fog or nocturnal dew. Associated with climatic aridity is low atmospheric humidity, except in the case of fogs which characterise the cool coastal deserts of South America and south west Africa. Dry air and the absence of cloud result in intense insolation during the day and rapid loss of heat by out-going radiation at night. In hot deserts daily maxima of over 60°C have been recorded, while diurnal ranges over 53°C in the Sahara are not uncommon. In other cases these violent daily fluctuations may be accompanied by as marked a seasonal range of temperature.

As in the cold deserts, climatic conditions which inhibit a closed vegetation cover are associated with harsh edaphic habitats. The mineral surface is exposed to intense insolation (soil-surface temperatures can exceed 65°C; those of 40 to 50°C are lethal for the majority of organisms) and evaporation. The latter, combined with the absence of leaching, results in the accumulation of soluble salts of sodium, calcium and magnesium in particular, on or just beneath the surface of the ground. In addition, the poverty of vegetation, the lack of soil organic-matter and aridity expose the unconsolidated and unprotected weathered rock-debris to erosion and transport by wind and flash-floods.

Survival of plants and animals in arid and particularly hot, arid zones is dependent, according to J. Cloudsley-Thompson and M. J. Chadwick, 'upon avoiding desiccation and keeping cool'. While the flora of arid zones contain some of the most drought-resistant and heat-tolerant species, those capable of withstanding a high degree of desiccation of their tissues or near-lethal temperatures are relatively few. Of the former the creosote bush (*Larrea divaricata*), native to the deserts of south-western U.S.A., is probably the most outstanding example. It can survive for over a year without rain. In contrast to many arid-zone shrubs it has a widely spreading, shallow root-system. During protracted periods of drought it loses all its leaves and a large number of its branches, as well as becoming generally reduced in stature. It can tolerate a dehydration of its tissues such that the osmotic pressure of the cell-sap exceeds fifty-five atmospheres. Small leaf-buds however remain dormant and are able to recommence growth as soon as moisture becomes available. The most heat-tolerant of species are the desert succulents which can remain active at temperatures between 58° and 65°C, which would be fatal for other higher plants. Since few plants can tolerate a high degree of desiccation, arid-zone species must be able either to evade or resist drought. Annual or ephemeral species—which account for fifty to sixty per cent. of arid-zone flora are examples of the former. Generally of small size and shallow-rooted they are able to complete their life cycles, on average, within six to eight weeks, or even within as short a period as eight to ten days in the case of certain ephemerals. They can therefore take advantage of periods when sufficient moisture is available, while their seeds remain quiescent during the intervening drought. Further, the seeds of arid-zone plants are characterised by a high degree of viability.

But whether annual or perennial an ability to germinate and

establish the new plant is critical to the survival of the species. There are a variety of mechanisms which would appear to aid the achievement of this goal. An ample supply of seeds and efficient means of distribution are even more essential than in humid environments. *Hygrochasy*, or the impedence of long-distance dispersal, is exclusive to plants of arid regions and ensures that the seed remains within the favourable habitat of the parent plant. In some cases dissemination is delayed until after the first few days of the rainy period, or the plant possesses mechanisms which actually limit the distance of dispersal. Also, in many desert plants the seeds in any one plant do not all germinate at the same time but do so in an 'orderly sequence' over a period that may cover several years. This would appear to ensure that, despite high mortality rates, there are always some seeds 'in reserve'. In another specific example seeds are released, a few at a time, whenever the fruit opens up after moistening.

In addition many seeds possess means of regulating the time and location of germination so as to ensure successful seedling establishment. Many have what has often been referred to as a 'built-in rain-gauge' and will not germinate until a given amount of precipitation has fallen. Inhibition of germination until conditions which will ensure seedling survival is thought to be due to the presence of chemicals which regulate the permeability of the seed coat; although water-soluble such substances require a certain amount of moisture for complete solution and removal. It has been further suggested that such growth inhibitors (or other toxic substances) may prevent the germination of seeds of the same or different species in close proximity to one another and so prevent excessive root competition for water. Some arid-zone species, such as the paloverde, iron-wood and smoke trees of south-western U.S.A., have seeds whose coats are so tough that germination can only be effected after severe mechanical abrasion such as occurs in the torrential flash-floods.

While annuals evade the unfavourable periods of arid-zone environments, perennials must resist drought and hence avoid the desiccation which they are unable to tolerate. This is dependent on the possession of a combination of anatomical, morphological and physiological characteristics which help the plant to cope with a deficiency or absence of water for varying periods of time. Some facilitate and increase water absorption. It is thought that many arid-zone plants can exert a greater suction pressure and hence extract more water from the soil than mesophytes. More obvious adaptations are the initial rapidity of root development and the low shoot: root ratio (1:3·5 to 1:6) characteristic of desert plants. Some species such as the mesquite (*Proposis spp.*) have long roots which may penetrate ten to thirty metres below the surface and so tap permanent sources of ground-water. Many others, including the creosote bush and cacti, have a large volume of shallow, spreading roots which can exploit very efficiently water supplied by brief downpours. In addition improved translocation of water through the plant is achieved by an increase in volume and width of water conducting cells, as well as by short stature.

Reduction of water loss is an acknowledged characteristic of xerophytic plants though the significance in this respect of many apparently xeromorphic features is debatable, as has already been discussed (see p. 45). A comparison of the morphology of *caatinga* and *cerradão* vegetation in South America reveals that the more xerophytic caatinga in fact possesses a smaller number of so-called 'xeromorphic structures' than the less xerophytic cerradão. The most important means of preventing excessive water loss would appear to be stomatal closure, particularly during the hottest period of the day; low to negligible cuticular transpiration; and the reduction of the transpiring surface to total plant volume. In these respects the most drought-resistant of the higher plants are the succulent Cacti and Euphorbiaceae, which possess well-developed storage tissues, a small surface-to-volume ratio and are characterised by low cuticular transpiration and rapid stomatal closure. In those particularly heat-tolerant species of cacti stomata, contrary to the normal reaction, remain closed during the day and open only at night. Most xerophytic perennials, however, are faced with the conflicting problem of avoiding desiccation and keeping cool at the same time. Those which cannot tolerate high temperatures must keep their stomata at least partially open throughout the day and maintain transpiration at a level sufficient to cool the leaves. P. Birot has pointed out that among the perennial xerophytes there are those with high assimilation and transpiration rates and whose leaves can withstand high temperatures. Although stomata open only between seven to nine and sixteen to seventeen hours, carbon assimilation can be of the order of 40 mg/dm²/hour—ten times that of most forest trees. On the other hand xerophytes which are unable to tolerate temperatures which would normally result in stomatal closure must maintain transpiration, though at a reduced rate, throughout the day and consequently their yield of dry matter remains constantly low.

In addition to physical drought and great heat, plants of

arid-zones must in a great many instances be able to tolerate 'physiological drought'. Permanent surface-water or ground-water sources may have salt concentrations much higher than that of the cell sap of normal plants, and this inhibits the uptake of water by osmosis. Some, particularly annual plants, may evade the detrimental effects of high salt-concentrations by the synchronisation of their life cycles with a rainy season sufficient to leach the superficial soil layers. Otherwise plants must have a range of tolerance such as to allow them to cope with the existing high degree of salinity but with reduced vigour. The true *halophytes* attain optimum performance under varying levels of salinity.

Arid-zone vegetation, like the tundra, is treeless, and is composed of a mosaic of communities which decrease in volume and density with increasing aridity. Not only is the vegetation cover discontinuous but communities tend to be 'open' with the characteristically even-spacing of individuals that is such a distinctive feature of many arid deserts. It has been suggested that the open, evenly spaced shoots is a result of extensive root development and competition. What is therefore apparently a structurally open community on the surface may, in fact, be biologically closed beneath the surface.

While all arid-zone vegetation is to a greater or lesser extent xerophytic it is much more diverse in composition and form than that of the tundra. The arid zone is, in contrast, spatially more discontinuous. Flora of discrete areas, derived from contiguous semi-arid regions, have tended to evolve in isolation one from the other. Despite the resulting diversity and variation however, E. S. Hills points to four main 'categories' of arid vegetation whose general form can be related to varying degrees of aridity. These are:

(1) *Frutescent* perennial 'scrubs' of desert and semi-deserts. These include the distinctive cacti communities of the American deserts and the low, woody, desert scrubs. The latter constitute the most highly organised *form* of desert vegetation and occur where precipitation and/or surface deposits give a particularly favourable soil-moisture regime.

(2) *Suffrutescent* perennial vegetation is the most widespread type of desert formation. It can be dominated by one or more species of dwarf (30–120 cm) succulents, small woody shrubs or perennial grasses. The dominance of the latter usually indicates soil conditions that allow some surface soil-moisture storage.

(3) *Ephemeral* or seasonal herbaceous vegetation is characteristic

of areas where there is a regime of recurrent annual rainfall but where soil storage is limited by low precipitation and high evaporation combined often with sub-surface indurated 'pans'. Composed of annuals and also perennials (with 'ephemeral' or seasonal vegetative development) the vegetation frequently takes the form of extensive 'grasslands' which are the basis of seasonal nomadic grazing.

(4) *Accidential* vegetation occurs in the most extreme of arid habitats where as little as ten mm, and not more than five mm may fall in one shower or cloud-burst every five to ten years. It tends to be localised on patches of loamy or finer-textured soils which retain sufficient moisture to allow the brief appearance of a carpet of ephemeral annuals.

Under the most rigorous conditions lack of precipitation, intensity of evaporation, or an absence, excessive mobility or salinity of soils may exclude vegetation completely. In the 'absolute' desert areas biologically inhabitable sites are extremely localised and limited in extent. This is the case in the most arid and sterile parts of the Sahara where permanent vegetation is confined to depressions—wadis—in which there is access to ground water. Particularly harsh desert conditions also occur along the littoral areas of Northern Chile and Southern Peru, and on the high plateaus of Central Asia. In the former practically the only source of moisture is that supplied by sea mists during winter. In addition, the high salt-content of this hygroscopic water can be sufficient to inhibit the growth of non-halophytic plants and to form indurated surface 'crusts'. Best adapted are halophytes and, more particularly, succulent epiphytes which can absorb moisture directly from the atmosphere. Local patches of more permanent shrubby evergreen vegetation or seasonal herbaceous annuals may occur where, as a result of aspect and relief, patches of particularly dense fog coincide with water-retentive soils.

On the high plateaus of Central Asia (3 500–4 500m) the worst features of cold and arid deserts coincide. The annual temperature regime is marked. Mean temperatures of the coldest month can be less than −10°C, while for three months of the summer mean temperatures exceed 13°C. In the later period, however, the diurnal range is great and, as a result, there is no frost-free month. The low summer precipitation is made the less effective by high evaporation. Winter cold and summer drought are exacerbated by exceptionally high wind speeds. The perpetual struggle against cold, drought, and exposure is intensified by a shortage of CO_2 which at these altitudes begins to become a significant ecological factor. The resulting landscape is particularly desolate—only small halophytic shrubs and xerophytic grasses can survive.

As in the tundra the extent to which vegetation can modify and ameliorate the physical habitat in the arid-zone environment is limited. Variations in the hydrological regime of the soil, as a result of relief, slope, texture and exposure are important factors determining the density and distribution of vegetation in any particular area. The mosaic of vegetation 'catenas' related to a variety of ecological gradients is perhaps even more striking than in the tundra. Among the commonest are those associated with mobility of parent material as in the case of shifting sand dunes; with playas and salt pans of enclosed depressions where there is a zonation or gradation of vegetation from less to more alkaline conditions; and with soil texture. In the latter instance a striking correlation has been noted by many authors between the dominance of shallow-rooting grasses and fine water-retentive material on the one hand and of deep-rooting shrubs and coarse debris on the other.

Animals which inhabit the arid zone have to contend with the same difficult living conditions as do plants. As in the tundra, species are fewer but more specialised than in humid environments. They must, above all else, be equipped to withstand lack of water, excessive heat, and limited and variable food supplies. Among the obvious advantages in these respects are low water requirements, tolerance of large water losses by body tissues and an independence of the need to drink at frequent and regular intervals. During periods of drought the only sources of water may be nocturnal dew, moisture absorbed hygroscopically by dried vegetation and that contained in the plant and/or animal food available. The ability to exist on the latter source of water and to conserve metabolic water by various physiological means is characteristic of many arid-zone animals.

Protection against water loss and dangerous levels of body heat are closely inter-related and pose similar dilemma as in plants. In the absence of surface shade, large animals must tolerate these extreme conditions; smaller organisms can avoid or evade them in a variety of ways. Among the morphological features which are thought to give protection against the detrimental effects of high temperatures is an impermeable body covering. Where this is absent a large perspiring surface to body-volume (achieved in many desert animals by the attenuation of peripheral organs) a sparse hairy rather than woolly coat, and a colouring which cuts down heat absorption are characteristic. But even in the classic case of the camel which exhibits these morphological features, survival is also dependent on physiological tolerance of high water-losses. The camel can endure a loss of water equal to over twenty-five per cent. of its body weight and is able to replace this by drinking large quantities at irregular intervals. Donkeys are almost as efficient in this respect. In contrast most mammals, including man are unable to survive losses of water greater than twelve to fifteen per cent. of their body weight.

Smaller animals, however, can avoid excessive heat by seeking shade or by reduction of activity. The former may be provided by boulders, stones and rock crevasses—and of particular importance unconsolidated mineral debris. The burrowing habit is widely developed, especially among desert insects as well as among a variety of smaller vertebrates. In areas where vegetation is discontinuous or absent direct evaporation can effect a complete desiccation of only the upper few inches of the soil. The atmosphere of the deeper soil layers remains more humid than that above. Also, as a result of the low heat-conductivity of the dry soil, temperatures decrease rapidly with increasing depth beneath the surface, where they are cooler and more equable than above. Hence the soil provides a more humid, cooler and more constant micro-climate than either the surface or the atmosphere above it (see Fig. 73). The distribution of sand inhabitants (or *sabulicoles*) is determined, on the one hand, by their tolerance of heat and aridity, on the other by edaphic factors. Of the latter texture is probably the most important since it affects ease of penetration as well as air, water and temperature conditions in the soil. Other

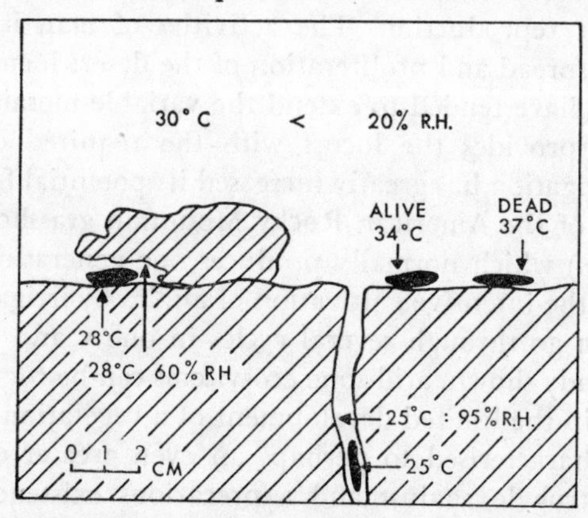

Fig. 73 A set of micro-climatic measurements in a habitat of the desert woodlouse (*Hemil epistus*) in Algeria. (From: Hills, E. S. 1967)

animals avoid heat-stress by resting and thereby reducing physical and physiological activity during the hottest period of the day or year. Nocturnal activity and summer dormancy (aestivation) are both distinctive features of invertebrate species of arid zones, and the larvae of insects can remain viable for long periods in a desiccated state.

The poverty and marked seasonal variation of arid-zone vegetation is reflected in small but often widely fluctuating populations of a few animal species. Mammals, in particular, are limited by the scarcity of fodder and slow reproduction rates. They respond to seasonal variation in food and water by migration, which contributes to the marked variation in animal populations in a particular habitat from one period of the year to another. Synchronisation of breeding cycles with periods favourable to vegetation growth is also characteristic of many arid-zone animals. It contributes to what has been referred to as the 'explosive' nature of reproduction and those marked periodic variations in populations which can be related to the variable character of precipitation. It also results in many, particularly insect, species becoming prolific pests when, by means of irrigation, man provides an abundant source of food in the absence of predators. In this respect none has been more spectacular or of such economic significance as the desert locust (*Schistocerca sp.*) This species of 'grasshopper' needs a variety of habitats dependent on its changing requirements at different stages of its life history. The eggs are laid in bare sand, the young 'hoppers' require abundant vegetation for food and shelter, while the adults can tolerate a scarcity of food and water; the latter conditions, however, inhibit reproduction. The activities of man have greatly assisted the spread and proliferation of the desert locust. Burning and grazing have tended to extend the variable mosaic of vegetation which provides the locust with the required diversity of habitats. Irrigation has greatly increased its potential food supply. In the case of the American Rocky Mountain grasshopper (*Melanopolus sp.*) which normally produces one generation annually in Arizona, the intensive cultivation of alfalfa by irrigation allows the insect to go through several cycles in one year.

As already shown, arid-zone ecosystems can be compared with the tundra in the limited development of a vegetation cover. The former is characterised to perhaps an even greater extent by a perpetual struggle against and a precarious existence in harsh environmental conditions. There is a small amount of animal biomass, trophic levels are few, and animal populations are susceptible to wide fluctuations in time and place. Under these circumstances any disturbance or modification of either the physical or biotic components can have drastic repercussions and set in train changes which may be irreversible. Archaeological evidence would suggest that man's use of arid and semi-arid regions has been more intensive and of longer duration than that of the tundra. Originally focused on the desert margins his effects have extended into the desert proper on the one hand and into the semi-desert and non-desert areas on the other. Long continued pastoral activities have led to over-grazing—to which the vegetation is particularly susceptible in dry years—and soil erosion. Mobile surface material has been transported into non-desert areas. Over-grazing and the destruction of the protective vegetation cover in the desert/non-desert ecotone has resulted in the expansion of desert conditions. Removal of vegetation has brought about changes in micro-climate, particularly of the soil, that have made the natural re-establishment of a continuous cover difficult and have created the so-called 'man-made' deserts.

On the other hand efforts to exploit the climatic potential of arid zones by means of irrigation have led to soil sterilisation and the extension of biological deserts. Because of aridity and the absence of leaching, soils of arid zones are naturally rich in the soluble products of weathering. These, including mainly soluble sulphates and chlorides of sodium, calcium and magnesium remain or accumulate in the upper part of the soil profile and impart a high degree of salinity. Inherent in the process of irrigation is the danger of increasing salinity and, even worse, alkalinity to levels which not only lower yields but may render the soil sterile. Salinity tends to be most pronounced where ground water is at or near ground surface. Irrigation by adding water, particularly where natural drainage conditions are poor, tends to raise the water-table and as a result of high evapo-transpiration soluble salts are continually concentrated and precipitated at or just below the surface. They often reveal their presence in a whitish 'bloom'. Above a critical level of salinity, yields of salt-tolerant crops tend to decrease. A concomitant of successful irrigation, however, is the need for adequate drainage to maintain the water-table at an optimum level and to counteract the tendency for the salt concentration of the soil-water to increase. Unfortunately drainage—the artificial equivalent of leaching—can have the deleterious effect of converting saline into even more intractable alkaline soils. Drainage may remove the excess of soluble salts but not the exchangeable sodium or insoluble carbonates. The latter, as well as producing high alkaline conditions (pH ten or more) causes a dispersion of the clay particles and a consequent loss of structure and permeability. Such conditions may render the soil completely sterile and extremely difficult to rehabilitate. The potential biological resources of the arid zones are, as far as can be judged, much superior to those of the tundra. The fact, however, that as much land has been sterilised as has been made to bloom by irrigation is a salutary reminder that the problems of exploiting successfully such an unstable and precariously balanced ecosystem have not yet been fully solved.

References

TUNDRA

BIROT, P. 1965. *Les formations végétales du globe*, S.E.D.E.S., Paris.

BENNINGHOFF, W. S. 1952. Interaction of vegetation and soil frost phenomena. *Arctic*, **5** (1): 34–44.

BILLINGS, W. D. and BLISS, L. C. 1959. An alpine snowbank environment and its effects on vegetation, plant development and productivity. *Ecology*, **40** (3): 388–397.

CANTOR, L. M. 1967. *A world geography of irrigation*. Oliver & Boyd, Edinburgh.

CHURCHILL, E. D. and HANSON, H. C. 1958. The concept of climax in arctic and alpine vegetation. *Bot. Rev.*, **24** (2/3): 127–191.

CROCKER, R. L. and MAJOR, J. 1955. Soil development in relation to vegetation and surface age at Glacier Bay, Alaska. *Ecology*, **43** (3): 427–448.

HANSON, H. C. 1950. Vegetation and soil profiles in some solifluction and mound areas in Alaska. *Ecology*, **31** (4): 606–630.

HOPKINS, D. M. and SIGAFOOS, R. S. 1951. Frost action and vegetation on Seward Peninsula, Alaska. *Bull. U.S. geol. Surv.*, 974C: 51–101.

LEMÉE, G. 1967. *Précis de biogéographie*. Masson et Cie, Paris.

PEARSALL, W. H. and NEWBOULD, P. J. Production ecology. IV. Standing crops of natural vegetation in the Sub-arctic. *Journal of Ecology*. **45**: 593–599.

PORSILD, A. E. 1951. Plant life in the Arctic. *Can. Geogr. J.*, March. **42** (3): 121–145.

SIGAFOOS, R. S. 1949. The effect of frost action and processes of cryoturbation upon the development of tundra vegetation. *Am. J. Bot.*, **36**: 832.

TRICART, J. 1952. Cours de géomorphologie. Part 2: Géomorphologie climatique, Section 1. *Le modlè des pays froids*. 1° Le modlè périglaciaire. Centre Documentation Universitaire, Paris.

WILSON, J. W. 1952. Vegetation patterns associated with soil movement on Jan Mays Island. *J. Ecol.*, **40** (2): 249–264.

WILSON, J. W. 1959. Arctic plant growth. *Adv. Sci.*, **13** (52): 383–388.

ARID ZONES

Amiran, D. H. K. 1965. Arid zone development: a reappraisal under modern technological conditions. *Econ. Geogr.*, **41** (3): 189–210.

Birot, P. 1965. *Op., cit.*

Cloudsley-Thompson, J. L. (Ed.) 1954. *The biology of deserts.* Institute of Biology Symposium. Oliver & Boyd, Edinburgh.

Cloudsley-Thompson, J. L. and Chadwick, M. J. 1964. *Life in deserts.* Foulis & Co., London.

Dickson. B. T. (Ed.) 1957. *Guide-book to research data for arid zone development:* Pt. I. *Physical and biological factors.* (*Arid Zone Res.* **9**), U.N.E.S.C.O., Paris, 1957.

Dresch, J. 1966. Utilisation and human geography of the deserts. *Trans. Inst. Br. Geogr.*, **40**: 1–10.

Highsmith, R. M. 1965. Irrigated lands of the world. *Geogr. Rev.* **55** (3): 382–389.

Hills, E. S. (Ed.) 1967. *Arid lands: a geographical appraisal.* Methuen, London.

Hudson, J. P. 1965–6. Irrigation problems under hot arid conditions. *Adv. Sci.*, **22** (98): 218–226.

Lemée, G. 1967. *Op., cit.*

Lowdermilk, W. C. 1935. Man-made deserts. *Pacific Affairs*, **8** (4).

Macfadyen, W. A. 1950. Vegetation patterns in the semi-arid desert plains of British Somaliland. *Geogr. J.*, **116** (4–6): 199–210.

Meigs, P. 1953. World distribution of arid and semi-arid homoclimates. In *Arid Zone Progm.*, I. *Reviews of research on arid zone hydrology.* U.N.E.S.C.O., Paris.

Meigs, P. 1966. *The geography of coastal deserts.* U.N.E.S.C.O., Paris.

Peel, R. F. 1966. The landscape in aridity. *Trans. Inst. Br. Geogr.*, **38**: 1–23.

Pond, A. W. 1962. *The desert world.* Nelson, London.

Reynolds, H. G. and Bohning, J. W. 1956. The effects of burning on a desert grass-shrub range in Southern Arizona. *Ecology*, **37** (4): 769–777.

Sears, P. B. 1949. *Deserts on the march.* Routledge, Kegan Paul, New York.

Shreve, F. 1942. The desert vegetation of North America. *Bot. Rev.*, **8** (4): 195–246.

Simons, M. 1967. *Deserts: the problems of water in arid lands.* Oxford University Press, London.

U.N.E.S.C.O. 1958. *Climatology: reviews of research. Arid Zone Res.*, **10**. Paris.

U.N.E.S.C.O. 1958. *Climatology and microclimatology.* Proceedings of the Canberra Symposium. *Arid Zone Res* **11**. Paris.

U.N.E.S.C.O. 1960. *Plant -water relationship in arid and semi-arid conditions: reviews of research. Arid Zone Res.*, **15**. Paris.

Wallén, C. C. 1967. Aridity definitions and their applicability. *Geogr. Annlr.* (Stockholm), **49**A (2/4): 367–384.

Walter, H. 1955. Le facteur eau dans les régions arides et sa signification pour l'organisation de la végétation dans les contrées sub-tropicales. *Année Biol.*, (Ser. 3): 271–283.

WENT, F. W. 1948 and 1950. Ecology of desert plants. I. Observations on germination in the Joshua Tree National Monument, California. *Ecology*, **29** (3): 242–253. II. The effect of rain and temperature on germination and growth. *Ecology*, **30** (1): 1–13.

WENT, F. W. and WESTERGAARD, W. 1950. III. Development of plants in the Death Valley National Monument, California. *Ecology*, **30** (1): 26–38.

WENT, F. W. 1955. The ecology of desert plants. *Scient. Am.* April. **192** (4): 68–76.

WHITE, G. F. (Ed.). 1956. *The future of arid lands*. Amer. Ass. Adv. Sc. Publ. No. 43, Washington, D.C.

15 Exploitation of Organic Resources

In comparison to many other forms of life the human population is still relatively small. Man, in fact, accounts for only a minute fraction of the earth's biomass. Ecologically, however, he is the dominant species. Not only is he omnivorous, but his diet is more varied than that of other animals. Also the uses he makes of the biosphere are more diverse. As a predator man is without equal; neither lack of food nor the presence of a powerful enough human-predator has been able, as yet, to set a limit to his numbers. Because of man's superior ability to exploit and modify both the organic and inorganic environment his influence on the biosphere has been greater than that of any other organism. He has, as a result, disrupted what has often been referred to as 'the balance of nature'—that complex, delicately and naturally regulated relationship between food producers and food consumers, and between these and their physical environment. As a result of his selective, uncontrolled use of organic resources he has altered drastically the relative proportions of species populations. He has depleted 'wild', at the expense of domesticated plant and animal populations. Many, formerly more prolific and widespread, are now rare, restricted in extent and/or reduced in numbers, while others have become, or are in danger of becoming extinct. This process has been accelerated particularly during the past four hundred years consequent upon the rapid increase and spread of human population, together with the technical developments which either facilitate the exploitation of resources or the deliberate modification of the physical environment.

In the first instance the reduction and elimination of species have been the result of the direct slaughter or removal of animals and plants at a rate greater than they can reproduce themselves. This unregulated 'robber-exploitation', whose inevitable result is the depletion of organic 'capital resources', is an intensification of man's early methods of acquiring food by hunting and collecting. It is characteristic of technically less-developed societies and of the more recently colonised areas of the world. It is still basic to the use of marine resources, since the extent of man's control of the seas is much less than in any other ecosystem. The effects of uncontrolled exploitation have been most obvious and drastic on animal populations whose terrestrial biomass is less than one per cent. that of plants. Species of 'higher' animals such as mammals, birds, and fish have been most severely affected. This is a result not only of their size but of their smaller populations and slower reproductive rates in comparison with other species. Further, in comparison to plants fewer animal species have been domesticated. A recent inventory, *The Red Book,* by James Fisher and others, of animal populations which have become extinct or have declined drastically since 1 600, records the loss of thirty-six species of mammals, with a further hundred and twenty so rare as to be on the verge of elimination; and of ninety-four and 187 species of birds similarly placed. It has been estimated that at present one species or sub-species of either mammal or bird becomes extinct every year.

This 'over-killing' of animals has, to a large degree, been for food; wild game is still the most important though, as yet, not fully developed or rationally exploited source of protein in Africa south of the Sahara. Much indiscriminate slaughter, however, has been motivated by a particular demand for a limited and highly priced animal-product. The late nineteenth century market for whale oil and whale bone resulted in the virtual extermination of the Arctic White Whale. Such potentially valuable marine animals of low reproductive rates as the seal and the turtle have suffered serious inroads into their populations. Indeed at a meeting of the International Union for the Conservation of Nature in 1969 it was reported that despite a high degree of 'official protection' the survival of all seven species of marine turtles was in jeopardy. Furs have long been a coveted luxury-product and an important item of trade. They provided an early and lucrative incentive to explore and colonise the North American continent. It is hardly surprising to learn that by the end of the nineteenth century, after only three hundred years of exploitation, the once prolific beaver had become extremely rare. The demand for more exotic products such as ivory and aphrodisiacs severely reduced the smaller populations of large mammals, such as the elephant and rhino. More recently the primates—and not least the Rhesus monkey—have acquired a high value for the purpose of scientific research.

In other cases animals have been deliberately and systematically slaughtered to protect human life, domestic and desirable game animals, as well as cultivated crops, from pests and predators. Predatory birds and carnivorous mammals have experienced the greatest losses. For instance the wolf had disappeared from Scotland by the beginning of the eighteenth century and the largest of our predatory birds, the golden eagle, is now a protected rarity. The systematic extermination of the North American bison initially for profit, but more particularly as a means of subjugating the Plains Indians must represent one of the most rapid and wholesale destructions of a large species-population in the world. In the early eighteenth century it has been estimated that there were at least sixty million bison 'on the hoof'; by 1913 only twenty-one individuals remained!

Until the eighteenth century the killing of animals for pleasure was subordinate to more practical ends. 'Game' in many European countries was the preserve of the aristocracy and hunting, although 'the sport of kings', made an important contribution to the table. Hunting primarily for sport developed during the eighteenth century and 'huntin' shootin' and fishin' became a respectable and fashionable pursuit of the more affluent and leisured classes. The profligacy of the 'strange', either dangerous or elusive, wild animals in areas subject to European colonisation saw the development of 'big game hunting', particularly in Africa and India. In the latter country it has been estimated that the numbers of the better-known wild animals, (e.g. tiger, leopard, deer, black-buck) have declined by a tenth of what they were fifty years ago. The Indian lion and rhino have been reduced to negligible numbers, while the Indian cheetah became extinct in the late nineteen-forties. Wanton poaching either to obtain a quick profit or for pleasure still continues. A recent report, for example, of the Fauna Preservation Society notes that the last refuge of the pampas deer in the Argentine has been invaded by hooligans with guns who are 'beating-up' a species now reduced to fifty individuals.

The reduction and elimination of plant species and populations has, in terms of absolute numbers, probably been even greater than that of animals. But in proportion to the total volume and diversity of the plant biomass the effect has been less spectacular. Direct depletion of species populations by overcropping, cutting or grazing has had its greatest and most obvious effect on trees and shrubs whose long reproductive cycle, combined with the vulnerability of their seedlings, makes them particularly susceptible to destruction. Those in greatest danger of extinction must be the restricted populations of 'relict' species such as the *Sequoias* (indigenous to California), the *Metasequoias* (a recently discovered 'living' fossil in Asia) and *Ginkgo bilboa,* the maiden-hair tree, only limited populations of which occur in the wilds of Japan. One species which, in this respect, has attracted much attention and generated a great deal of controversy is the Californian redwood (*Sequoia sempervirens*). This, the tallest tree in the world, is confined to a narrow coastal strip of land in northern California and southern Oregon. Since the mid-nineteenth century it has been losing ground rapidly in face of uncontrolled logging; and today it is further threatened by proposed highway and industrial developments. Similarly vulnerable are the relatively sparse and also geographically restricted neo-endemics, the most recent products of evolution.

More recently, however, annual and perennial herbaceous plants have been subjected to increasingly intensive and efficient destruction by *herbicides*—chemical and now usually selective weed-killers which have replaced the more laborious mechanical methods of weeding. The significance of the hormone herbicides, developed since 1940, lies not only in the efficiency with which they can destroy almost all broad-leaved plants but in the consequent reduction of species of animals formerly dependent on them. Their indirect effect on wild-life, through habitat modification, can be as great as their direct effect on plants. Plant populations, however, have probably suffered greater depletion as a result of man's modification and/or destruction of former habitats. In this respect the time-honoured and closely inter-related pastoral activities of grazing and burning have given advantage to certain species of plants (and consequently of animals) better adapted to survival under these conditions. Most obvious and widespread has been the clearance or submergence of a pre-existing 'wild or semi-wild' vegetation cover and its replacement by another form of land use—be it arable crop, forest plantation, reservoir, highway, airport, buildings or industrial waste. On the one hand, drainage has been accompanied by the disappearance of 'wet-land' habitats and their associated flora and fauna; on the other irrigation has replaced a xerophytic by a hydrophytic condition. To a marked extent, however, all these habitat changes have been effected deliberately in the interests of a particular end. Man has sufficient, if not always definitive, *technical* ability to control the processes involved should he so wish. In contrast, the more recent and rapidly increasing *pollution* of the physical environment of air, soil and water is 'accidental'—in that it is the result of actions not originally intended to modify habitats—and is still largely uncontrolled. Already however it is becoming clear that it is having a seriously disruptive, though not as yet completely understood, effect on species populations. Habitat pollution is the result of two main factors; first, the discharge of industrial and domestic waste products and second, the use of herbicides and more particularly insecticides. Both have attained alarming proportions only within the last thirty years.

While much domestic and industrial waste material, from coal bings to used-car 'cemeteries' is dumped on the surface of the ground and may locally sterilise or poison the area concerned, a very high proportion is discharged either into the air or in a suspended or soluble form into rivers, lakes and the sea. Air pollution is the result primarily of the physical and chemical by-products of coal and oil combustion. In so far as these are concentrated over urban industrial areas, it would appear to have a greater direct effect on man than on either wild or domesticated plant or animal life. Nevertheless it has been suggested that plants are more sensitive than animals (including man) because of their ability to absorb and concentrate in their tissues gaseous substances which normally only occur in small amounts in the atmosphere. Air pollution can have a direct physical effect on plant activity through the reduction of light intensity and the deposition of 'soot' on leaves. The former reduces photosynthesis, the latter both photosynthesis and transpiration. Evergreens are particularly susceptible because they are subjected to a longer and, particularly in winter, more intense period of pollution; in heavily polluted areas it has been discovered that they may have a rate of transpiration only one-tenth of normal leaves. Leaf-fall in coniferous species is accelerated, and growth can be inhibited. More significant, however, are the chemical effects of increased amount of gaseous or solid substances. Of the former the most important include carbon monoxide and dioxide, the commonest constituents of the exhausts from coal fires and motor vehicles. Sulphur dioxide is the main product of heavy-oil combustion and is thought to be the cause of the lethal character of the thick yellow smogs of urban industrial areas; and ozone, the product of a photo-chemical reaction between strong sunlight and motor exhaust which is responsible for the notorious 'blue' smog of Los

Angeles and other urban areas in Southern California (see Fig. 74). Among the effects of these pollutants are loss of flower buds, the acceleration of autumn coloration and leaf fall in deciduous trees and other herbaceous perennials; and in the case of ozone, the bleaching of leaves as a result of the reduction and eventual

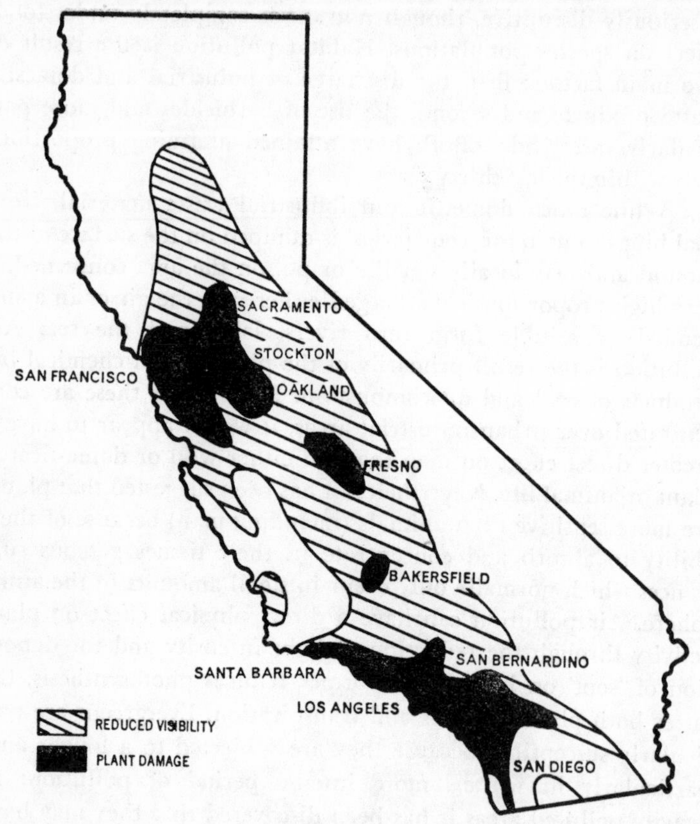

Fig. 74 Extent and effect of smog in California. (From: Haagen-Smith, A. J. 1964)

cessation of chlorophyll formation. In other cases, the concentration of substances, such as lead and fluorine absorbed either directly from the air or from the soil may be concentrated at levels toxic either to the plants or the animals which eat them. The influence of air pollution on soil reaction as a result of the acidulation of rain water by such substances as carbon dioxide and sulphur dioxide, still awaits more detailed investigation.

Plants vary in their tolerance of air pollution which may either reduce yields or inhibit growth of certain species. Because of the geographical location air pollution affects mainly the floral composition of parks and gardens and agricultural land around urban areas. Its effect on wild life is, as far as is known, limited. There are however indications that certain species of lichen are particularly sensitive to air pollution—a factor now thought to contribute to their absence from the trunks of urban trees! Attempts have been made to use the percentage of lichen cover as an index of the degree of air pollution.

Much industrial and even more domestic waste is discharged as effluent. The resulting pollution of water bodies is not only more widely distributed but has, as far as is known, a more pronounced effect on plant and animal populations than air pollution. These effects may be direct because of the high toxicity of the substances concerned. This is the case where high concentrations of copper, lead, zinc, or certain insecticides poison and kill living organisms. But the ecological effect of effluent on the oxygen content of the water into which it is discharged is even more important. Synthetic detergents—which within the last thirty years have all but replaced soap—are among the most abundant by-products of a wide range of industrial processes. Heavy concentrations in lakes and slow-flowing rivers or canals often accumulate and form a persistent foam ('detergent swans') whose exact biological repercussions are not yet fully understood. They are, however, known to decrease oxygen absorption and are thought in some instances to cause the death of fish through asphyxiation. More important however is the progressive de-oxygenation of water as a result of increased primary biological productivity together with acceleration of decomposition. This is the direct result of the increasing discharge of plant nutrients leached from the soil and of organic matter in the form of human and animal sewage. In the first instance heavy 'fertilisation', with nitrates and phosphates, particularly of shallow, sheltered and relatively calm water bodies causes increased plant production. This however may be so prolific as to form an algal scum which prevents re-oxygenation of the water below to such an extent that the existence of all but the most anaerobic forms of life is inhibited. In addition, the amount of organic matter produced or discharged in the form of untreated or incompletely treated sewage may be such that its decomposition results in de-oxygenation and the eventual elimination of most plant and animal life. As has been found in southern Sweden such lakes, 'dead' or 'dying' as a result of pollution by raw sewage, become choked with partially decomposed organic matter which accumulates on the bottom of the lake. Unless resuscitated the former lake will finally be completely filled in and disappear. In most urban/industrial countries of the world the extent and degree of water pollution have reached alarming proportions. It is particularly bad in rivers, lakes and, above all, estuaries near to large urban populations. The latter are particularly important as marine animal breeding grounds and their populations are all the more vulnerable to destruction by either the direct or indirect effects of effluent.

By far the most lethal pollutants of air, water and soil are pesticides and the by-products of nuclear explosions. The former include the organo-chlorine (D.D.T., dieldrin, BHC) and organo-phosphorus (parathion etc.) compounds—'insecticides' of varying toxicity to man and animals. The latter are the radio-active isotopes of which, as far as is known, the biologically most significant are Strontium-90, Caesium-137 and Iodine-131; these together with others such as Cerium-144 and 141, Ruthenium-106 and 103 and Baryum-140, reach the earth's surface as dust or 'fall-out'. The already known or possible effects of these two groups of pollutants are attracting most attention and giving rise to growing concern throughout the world to-day. Produced only since the Second World War, both are easily transported and widely distributed. They have in certain areas and organisms already reached dangerously high levels of concentration.

The ecological significance of both the insecticides and radioactive fall-out is related, *first,* to the fact that some of these substances are chemically very stable, breaking down or 'decomposing' only slowly. Once discharged into the physical habitat they can, therefore, persist for a long time in their original or even more toxic form. This, for instance, is a characteristic of the more popular organo-chlorine than of the organo-phosphorus insecticides which, in comparison, are very unstable. While the latter are highly poisonous to birds and mammals, including man, they break down so rapidly that they lose their toxicity only a few hours after application. Conversely, some of the radioactive isotopes such as Strontium-90 and Caesium-137 are 'long-lived' and decay much more slowly than Iodine-131 which has a 'half-life' of only eight days. *Second,* these substances which may be present in very small quantities in the soil, air or water can become highly concentrated once they enter the biological cycle. They are absorbed directly by the tissues of invertebrate animals or in the case of some of the particularly poisonous insecticides, those of higher animals including man. Others, including the radioactive isotopes, can be absorbed directly from the air, water and soil by plants. If

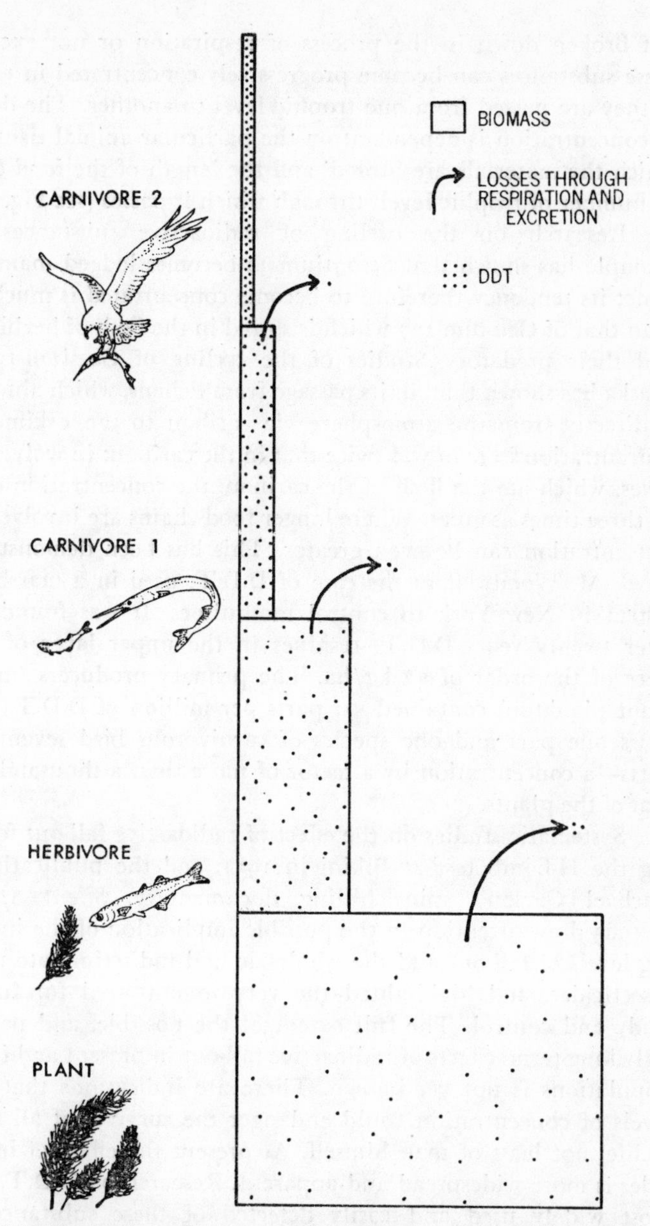

Fig. 75 Diagram illustrating concentration of DDT residues in their passage along a simple food chain. As 'biomass' is transferred from one link to another along such a chain, usually more than half of it is consumed in respiration or excreta. Losses of DDT residues, on the other hand, are small in proportion to the amount that is transferred from one link to another. (From: Woodwell, George, M. 1967)

not broken down in the process of respiration or not excreted these substances can become progressively concentrated in tissues as they are passed from one trophic level to another. The degree of concentration is dependent on the particular animal tissues in which the materials are 'stored' and the length of the food chain or number of trophic levels through which it passes (see Fig. 75).

Research on the cycling of radioactive substances, for example has shown that Strontium-90 becomes lodged mainly in bone; its tendency therefore to become concentrated is much less than that of Caesium-137 which is stored in the flesh of herbivores and their predators. Studies of the cycling of Caesium-137 in Alaska has shown that, in its passage from lichens, which absorbed it directly from the atmosphere via caribou to the Eskimo the concentration in man was twice that in the caribou; in wolves and foxes, which ate the flesh of the caribou, the concentration could be three times as great. Where longer food-chains are involved the concentration can be even greater. This has been demonstrated by G. M. Woodwell in the case of D.D.T. used in a marsh ecosystem in New York to control mosquitoes. It was found that after twenty years D.D.T. residues in the upper layer of mud were of the order of 32 kg/ha. The primary producers (mainly plant plankton) contained ·04 parts per million of D.D.T., minnows one part and one species of carnivorous bird seventy-five parts—a concentration by a factor of more than a thousand over that of the plants.

Systematic studies on the effect of radioactive fall-out following the H-bomb test at Bikini in 1954, and the publication of Rachael Carson's spine-chilling documentary, *Silent Springs* in 1963 drew attention to the possible implication of the increasing levels of fall-out and the wholesale and indiscriminate use of insecticides, and highlighted the very urgent need for further study and control. The full extent of the possible, and particularly long-term, effects of radioactive fall-out in present and future populations is not yet known. There are indications that high levels of concentration could endanger the survival of all forms of life, not least of man himself. At present the effect of insecticides is more widespread and apparent. Research on D.D.T.—the most widely used and easily detected of these substances—is beginning to reveal the full effect of insecticides on animal populations. Concentrations in the tissues of higher animals who feed on plants treated with these preparations or who prey on such herbivores may build up to levels which can cause sterility or death. In either case this could lead to the decimation and eventual elimination of a species. There is increasing evidence that among the most serious casualties are the higher carnivores—particularly carnivorous mammals, birds and fishes. As George Woodwell remarks, 'Because of the wide distribution of D.D.T. the effects of the substance on a species of animal can be more damaging than hunting or the elimination of a habitat (through an operation such as dredging marshes). D.D.T. affects the entire species rather than a single population and may well wipe out the species by eliminating reproduction'. There are, further, indications that the use of insecticides is in the process of intensifying the very problem they were designed to solve. Insects, and other invertebrate pests have the ability because of their large populations. In doing so he has, consciously or unconsciously, increased tively rapidly new strains resistant to chemical attack.

Intentionally or accidentally, by direct or indirect means, man has reduced the number of certain species and their populations. In doing so he has, consciously or unconsciously increased the numbers and range of others. This process has been further intensified by his deliberate selection, protection and propagation of others for particular uses. In view of the tremendous diversity of his potential organic resources and his technical developments, man has been peculiarly selective and conservative in his choice and use of plant, and more particularly, of animal species. Despite the greater diversity of the latter the number of species exploited is very small in relation to the total available. Several factors account for his apparent fastidiousness. The mobility of all animals obviously makes them more difficult to collect and tend than plants! In addition a very high proportion are of a size and/or habit of life which would make their exploitation both difficult and uneconomic; in most ecosystems, over ninety per cent. of the zoo biomass may be composed of invertebrates. The large, though less abundant, herbivores are from man's point of view the most efficient converters of plant carbohydrates into animal protein. They are, hence, the most valuable and easily exploited. Of the dozen or so species of animals which man has successfully domesticated for food or draught purposes all are herbivores; and three—the cow, sheep and pig—make the largest contribution to his protein supply. The increase in the numbers of domesticated animals has been accompanied by that of wild herbivores as a result of the deliberate reduction of carnivorous predators. There have been three important consequences of the greatly increased numbers of herbivorous animals. First, formerly more mixed, diverse animal communities have been replaced, more especially above ground, by less diverse and more uniform populations. Secondly, because of the reduction in the number of species of herbivores grazing in many ecosystems is more selective than formerly. Thirdly, grazing pressures have been greatly intensified, both directly and indirectly.

The domesticated herbivore is one of man's principal means of exploiting organic resources. Pastoralism is characteristic of those areas where a deficiency of either heat or moisture tends to make cultivation (in the absence of irrigation) unreliable, but where there is sufficient 'wild' vegetation to provide fodder for stock. The 'carrying capacity' of such land is often extremely low (tens or sometimes hundreds of hectares per breeding animal are normally required) and is dependent on the vegetation available. To this extent, the amount and type of plant growth 'controls' the intensity of use that can be made of a particular area. However, a pastoral economy is not a 'closed' system; animal products are constantly moving out of the ecosystem. Extraction of an animal crop is therefore accompanied by a continuous loss of nutrients derived from the soil via the vegetation. This depletion combined with, and intensified by, selective grazing results in a gradually changing balance of plant species and a progressive decline in primary biological productivity. Grazing, which is so often accompanied by burning, tends initially to give herbaceous species an advantage over woody ones (see pp. 91–94). In preventing tree and shrub regeneration it favours the spread and maintenance of 'pasture land'. Selective grazing of the more palatable, nutritious and, usually, more nutrient-demanding species accompanied by nutrient depletion results in a decline in soil fertility. This is usually reflected in an increase in less nutritious and nutrient-demanding species, the replacement of more valuable perennial by annual grasses and herbaceous weeds, and, in semi-arid areas, an increasing number of xerophytes. The inevitable concomitant is a decrease in primary productivity and a reduction in the carrying capacity of the pasture. Such are the outward and visible signs of *overgrazing* which, if maintained at constant stocking densities, will eventually result in the removal of the entire vegetation cover at a rate greater than it can be replaced. This combined with the loss of soil organic-matter must terminate in soil erosion.

The continued use of 'natural' range land without replacement of the nutrients removed by stock is characteristic of most pastoral economies. It must inevitably result in overgrazing. This has, however, been accelerated and intensified by the rainfall variability which is a climatic hazard of many of the world's traditional pastoral areas. The occurrence of often protracted periods of drought can, and does, reduce productivity and carrying capacity without any change in the number of stock. Overgrazing, however, is as often the result of overstocking in response to increasing demand and the incentive to make a quick short-term profit without regard for the long-term consequences on the pasture resources. Many temperate range lands have suffered as a result of the deliberate mining of an irreplaceable capital resource. In the U.S.A. it was estimated that in the short space of fifty years, between 1880 and 1930, overstocking of the western range reduced its carrying capacity by about half its original potential. More recently concern has been expressed about the rapid deterioration of the semi-arid range land of Australia. Occupying about three-quarters of the total area it supports about a third of the country's sheep and cattle stock. Since it was originally opened up just over a hundred years ago, not only have many of the most valuable native perennial grasses disappeared but the productivity of the range has declined to a third of its early level.

Overgrazing probably constitutes a much more serious problem in underdeveloped inter-tropical areas of the world. There is an even greater lack of integration of pastoral and arable farming economies than in temperate countries. Stock rearing and crop production tend to be more distinct and geographically separate activities. Stocking rates are exceptionally high and are increasing—consequent upon growing rural population densities —combined with static or in some cases declining amounts of rough-grazing land. Overstocking is exacerbated by the fact that cattle, in particular, often have a social or religious significance which far outweigh and indeed tend to detract from their value as a source of food. Among many pastoral communities in Africa a man's importance is assessed by the number of cattle, irrespective of their physical condition, which he possesses. In India there are religious taboos which forbid the Hindu to kill the 'sacred' cow; and as a result one of the largest cow populations in the world makes a negligible contribution to the human food supply. Because of a shortage of domestic fuel, animal dung, which might otherwise replace the drain on soil fertility, is used as a substitute for wood.

Also, as has already been noted, the effects of overgrazing and the destruction of the vegetation cover are more drastic in the tropics than elsewhere. Depletion of soils whose nutrient and organic capital is naturally low and exposure to climatic extremes of high temperature, torrential rainfall or severe drought can effect rapid and often irreversible soil erosion. The essential tragedy is the exploitation and depletion of a capital organic resource to so little effect, particularly in face of the urgent need to increase food and not least protein production in these areas. In view of the problems attendant upon the introduction of improved high-yielding breeds of cattle, and the type of pasture necessary to maintain them, it has been suggested that an effort should be made to make a more rational use of 'wild' rather than 'domesticated' animals. Game has for a long time been an important source of food in savanna regions where it has, unfortunately, suffered the effects of uncontrolled exploitation. However, because of the greater diversity of species and their varying feeding habits game animals make fuller use of wild fodder resources. By means of regulated breeding stocks and carefully controlled grazing and browsing, savanna vegetation could produce a much greater quantity of human food than it does at present without the inevitable consequence of selective overgrazing.

The exploitation of organic resources has been characterised by a progressive development from the uncontrolled depletion of capital and the consequent decrease in biological productivity to the regulated build-up and increase in fertility. This process has proceeded further in some types of resources, some forms of exploitation and some areas of the world than others. It has attained its greatest intensity in crop production under both commercial agricultural and silvicultural systems in, particularly, the temperate regions of the world. In the development of what have in modern jargon been termed 'agro-systems' three stages can be identified: first, extraction or mining accompanied by decline in primary biological productivity or yield; second, sustained yield; and third, increasing yield.

As far as is known the earliest method of crop-production was under a system of *shifting agriculture* similar to that still practised today in subsistence economies in inter-tropical areas of the world. It is particularly well-suited to wooded areas and no doubt originally evolved there. A patch of ground is cleared by felling or ringing the trees which are burned in situ. The minerals released in the ash provide an additional though temporary source of plant nutrients. The cleared area is planted often with a considerable variety of crops which give maximum ground cover and competition to weeds. Cropping is purely exploitive and is continued for a number of years until yields begin to decline. At this stage the area is abandoned and a new site cleared. Regeneration of the wild vegetation serves to rehabilitate and restore the fertility of the cropped area. The basis of shifting agriculture, therefore, is one of a long-term rotation of a few years cropping—probably no more than three at most—followed by a period of 'fallow' long enough in humid tropical areas for secondary forest to re-establish itself. Although 'primitive' in the sense of being very ancient, it is a system peculiarly well-adapted to tropical conditions where the potential soil nutrient-capital is small and where the complete destruction of the protective forest cover can result in soil sterility and erosion.

Archaeological evidence has revealed that 'shifting agriculture' on the basis of 'slash-and-burn' of the once more extensive forest was characteristic of early Neolithic farming in Western Europe. It was maintained for a long time in a more sophisticated form in later sedentary systems of agriculture in medieval times. For long the 'fallow period' was the only means of restoring soils depleted of nutrients. Under the old infield-outfield system of farming in Scotland, while the former area was manured, the latter was worked on the basis of shifting cultivation. In the plantation agriculture of seventeenth- and eighteenth-century North America the production of particularly exhausting crops, such as tobacco and cotton, was initially accompanied by a form of shifting agriculture made possible by the abundance of land, and soils exhausted by over-cropping were readily abandoned in favour of virgin soils. However the system can degenerate into one involving the progressive decline of productivity and the final reduction of areas to biological deserts. This may, paradoxically, be the result of too little or too much land. In the first instance, increasing population pressure (such as is being experienced by many underdeveloped tropical countries today) on often dwindling land resources results in a reduction of the 'recuperative fallow' period necessary after cropping. If this becomes too short to allow 'nature' to make good the previous losses a gradual decline in biological productivity will inevitably follow. In the second case, the opening up of relatively virgin territory in North America, South Africa and Australasia was accompanied, particularly in the late nineteenth and early twentieth century, by a ruthless exploitation and mining of soils. The sheer abundance of land often associated, as in the 'grasslands', with soils of seemingly unlimited fertility undoubtedly led to the profligate use of organic resources. It was not, however, until the nineteen-thirties, when a period of protracted drought in the south western

Great Plains of the U.S.A. resulted in severe wind-blowing of soil deprived of a protective vegetation cover and depleted of their organic content, that the extent and magnitude of soil erosion throughout the world was fully realised. The notorious 'Dust Bowl' was but one—albeit spectacular and highly publicised—of the results of 'over-cultivation'.

Sustained cultivation of the same land is dependent upon the return of plant nutrients taken out by the crop. The effect of animal dung on yield was early recognised. The systematic application of human excreta or 'night soil' has long been characteristic of the intensive, double- or treble-cropping of rice in China. The first European colonists in North America were quick to observe the Indian technique of burying a fish with each seed of corn (maize) planted. In other cases, annual flooding—as in the flood-plain of the Nile or the Ganges—is accompanied by renewal of nutrients in the form of fresh alluvial deposits. If long-term cultivation of a given area allowed the development of 'sedentary' as distinct from 'shifting' methods of agriculture, the concomitant concentration of people and animals provided a larger and more reliable source of dung. Under the medieval systems of farming in Britain the folding of 'community' herds of stock on arable land after the harvest was a common practice. But the amount of dung available was limited and much was lost to the cultivated fields because of unenclosed land over which animals could range widely for much of the year. In addition, the emphasis remained on the continuous cultivation of a grain crop interrupted only by the periodic fallow or rest year. Productivity was low and yields were sustained at no more than minimum levels of fertility.

Improved methods of husbandry had, however, to await the evolution of agricultural techniques, *and* the necessary reformation of systems of land management and tenure. These began to take effect during the eighteenth century in Britain. The introduction of systematic crop rotations permitted a more efficient use of soil resources by a carefully selected series of crops differing in their nutrient requirements, potential uses and methods of cultivation, than did the older monocultural system. The introduction of the clover and/or grass-ley provided a 'break' in crop production and source of grazing much superior to the weed-infested fallow. The inclusion of the ley and other fodder crops, not least root crops, in the rotation finally effected the integration of stock-rearing and fattening with crop-production to the mutual benefit of both. Eventually, growing scientific knowledge, a rapid expansion of urban/industrial populations and a mounting demand for food brought in their wake the greatly increased use of *organic fertilisers.* Animal manures, blood and bones (waste products of the urban abattoirs and it has been said, of battle-fields!) and fish meal together with human sewage became more abundant and readily available. These were supplemented by the use of recently discovered sources of phosphorus-rich 'guano' deposits in Peru. By the beginning of the nineteenth century, when the early development of an urban/industrial society (as in Britain) had provided the impetus and the means of agricultural improvement, methods of conserving fertility were being developed and established.

It was not, however, until the mid-nineteenth century that the action of organic fertilisers was completely understood. Increasing knowledge about the chemical characteristics and the rôle of plant nutrients, together with advances in plant genetics, initiated what has been referred to as the 'fertility or biological revolution' of the present century. This has been reflected in rapidly-rising crop yields (particularly since 1939) consequent upon the substitution of inorganic for organic fertilisers, the chemical control of competition from weeds and animal pests, and the breeding of high-yielding strains of crops and breeds of animals. Increasing use of what are called 'artificial' fertilisers has been second only to irrigation in achieving and sustaining maximum biological productivity. Of the 'Big Three' plant nutrients potassium was the first to become available in an inorganic form. Initially the main source of the water soluble compound, potassium-oxide (K_2O) was wood-ash, hence the name potash. By the mid-nineteenth century this had been replaced by that derived from extensive deposits of potassium chloride, first in Germany and France, later in North America, the U.S.S.R. and the Middle East. The significance of phosphorus in bone and fish meal was not, however, realised until 1840. The subsequent discovery that the treatment of bones with sulphuric acid made phosphorus more readily available to plants, in fact, initiated the modern chemical fertiliser industry. The use of bone as the main source of phosphates was superseded towards the end of the nineteenth century by basic slag (the by-product of the smelting of highly phosphatic iron-ores) and mineral deposits of which the richest known reserves are in Florida, Western U.S.A., North Africa and the U.S.S.R.; this latter source now supplies about ninety per cent. of the world's phosphorus requirements.

Nitrogen, because of its deficiency in soils (particularly of inter-tropical regions) and its relative scarcity in a readily available mineral form, is the most vital of plant nutrients. The successful economic production of inorganic nitrogen was not achieved until the twentieth century. It was initially a by-product of the rapidly-developing chemical industry. The two early sources were ammonium sulphate, formed during the manufacture of coal-gas, and somewhat limited mineral deposits of sodium nitrate originally exploited as a base for explosives and other chemicals. The synthesis of ammonium from a combination of nitrogen and hydrogen in 1910 resulted in a major advance in the fertiliser industry; and the economic production of nitrates was stimulated by the technical developments and demands (for both explosives and fertilisers) created by two world wars. The spectacular growth of the fertiliser industry within the last thirty years is symptomatic of the increasing use of nitrates. Nowhere have the results been as dramatic as in the U.S.A. where, until the mid-thirties, agriculture was still largely exploiting capital soil-organic resources. Between 1940 and 1950 agricultural production increased by fifteen per cent. and between 1950 and 1960 by forty per cent., largely as a result of an increase in the use of artificial fertilisers.

There have been, then, two outstanding and far-reaching results of the 'fertility revolution'. One is the very great, though by no means evenly distributed, increase in soil nutrient levels. As C. J. Pratt points out in his excellent account of developments in the use of chemical fertilisers, 'Of all the short-range factors capable of increasing agriculture readily . . . the largest yields and most substantial returns on invested capital come from chemical fertilisers'. The other is the regulation of the optimum, or a 'better' nutrient-balance in soils where formerly an excess or deficiency of either macro- or micro-nutrients had limited yields either directly or as a result of their effect on the availability of other essential minerals. However, while the use of chemical fertilisers has gone far to solve that problem basic to the rational use of biological resources—of increasing and sustaining maximum yields—it has created others.

Of these probably the most important is that of minimising losses or damage due to excessive use of inorganic nutrients. The former results from the ease with which inorganic compounds applied to the soil in a readily-soluble form can be removed by leaching. Among the most soluble, and therefore most susceptible, are the nitrates. Leaching involves the removal of these and other nutrients in surface and ground-water flow from terrestrial to, ultimately, lake or marine ecosystems. In shallow waters they may be recovered fairly rapidly by water plants. An important amount however may be deposited in habitats such as de-oxygenated lakes or ocean bottoms where their return to the nutrient cycle is retarded or prevented. In other cases a proportion of the mineral elements may be rendered less available or unavailable either because of the base-status of the soil or because they are converted into insoluble forms. In the latter case certain nitrogenous fertilisers can increase soil acidity and thereby reduce the availability of other essential nutrients. Under acid soil conditions the ease with which phosphorus is 'fixed' in insoluble compounds with iron, aluminium, etc., is even more serious in view of its relatively limited supply in nature. It has been estimated that as a result of such losses the proportion of soil nitrogen and phosphorus utilised may rarely exceed seventy-five per cent., while that of phosphorus can be as low as ten per cent. On the other hand, the addition of increasing amounts of fertilisers does not result in an indefinite increase in productivity. Above an optimum (which will be dependent on the type of plant, nutrient and soil) yields may become static or even create conditions of soil toxicity.

The major problem, therefore, which the use of artificial fertilisers has created is that of controlling their application in such a way as to give maximum availability of nutrients. Various methods have been developed although they are, as yet, in the experimental stage. These include the use of inorganic compounds which 'decompose' slowly or even of 'packaged soil-granules' containing the seed and the correct mixture of substances required for maximum growth. Chemical ploughing has been tried on a small scale. This involves the elimination of stubble and cover crops by spraying with herbicides instead of ploughing; new seed and fertilisers are injected together directly into the soil. The aim of this system is to combine the advantages or inorganic and organic nutrients while reducing their respective disadvantages. Retention of nutrients and the degree of leaching is influenced by soil texture and structure. The value of organic matter lies in the provision of a means of protection against the effects of frost, drought and erosion; and also in the creation of soil tilth which provides optimum physical conditions for plant growth, as well as increasing its adsorptive capacity. The main disadvantage, from an agronomic point of view, of organic matter as a source of nutrients is the slow rate of mineralisation of the most nutrient-rich humus. The advantage of 'artificial' fertilisers is their immediate availability; their obvious disadvantage is that this may be greater than the absorptive capacity of the plants.

One of the most important and widespread results of the predominant use of inorganic fertilisers in all except certain subsistence or specialised systems of agriculture is the increasing abandonment of the older traditional crop rotations. There has been an increasing tendency to return to continuous monocultures even in Britain. This has been intensified by the segregation of fodder-production and animal-feeding characteristic of modern methods of intensive 'meat or milk' production: it has been further exacerbated by financial incentives to extend the acreages of grain crops. For example in Britain barley acreage alone has doubled; the yield has all but trebled within the last ten years. Such intensive monoculture is already creating a new set of problems in the parallel multiplication of soil and airborne diseases. If not controlled losses could negate the gains achieved at the expense of fertilisers, insecticides, herbicides and mechanisation and reverse the cycle of productivity.

References

AGRICULTURAL RESEARCH COUNCIL. 1967. *The effects of air pollution on plants and soil.* London.

ANDERSON, M. 1951. *A geography of living things.* English University Press, London.

AVRILL, R. 1967. *Man and environment.* Pelican Books.

BARR, J. 1969. *Derelict Britain.* Pelican Books.

BENNETT, H. H. 1944. Food comes from soil. *Geogr. Rev.*, **34** (1): 57–76.

BROOK, M. (Ed.) 1967. *Biology and the manufacturing industries.* Symposia of the Institute of Biology, No. 16. Academic Press, London.

CAREFOOT, G. L. and SPROTT, E. R. 1968. *Famine on the wind: plant diseases and human history.* Angus and Robertson, London.

CARSON, R. 1963. *Silent Spring.* Hamish Hamilton, London.

DARLING, F. F. 1960. Wild life husbandry in Africa. *Scient. Am.* November, **203** (5): 123–137.

DAVIES, W. 1960. Pastoral systems in relation to world food supplies. *Adv. Sci.* **17** (67): 272–280.

ENGLER, F. E. 1964. Pesticides—in our ecosystem. *Science, N.Y.*, **52** (1): 110–36.

ELIASSEN, R. 1952. Stream pollution. *Scient. Am.* March. **186** (3): 17–20.

FISHER, J., SIMON, N. and VINCENT, J. 1968. *The red book: wild life in danger.* Collins, London.

GOODMAN, G. T., EDWARDS, R. W. and LAMBERT, J. M. (Eds.) 1965. *Ecology and the industrial society.* Symposium No. 5 of the British Ecological Society. Blackwell, Oxford.

GRAHAM, E. G. 1944. *Natural principles of land use.* Oxford University Press.

HAAGEN-SMITH, A. J. The control of air pollution. *Scient. Am.* January, **210** (1): 24–31.

HEIM, R. 1952. *Destruction et protection de la nature.* Armand Colin, Paris.

HUTCHINSON, Sir J. 1966–7. Land and human populations. *Adv. Sci.*, **23** (111): 241–254.

JACKS, G. V. and WHITE, R. O. 1939. *The rape of the Earth: a world survey of soil erosion.* Thomas Nelson & Son Ltd. London.

LE CREN, E. D. and HOLDGATE, M. W. 1962. *The exploitation of natural animal populations*: Symposium No. 2 of the British Ecological Society, Blackwell, Oxford.

MCNEIL, M. 1964. Lateritic soils. *Scient. Am.* November. **211** (5): 96–106.

MELLANBY, K. 1967. *Pesticides and pollution.* Collins, London. (Fontana paperback, 1969).

MOORE, N. W. 1967. A synopsis of the pesticide problem. *Adv. ecol. Res.*, **4**: 75–126.

OSBORN, F. 1948. *Our plundered planet.* Little and Brown, Boston.

OVINGTON, J. D. 1963. *Better use of the world's fauna for food.* Symposia of the Institute of Biology, No. 11. London.

PIEMERSEL, R. L. 1954. Replacement control: changes in vegetation in relation to control of pests and diseases. *Bot. Rev.*, **20** (1): 1–32.

PRATT, C. J. 1965. Chemical fertilisers. *Scient. Am.* June: 62–72.

RUSSELL, Sir E. J. 1961–2. Food production for the expanding world population. *Advmt. Sci.*, **18** (75): 427–435.

RYDER, M. L. 1966. The exploitation of animals by man. *Adv. Sci.*, **23** (107): 9–18.

SEARS, P. B. 1959. *Deserts on the march* (3rd ed.). Univ. Oklahoma Press, Norman.

SHARPE, C. F. 1938. *What is soil erosion?* U.S. Dept. Agric. Misc. Publ. No. 286. Washington, D.C.

SIMMONDS, I. 1966. Ecology and land use. *Trans. Inst. Br. Geogr.*, **38**, 59–72.

STAMP, L. D. 1958. The measurement of land resources. *Geogr. Rev.*, **48** (1): 1–15.

STAMP, L. D. 1965. Man and his environment. K.C.S. *Sci., J.* May: 76–80.

THOMAS, W. L. (Ed.) 1958. *Man's role in changing the face of the earth.* Chicago University Press.

VOGT, W. 1948. *The road to survival.* Sloane, New York.

WOODWELL, G. M. 1967. Toxic substances and ecological cycles. *Scient. Am.*, March. **216** (3): 23–31.

YAPP, W. B. and WATSON, D. J. (Eds.) 1958. *The biological productivity of Britain.* Symposia of the Institute of Biology, No. 7. London.

16
Conservation of Organic Resources

The concepts, aims and methods of conservation developed from a concern for the way in which man was misusing and depleting his natural resources. Over exploitation leading to the simplification and impoverishment of plant and animal life and accompanied by reduction of biological production (and, in some cases, the destruction of organic resources) is not a recent process. It has been in operation for many centuries. Although initially limited in extent, it had left an early indelible mark on long-settled and formerly more densely populated areas of the world. Deforestation and the consequent erosion of hillsides, particularly in such semi-arid regions as Mediterranean Europe, North Africa, the Middle East and parts of Monsoon Asia are thought to have intensified soil aridity and created 'man-made' deserts. The consequent accelerated run-off together with flooding and increased silting in adjacent lowlands may well have contributed to the deterioration of ancient irrigation schemes. Pierre Gourou records that the decline of the once highly-developed Mayan civilisation in Central America has been attributed to the depletion and destruction of fragile tropical soils as a result of increasing pressure of population. However during the last two hundred years the rate and scale of exploitation has accelerated rapidly following the increase of world population, the colonisation of new territories and the use of more sophisticated techniques. But it was not until the end of nineteenth century that man began to comprehend fully the nature and extent of his actions.

The wasteful exploitation of the recently colonised 'virgin lands', particularly in North America and other temperate regions in the southern hemisphere, was stimulated by an abundance of land and expanding world markets for agricultural products. At the same time, however, knowledge about biological principles and a greater understanding of the ecological implications of organic resource-use were beginning to emerge. To this end the publication of George Marsh's now-classic book, *Man and Nature: Physical Geography as modified by Human Action*, in 1864, had a profound influence, not least in the U.S.A. What was to become known as the 'Conservation Movement' was born of the realisation of the extent to which ruthless, uncontrolled clear-felling was destroying the last great stands of virgin forest in the western part of the United States. Drought and agricultural depression in the nineteen-thirties had an even more dramatic and universal impact. They revealed the appalling degree and extent to which soils had been ruthlessly over-cultivated. The 'Dust Bowl' of Texas and Oklahoma and the gully-ravaged hillsides of the Tennessee Valley became the classic case-studies of the disease of soil erosion. These served to focus attention on the extent of the problem in other parts of the world. The establishment of the U.S. Soil Conservation Service in 1934 was a major land-mark in the history of the conservation movement. Until the Second World War conservation policies and techniques had progressed more rapidly in the United States than elsewhere and were mainly concerned with problems of forest and soil use. Since the war, increasing pressure on land in western urban/industrial countries and the growing populations and needs of the under-developed tropical countries has intensified the problems attendant on the exploitation of organic resources. Concurrently attention has become focused on the need to control and regulate the use of organic resources; and the concept of conservation as a method of 'applied ecology' or 'resource-use planning' is now universally accepted, if not always applied.

However, since the inception of the conservation movement there has been an evolution—indeed one might say revolution—in conservation principles and policies. The problem of *what* to conserve, *how* to conserve and for what *purpose* or *to what end* conservation should be directed have multiplied in complexity as man's knowledge of and his demands on the biosphere have increased. 'Conservation', it has been said, 'is now a protean omnibus term' which implies a way of thinking as much as a process; it is both a philosophy and a technique. It was originally defined by Pinchot, the first Director of the U.S. Forest Service (and often regarded as 'the father of the conservation movement') as resource use designed to ensure the greatest good for the greatest number for the longest time or, in the words of the late J. F. Kennedy '. . . the prevention of waste and despoilment while preserving, improving and renewing the quality and usefulness of our resources'. In meaning and scope conservation can range from the protection of birds' eggs to the regulation of the quality of man's environment.

Inherent in the concept of conservation, as applied to organic resources, are two extreme views or aims which, interpreted literally, would appear to be so contradictory as to defy reconciliation. The first is that of *preservation* or *protection* of 'wild' life or of 'natural' habitats for modification, depletion or destruction by man. This was the initial objective of conservation policies prompted by an understandable, if somewhat quixotic, desire to 'protect nature from man'. Attempts to preserve and protect nature—as Robert Avrill points out—pre-dated the modern conservation movement. The protection of royal hunting-grounds and game for sport no doubt helped to preserve species and habitats that might otherwise have disappeared Also the protection of animal life is basic to certain religious beliefs not least among the most orthodox Jains, an extreme sect of Hinduism, who regard all life as sacred and will not deliberately kill any animal. One of the earliest pieces of conservation legislation must have been the Act passed in 1869 in Britain to ensure the protection of certain sea-birds. The need for preservation *per se* has been supported on several grounds. For some it is an *ethical* problem; it is considered morally wrong that man should deliberately destroy any part of what is the common heritage of present and future generations. Preservation in the interests of *scientific* knowledge and research can be justified for either purely academic or more practical ends. Knowledge of all forms of life and their ecological relationships is essential to the rational management of organic resources. This can only be successfully pursued in actual habitats; and the more varied and diverse that are available for study and experiment the better. Further, it has been pointed out that it is in man's best interests to preserve as varied a 'genetic pool' as possible which could, with advances in biological knowledge, provide new resources in the future.

Thirdly, the *aesthetic* reasons for preservation of wild life cannot be ignored; indeed they have, of late, been given increasingly serious consideration. Diversity of life and variety of habitats are generally accepted as intellectually more satisfying and emotionally more pleasing than uniformity and monotony of landscape. Indeed some of our most beautiful and seemingly natural landscapes in Britain, in which a variety of wild and cultivated habitats are combined, were deliberately created for the aesthetic satisfaction (as well as the social prestige) of former landowners. Today, as the demand for 'open spaces' for recreational activities increases, organic variety and diversity enhances the value of land for this purpose. Scenic 'beauty' and/or interest, to which wild life makes an important contribution, has become a valuable and, in over-crowded countries such as Britain, a scarce resource. Public concern (highly developed in the United States and now beginning to gain momentum in the U.K.) for wild life has been stimulated by an increasing number of people interested in 'outdoor' activities and equipped with a greater curiosity and knowledge about 'nature' than ever before. As a result preservation has acquired a more readily acceptable and comprehensive *economic* justification. It has also been expressed, both explicitly and implicitly, that the preservation of maximum organic diversity is not only ecologically preferable but, *in the long-term,* economically desirable and profitable. The biological productivity of complex and mixed communities is higher than that of simple monocultures; and although the *short-term* economic return of a particular product from the former may be less this is compensated by the greater stability and sustained yield with smaller 'inputs' of fertiliser etc., over a longer period of time than in the latter.

Preservation and protection is dependent on legislation specifically designed to protect species or habitats from destruction. In the former case the aim has been to ensure reproduction at a rate sufficient to make good losses by natural and other causes. It has involved the outlawing of the collection of birds' eggs and the picking of rare plants, together with the legalisation of 'closed seasons' for hunting and fishing etc., during the vulnerable periods of mating and reproduction. In this respect the 'ornithological lobby' has been most successful; and the protection of wild-fowl has perhaps influenced conservation policies to a disproportionate extent in many countries—not least in Britain. The enforcement of legislation designed to protect such mobile and elusive species as birds and fishes, however, is not easy at either national or international levels. The *Nature Reserve,* the *Wild Life Refuge* and the *National Park,* designed to protect a particular habitat and its associated plant and animal communities, are more effective and easily implemented methods of preservation. There is little point protecting a species if the area in which it breeds and/or feeds is destroyed. As has already been noted, habitat modification can be as effective in the reduction or elimination of a species population as direct over-exploitation.

Unfortunately the protagonists of preservation have often, unwittingly, it must be admitted, defeated their own aims, and have hindered rather than promoted development of conservation. In the first instance, conservation in its simplest and earliest form was to protect, indeed prevent access of man to areas thought to be pristine in order to 'let nature alone' and, it was assumed, to maintain the *status quo*. Unfortunately, the early, rigid preservationists tended to forget or failed to take into account two important facts. First, that any ecosystem is *dynamic* in character and second, that *man* is, and has indeed for a long time been, an almost universal habitat factor; there are few areas into which his direct or indirect effects have not penetrated to some extent. As a result areas set aside as nature reserves or parks tended to suffer from the protection they were given. In an address to a meeting of the International Union for the Protection of Nature, in 1956, Professor Romell, discussing the man-made 'nature' of the northern lands of Scandinavia and Finland, said: 'Some of them have changed to the point of being unrecognisable. Rare plants have disappeared from their specially protected habitats because the latter changed under protection. There has been 'a conservation to death'. The former landscape and habitat directly created by man was fast disappearing as birch scrub, the precursor of the formerly extensive and more monotonous coniferous forests, quickly established itself. J. Morton Boyd notes that the basic mistake in the establishment of National Parks in East Africa—designed originally to protect the game animals—was the exclusion of man other than as a sightseer and the policy of letting 'nature take its course'. Here the results were even more disastrous; animals such as elephant, hippo, and buffalo, whose populations had formerly been kept in check by hunting, increased to the extent that widespread devastation of their habitat resulted. In fact, in many seemingly wild habitats man has already created an imbalance, the effects of which may become worse if he is excluded. The ecological 'niche' which man has occupied for so long cannot be left vacant. In most areas successful preservation of a rich, and varied habitat depends on successful management towards this end!

Secondly, preservationists in an excess of zeal to protect species or habitats tend, at times, to be primarily concerned with the 'rare' species whose populations are particularly small and whose habitats very restricted in extent. Unfortunately assessment of rarity may be on a local rather than a national or international basis. The extinction of species is part of the evolutionary process in which man has, admittedly, become a very powerful force. He can, however, preserve both species and their habitats to an extent unknown before. Justification for the preservation of an area from exploitation or development can only be made on the basis of what is the most desirable use of the resource or area for the greatest number of people. Merely to use the argument of preservation in order to alienate an area in the interests of a small section of the population is, in areas where pressure on land resources is extreme, intellectually dishonest. It also does the cause of conservation a grave disservice. If the main reason for preservation is scientific it may, in the final assessment, be necessary either to 'transplant' the species to another suitable habitat or re-create a suitable habitat elsewhere. In the interests of preservation neither technique is impossible nor necessarily undesirable. Both, as in the case of the transplantation of rare species to existing reserves or the creation of particular habitats in parks and gardens, are becoming academically and ecologically acceptable! The need for 'natural' ecological testing-grounds or experimental areas relatively undisturbed by man is real and indeed urgent. Understanding of such habitats is essential if man is to learn how to manipulate his organic resources efficiently. From this point of view small, relatively isolated islands provide ideal conditions. Not only are they often rich in endemic species but human population and exploitation tend to be limited. Also they are well-defined ecosystems whose isolation makes study and control easier than in a land-locked reserve.

Unfortunately, conservation is still regarded by many as synonymous with the preservation of wild life and as implying a static rather than dynamic approach to the problem. The second aim, however, implicit in the principles of modern conservation is that of the *use* or *production* of organic resources. This was originally based on the concept of *sustained yield* of those products for which there is the greatest demand and which would give the maximum economic return from capital invested. As the significance of the *renewable* (or in modern terms 'flow') character of organic—as distinct from inorganic—resources became fully understood so it was realised that rational use depended on conservation of organic capital. The idea of sustained yield or conservation use was first deliberately applied to forestry, to the maintenance of the breeding stocks of marine animals which had suffered from over-fishing, and was implicit in the principles of soil conservation, the aim of which was to sustain, in particular, agricultural fertility. Conservation in the interests of the production of food, fodder, timber, etc., needs little justification. Many would maintain that, in face of growing pressure of world population on food resources, the preservation and protection of wild life for non-productive reasons is a luxury which the world cannot afford. However to achieve maximum sustained yields of 'consumable organic products' necessitates the selection of plants and animals, the shortening of food-chains with a consequent reduction in the diversity of wild life and intense specialisation in land use. It assumes a continuing demand for a particular product and minimises the possibility of alternative uses of either the resources or the area it occupies.

The problem in conservation policies therefore, is how to reconcile the too extreme and apparently conflicting aims of sustaining maximum productivity of a particular resource while at the same time preserving the greatest biological variety possible. This has created the dilemma of attempting to reconcile a course of action which is economically most profitable with one that is ecologically most desirable in the management and exploitation of organic resources. The evolution of conservation principles has been characterised by a convergence of these two originally divergent aims in the concept of *multi-purpose use* of either a particular resource or area. This concept was, as has been noted previously, initially applied to forest resources, as their potential for uses other than timber production came to be more fully appreciated and understood. The aim of multi-purpose use is the development of the maximum potential of forest resources by management designed to maintain a balance between demands for timber, water, recreation, wild-life conservation, etc. It is now the accepted policy in the United States Forest Service's management of their National Forests—the successors of the early Forest Reserves. Although much slower to develop in Great Britain, the principle of multi-purpose use is accepted, if not always fully implemented as it might be, in the National Forest Parks established in 1935 under the auspices of the Forestry Commission. In addition the creation of Forest Nature Reserves is a result of the recognition of the importance of forest management in the increase and conservation of wild-life. Forests, because of the size and complexity of their biomass, are particularly well-suited to multi-purpose use. It is also applicable, although it has not been employed to the same extent in other ecosystems. The combined use of moorland or rangeland for stock, game, recreation and military exercises is a problem particularly relevant to the use of Britain's dwindling open spaces. The need to reconcile the, at present, conflicting uses of estuaries, lake and other wetland habitats needs little emphasis in the light of current developments in Europe and America.

The concept of multi-purpose use has already effected a re-assessment of the rôle of areas originally designated as Nature Reserves, Wildfowl Refuges, National Parks and the like. The management of National Nature Reserves as multiple resource units is now fundamental to conservation policies in Britain. The Loch Druidibeg Reserve in South Uist (Outer Hebrides) is a classic example; in the words of J. Morton Boyd 'it contains land owned by the Nature Conservancy and the property of a private landowner. It includes sanctuary breeding sites for wild geese, agricultural land, common grazings, fresh-water fisheries and examples of most of the wild and man-made habitats of the Outer Hebrides. It is managed as a partnership between the Conservancy, the landowner, the sporting tenant and the crofters. This is a multiple-resource project incorporating all the main land use practices of the district and has resulted in a very successful nature reserve in the modern image'. The recent extension of the Loch Lomond Nature Reserve to incorporate (with the already designated islands) the important wild-fowl breeding area of the Endrick Marshes also involves a similar combination of marsh, wood and agricultural land, and of public and private ownership and management. In the areas designated National Parks in Britain under the Act of 1949 the twin aims of preserving and enhancing natural beauty and providing facilities for public recreation were directed, within the limited powers of the Act, towards controlling already existing types of land use and further developments to these ends. More recently the French Government has decided to establish Regional Parks (of which the Camargue area of the Rhône delta was the first example), as distinct from the National Parks instituted in 1960. The latter are confined to large areas of 'wild' countryside in which there is little human activity. The former, on the other hand, are areas or *'pays'* characterised by the distinctive nature of their wild life, types of land-use and recreational potential. All local interests are combined and represented in the formulation of a management policy. The aim is to link the maximum development of the local rural economy with the conservation of the unique character of the landscape. Others are envisaged in such contrasting regions as Brittany, the Vosges, the Jura and the Landes coast of Aquitaine. In contrast the great National Parks of the U.S.A. (the first of which, Yellowstone, was established as early as 1877) and those created by colonial administration in such tropical countries as

Africa, are of more specialised use. Recreation is the sole use of the former and wild life is from this point of view one of their most important resources. The U.S.A. is fortunate that the foresight of her early conservationists and a wealth of land, and particularly *public land,* resources has allowed the creation and controlled use of such a large area for this purpose. The National Parks of Africa, on the other hand, originally instituted as game preserves, are economic and ecological anachronisms. In view of the disastrous effects of excluding man from these areas and the effects of misuse or over-use of tropical vegetation and soils a more realistic and rational approach to conservation is urgently required. Boyd sees a solution not merely in multiple resource use, which is beginning to be recognised in many of the National Parks of East Africa, but in the creation of *conservation units.* Larger than other types of parks, reserves or areas of controlled land use, these would combine (as is already being done in the Ngorongoro Conservation Area) all types of use—including wild-life conservation—which might, if a large enough area were involved, be managed on a rotational basis.

Fundamental to the modern approach to conservation is management which permits the maximum use of biological resources consistent with the maintenance of the greatest quantity and diversity of organic capital possible. The use least easy or amenable to control today is probably that for recreational purposes. It is made difficult by the still very incomplete knowledge of the 'carrying capacity' or optimum population pressure which plants and animals can sustain. It is exacerbated by increasing demands for 'open space' by growing urban/industrial populations and by even more sophisticated techniques—from the motor car to the impedimenta necessary to maintain an urban way of life in the countryside. Greater mobility and affluence since the Second World War has been paralleled by a phenomenal increase in the numbers using the National Forests and Parks in the U.S.A. Concern for the resources of these areas was instrumental in the decision by Congress to establish *Wilderness Areas* within the existing National Parks. The object is to protect areas of high scenic value, not so much from man *per se* as from deterioration consequent upon over-use. To this end access to the Wilderness Areas is confined to those on foot, horseback, or canoe, and only shooting and fishing of wild-life necessary to maintain the ecological balance will be permitted. Not all countries, however, have the land resources or can afford this modern form of preservation. However, it is not beyond the bounds of possibility that wilderness conservation may require some form of international co-operation and agreement such as has had to be applied to the exploitation of certain extra-territorial marine resources. In the final chapter of his book *Man and Environment,* Robert Avrill suggests that 'In a Europe planned as a physical entity, the Scandinavian coastline, much of Scotland, the Black Forest, the Alps and many similar areas would receive priority for conservation and enhancement.' A logical extension of this idea must eventually be the assessment of conservation priorities on a universal basis.

Economic and technical developments in agriculture are (as in urban areas) tending to produce extensive landscapes of a uniform monotony. High capital investment and the subsequent economies of scale are reflected to an ever-increasing extent (particularly in the temperate regions of the world) in amalgamation of farms and fields into larger units and, as has been previously noted, a return to monocultural methods. This has been achieved by the eradication of hedges, embankments and smaller woodlands as well as by the increasing application of herbicides, pesticides and artificial fertilisers. It has been accompanied by increasing soil pollution and, in some cases erosion, a drastic impoverishment of 'wild life' and the creation of ecosystems particularly vulnerable to disease and the vagaries of weather. In face of the difficulty of reconciling a high degree of *technological* efficiency with a diverse, and hence more stable and aesthetically satisfying landscape, there are those who would advocate the development of conservation farming. The ecological justification for the greater use of organic manures to maintain soil structure, for the retention of hedge and copse as sanctuaries for wild-life and for a return to systems of mixed rotation farming is sound. Politically and economically it is, at present, an unrealistic pipe-dream. Even those incentives designed to counteract the effects of over-production and to encourage soil conservation have tended to defeat their own ends. The Soil Bank programme instituted in the U.S. after the Second World War is a classic example. Combined with acreage restrictions it was designed to check the surplus production of certain food and fodder crops, as much as to build-up soil fertility. However, guaranteed prices for crops plus Soil Bank payments merely served to stimulate increased yields on the best land and (in many cases) the deliberate cultivation of pasture land not previously cropped in order to qualify for the subsidy for converting it back to grassland!

However as the agricultural landscape has become more uniform and standardised, less varied in appearance and less rich in species, there has been a tendency for wild life to become relatively more abundant and diverse in urban areas. The significance of this somewhat paradoxical situation has recently been attracting considerable attention in Britain, as well as in America; Gottman, for instance, has drawn attention to the marked increase in woodland in the surburban areas of the Boston–New York–Washington metropolitan region (or Megalopolis). In Britain the increasing biological diversity of urban areas is undoubtedly a function of the peculiar variety of habitats available. Some are a direct product of existing urban functions, such as reservoirs, sewage-farms, railway embankments, gravel-pits and the multitude of 'rock-ledges' in the 'city-canyons'. Others are derelict or abandoned sites; industrial waste-ground, disused railway lines, yards and stations, old canals, bomb-sites or cleared spaces awaiting re-development. They are inhabited not only by now highly-urbanised though still, technically speaking 'wild' species, of which birds such as the pigeon, starling and sparrow, are so prolific as to be regarded as pests, as well as immigrant weeds and vermin.

More important and extensive, however, are the private gardens and public parks (including botanical and zoological museums). Recently, as a result of mounting pressure for recreational land the rôle of the municipal park is being re-assessed. Its function as a form of multi-purpose land use in which wild-life conservation, recreation and education can be profitably integrated is attracting increasing attention. Investigations in the Royal Parks have revealed the extent to which bird-life can be increased and diversified by judicious management of their plant-life and water-bodies. Even more exciting has been the revolutionary change in attitude towards the management of Municipal Parks—initiated largely by the inspiration and action of Mr Oldham, Director of City Parks in Glasgow. The image of the city park as an open space where appearance and use were subordinated to the achievement of tidiness and cleanliness is gradually being replaced by that of a semi-natural habitat, a rural enclave—a countryside oasis in the urban desert! The city park, together with such proposals as the Lea Valley Regional Park and the Country Parks (one of the first being the Wirral Way in Cheshire using a disused railway line), envisaged under the terms of the Countryside Act 1966, have been quaintly, though it is hoped not quixotically, regarded as potential 'honey-pots'. That is to say they would attract and concentrate those in search of the open spaces or merely escape from the back-garden or yard, and relieve the summer week-end congestion on roads out of the large cities. In such parks, however, the problem of management to allow for and to withstand exceptionally heavy recreational use is particularly difficult. The solution of this particular problem, as well as the wider one of the conservation of the intrinsic character and value of organic resources can only be achieved by the maximum development of the educational potential of every type of conservation unit. In this process, because of their location, city parks have a vital rôle to play as 'field' or 'outdoor' education centres, such as have been proposed and are being developed in Glasgow. The whole concept of conservation as initially envisaged depends ultimately on a universal recognition of man's rôle in the biosphere, and on informed public opinion and cooperation.

References

AVRILL, R. 1967. *Man and environment.* Penguin Books.

BLACK, J. N. 1964–5. The ecology of land use in Scotland. *Trans. Proc. bot. Soc., Edinb.*, **40** (1): 1–12.

BOYD, J. M. 1966. The changing image of the National Park. *New Scient.*, 28th April: 254–256.

CRAIG, D. 1962. Resource utilisation and the conservation concept. *Econ. Geogr.*, **38** (2): 113–121.

DARBY, H. C. 1963–4. British National Parks. *Adv. Sci.*, **20** (86): 307–18.

DARLING, F. F. and DURWARD, A. L. (Eds.) 1966. *Future environments of North America.* The Natural History Press, New York.

DARLING, F. F. 1964. Conservation and ecological theory. *British Ecological Society Jubilee Symposium*, 1963. A. Macfadyen and P. J. Newbould, (Eds.) Oxford.

DASMANN, R. F. 1959. *Environmental conservation.* Wiley, New York.

GOUROU P. 1961. *The tropical world.* Longmans, London.

JACKS, G. V. and WHYTE, R. O. 1939. *The rape of the Earth: a world survey of soil erosion.* London.

JARRETT, H. (Ed.) 1961. *Comparison in resource management: six notable programmes in other countries and their possible U.S. application.* Oxford University Press, London.

JARRETT, H. (Ed.) 1961. *Perspectives on conservation: essays on America's natural resources.* John Hopkins Press, Baltimore.

LOWDERMILK, W. C. 1960. The reclamation of a man-made desert. *Scient. Am.*, March, **202** (3): 54–63.

MARSH, G. P. 1864. *Man and nature: physical geography as modified by human action.* Sampson Low, Marston, Low and Searle, London; Scribner, Armstrong & Co., New York.

NICHOLSON, M. 1970. *The environmental revolution*. Hodder and Stoughton, London.

OVINGTON, J. D. 1956. Scientific research and nature reserve management. Paper presented at 6th Techn. Meeting Intern. Union for the Protection of Nature (UIPN/AG5/RT6/1/3). (Mimeo)

OVINGTON, J. D. 1964. The ecological basis of the management of woodland nature reserves in Great Britain. *British Ecological Society Jubilee Symposium, 1963*. Eds. A. Macfadyen and P. J. Newbould. Oxford.

PRICE, E. T. 1955. Values and concepts in conservation. *Ann. Ass., Am., Geogr.* **45** (1): 63–84.

RAUP, H. M. 1964. Some problems in ecological theory and their relation to conservation. *British Ecological Society Jubilee Symposium, 1963*. Eds. A. Macfadyen and P. J. Newbould, Oxford.

ROMMELL, L. G. 1956. Man-made 'nature' of northern lands. Paper presented at 7th Techn. Meeting Intern. Union for the Protection of Nature (UIPN/AG5/RT6/1/2). (Mimeo).

SIMMONS, I. G. 1965. The future of the redwoods. *New Scient.* 9th December: 746–748.

STAMP, Sir L. D. 1969. *Nature conservation in Britain*. Collins, London.

STANDING COMMITTEE ON NATIONAL PARKS and the CPRE and CPRW. *Afforestation in National Parks*, Study No. 1. *Future of the National Parks and the countryside*, Study No. 2, London. n.d.

STODDART, D. R. 1968. The conservation of Aldabra. *Geogrl. J.*, **134** (4): 471–486.

REPORT ROYAL SOCIETY OF ARTS AND THE NATURE CONSERVANCY 1966. The Countryside in 1970. Second Conference, November, 1965, Study Group No. 3. *Technology in Conservation*.

WAGNER, J. A. 1964. The carrying capacity of wild lands for recreation. *Forest Sci., Monogr.*, **7**.

WALTER, R. 1969. The case for conservation agriculture. *New Scient.*, 24th April: 179–181.

ZOBLER, L. 1962. An economic-historical view of natural resource use and conservation. *Econ. Geogr.*, *38* (3): 189–194.

Index